FUSION REACTOR PHYSICS

principles and technology

by

TERRY KAMMASH

Professor of Nuclear Engineering
The University of Michigan
Ann Arbor, Michigan

Second Printing, 1976
Third Printing, 1977

P. O. Box 1425, Ann Arbor, Michigan 48106

Library of Congress Catalog Card No. 74-14430
ISBN 0-250-40076-6

Manufactured in the United States of America

PREFACE

For a number of years the speculative assumption among physicists and engineers has been that nuclear fusion will eventually be proved and then established as a practical source of commercial energy, possibly on a scale sufficiently vast to meet a significant portion of man's energy needs.

This assumption has been repeatedly encouraged by new developments, and there is now perhaps more reason for optimism than ever before.

Impressive progress in the physics of confinement for high temperature plasma has led recently to world-wide hope for early demonstration that controlled thermonuclear reactions are scientifically feasible. Such a demonstration will encourage the further expectation that power production from controlled fusion is closer, though it cannot of course guarantee that result.

After controlled thermonuclear reactions are proved, the technological feasibility of fusion power will then have to be methodically established. Acceleration of activity in this area is already taking place. Many countries throughout the world have embarked on vigorous research efforts to advance the technological aspects of fusion reactors.

This book has been written to provide detailed coverage on the basic elements of fusion reactor physics and technology. It is intended to instruct and familiarize workers in the general field of nuclear energy, students, and others interested in the present state and future prospects of nuclear fusion, with comprehensive information in a single volume that was previously available only from a large number of sources.

The book is an outgrowth of class preparations and notes I have made for several years while teaching courses on this subject to seniors and graduate students in Nuclear Engineering at the University of Michigan.

The material is presented at the level of a student with senior standing in Engineering or the Physical Sciences. My own classes have confirmed that an undergraduate course in Modern Physics augmented by a course in Advanced Mathematics for engineers and physicists are adequate preparation for understanding the contents. An elementary course in Plasma Physics, if available, might also prove useful in deriving optimum benefit.

In addition to serving general scientific readers, it is expected that the work will provide an adequate text for courses in this area. The subject matter covered includes material sufficient for a two-semester course. However, with proper selectivity, a meaningful one-semester course can be extracted from it as well.

For example, if the reader wishes to examine the system characteristics of various approaches to fusion power without considering the interplay between design factors and confinement principles, he can turn directly to Chapter 16, bypassing such chapters as 9, 11 and 13.

A one-semester course based on this book could easily include the subject of radiation damage on fusion reactor first walls while omitting the first section of Chapter 14. Such a course could also include the subject matter of Chapters 5 and 6 on plasma heating by relativistic electrons and RF radiation by utilizing only the working formulas that appear at the conclusions of these chapters.

Various chapters may contain material beyond the immediate reach of some students without further preparation, but the subject is presented so that such students and other readers can move past the advanced material without loss of continuity or understanding.

A case in point is the section on Sputtering Theory found in Chapter 14. The sputtering yield formulas are derived from basic principles using transport theory, and the application of these formulas to fusion devices is presented in subsequent sections of Chapter 14. Readers can go directly to those sections without jeopardizing comprehension or appreciation.

Throughout the book, emphasis is on basic principles rather than specific technological problems. In the case of fusion reactors, such problems can be so numerous and complex that thorough immediate coverage is impracticable. I have attempted to choose those topics and concepts which are essential in understanding the overall power producing system. A further effort has been made to concentrate on basic and established information due to the likelihood of drastic change in the near future.

The criteria for selection of included materials have been based not only on the importance of a particular topic but also on its appropriateness and compatibility with the level of understanding I have tried to maintain. In those instances where for purposes of illustration or special clarity a subject matter is treated in an over-simplified manner, with potentially unrealistic results, I have attempted to alert readers to the reasons for simplification and to recommend following up with the more exact treatments cited in the references listed at the end of the chapters concerned.

References have been limited to the relevant ones. It should be noted, however, that most of these include many additional references which readers would find instructive and useful.

Since much of this book is devoted to engineering-related topics, it was felt logical to use the MKS system of units for their discussion. The CGS system has been used in those chapters devoted almost exclusively to physics. This choice was made for the convenience of both physicists and engineers. It is hoped the use of the two systems will not cause confusion. Conversion tables are provided at the end of the book to facilitate conversion.

Terry Kammash
January 1975

ACKNOWLEDGMENTS

The assistance I received from my students at the University of Michigan in the preparation of this book has been truly invaluable. I am very grateful to each of them.

I am indebted to the Princeton Plasma Physics Laboratory, the Thermonuclear Divisions of the Oak Ridge National Laboratory, the Los Alamos Scientific Laboratory, and the Lawrence Livermore Laboratory. These institutions made available many reports and documents before publication and have been consistently helpful.

I wish to express equal appreciation to the Culham Laboratory of the United Kingdom Atomic Energy Authority (UKAEA) for their permission to use data and figures from some of their reports.

I greatly appreciate the support I received from the Division of Controlled Thermonuclear Research of the U.S. Atomic Energy Commission for much of my research that appears in this book, and I am deeply grateful to its director, Dr. Robert L. Hirsch, and to Drs. Stephen O. Dean, Bennett Miller, Alvin W. Trivelpiece and George K. Hess, Jr., for their interest and valuable comments.

The constructive suggestions of Professors Hans H. Fleischmann of Cornell University and George H. Miley of the University of Illinois are also acknowledged with much gratitude.

My thanks are due and sincerely given to the many authors whose papers I followed closely in various parts of the book.

Most of all I wish to thank my wife Sophie and son Dean. Without their love, understanding, and patience, this undertaking might never have been attempted, and could never have been completed.

If this book succeeds in making a contribution to the future progress of nuclear fusion, these and many others, including my past teachers and present colleagues, deserve to share the credit.

CONTENTS

LIST OF SYMBOLS

P	Pressure, power, momentum
n,N	particle density
σ	microscopic cross section, electrical conductivity
v,V	velocity
W,E	energy
τ	equilibration time, slowing down time, confinement time
k	Boltzmann constant, wave number
T	temperature
Z	charge number
μ	magnetic moment, reduced mass, magnetic permeability
B	magnetic field
m	mass
ρ	Larmor radius, fluid density
Ω,ω_c	cyclotron frequency
e	electronic charge
c	velocity of light
D	diffusion coefficient
υ	collision frequency
λ	range, mean free path
Λ	logarithmic cutoff
Λ_D	Debye length
ω_p	plasma frequency
ϵ,ξ	efficiency
φ,Φ	neutron flux, magnetic flux, K.E. flux, electric potential, error function
Σ	macroscopic cross section
L	length, diffusion length, inductance
I	inventory, current
γ,Γ	relativistic parameters
β	v/c ratio of velocity to speed of light
T_{ij}	stress tensor
f(v)	distribution function
Q	fluid flow rate heat flow rate reactor figure of merit
A	area, aspect ratio
M	mass, energy multiplication, Hartmann number,modulation
θ	thermal energy
η	viscosity, efficiency
R_e	Reynolds number
R_m	magnetic Reynolds number
δ	boundary layer thickness
N_u	Nusselt number
S	neutron source, fueling rate, stress
Q_n,Q_f	fusion energy
f_b	burn-up fraction

CHAPTER 1

INTRODUCTORY REMARKS–FUSION POWER FEASIBILITY

For about a quarter of a century many countries throughout the world have been engaged in research aimed at producing power from fusion nuclear reactions. The primary motivation is the availability and easy accessibility of an inexhaustible source of fuel for use in fusion reactors. Although it only exists as 1 part in about 6500, there is enough deuterium in ocean waters to meet mankind's energy needs for millions of years. It is for this reason that nations are counting on controlled fusion as the long-range answer to the "energy crisis."

Although limitless fuel is the primary feature, there are other features of fusion reactors which make them especially attractive. Those most often cited are

A. Lack of radioactive waste products compared to those which are inherent in fission reactors.

B. Inherent safety against nuclear explosions. The conditions required for energy production in a fusion reactor depend on the delicate balance between certain parameters which, if destroyed, shuts the reaction off automatically.

C. Absence of a wasteheat problem. This is a consequence of the lack of radioactive products which could release energy after the fusion reactions ceased. In the case of fission reactors, radioactive waste disposal and associated problems are major considerations.

D. Low biological hazard in the event of sabotage or natural disaster. Since it does not occur in nature, tritium, needed in fusion reactors, must be produced; this can be accomplished in the "blanket" of the reactor via a nuclear reaction (see below) involving neutrons and lithium. Although the resulting tritium will be fed back into the reaction chamber, there is always the possibility of leakage or loss. If this happens we may find comfort in the fact that tritium is essentially biologically innocuous.

E. Reduced danger of diversion of weapons-grade material for clandestine purposes. This is to be contrasted with fission breeder reactors where weapon materials can be produced as a side benefit to power production.

F. Relatively low waste heat plus the potential for direct conversion. The direct conversion of charged particle energy to electrical energy will be examined in some detail later. All the heat produced in a fusion reactor can be used to produce steam; there are no hot stack gases as in the case of oil, gas, or coal power plants. A recent study by Galbraith and Kammash (1971) showed that a mirror reactor, for example, even in its most elementary form can be competitive with a conventional source (*e.g.*, coal-fired plant with 50% thermal efficiency) if current estimates of direct conversion and injection efficiencies prove realistic.

Even though some of the above features require further analysis, the fact remains that fusion reactors now appear "ecologically" superior to fission reactors.

In the remainder of this chapter the underlying principles of fusion reactors will be discussed (although some of them will be examined in detail later on), and the state-of-the-art in research leading to fusion power feasibility will be summarized.

1.1 FUEL CYCLES AND BREEDING REACTIONS

As pointed out earlier the nuclear reactions of interest to fusion reactors are those involving deuterium, namely

$$ {}_1D^2 + {}_1D^2 \longrightarrow {}_2He^3 + n + 3.2 \text{ MeV} \tag{1.1}$$

$$ {}_1D^2 + {}_1D^2 \longrightarrow {}_1T^3 + p + 4.00 \text{ MeV} \tag{1.2}$$

$$ {}_1D^2 + {}_1T^3 \longrightarrow {}_2He^4 + n + 17.6 \text{ MeV} \quad (1.3) $$

$$ {}_1D^2 + {}_2He^3 \longrightarrow {}_2He^4 + p + 18.3 \text{ MeV} \quad (1.4) $$

$$ {}_3Li^6 + \text{“slow” } n \longrightarrow T + He^4 + 4.8 \text{ MeV} \quad (1.5) $$

$$ {}_3Li^7 + \text{“fast” } n \longrightarrow T + He^4 - 2.5 \text{ MeV} + n'(\text{slow}) \quad (1.6) $$

The last two reactions are included to illustrate the breeding of tritium by the interaction of neutrons with lithium. Note that the last reaction is endothermic, *i.e.*, energy must be supplied to make it go.

The most likely fuel cycle for first generation fusion reactors is reaction (1.3), *i.e.*, the D-T reaction with an energy output of about 17.6 MeV. It has a relatively high cross section of about 5 barns at 100 keV, and most of the energy (about 14.1 MeV) is carried by the neutron. The alpha particle carries the remaining 3.5 MeV which is considerable and, as we shall see later, could constitute a major source of heating of fusion plasmas. Since the neutrons of this reaction will be the main cause of radiation damage and induced radioactivity in the reactor, many quickly point to reaction (1.4) as the “clean” fuel cycle since it does not produce neutrons. Instead, it produces charged particles which could readily be used in a “direct conversion” scheme. These are indeed attractive features but unfortunately do not dispense completely with the problem. As long as D is used the first two reactions will take place, even though to a lesser extent, and reaction (1.3) will automatically follow. Aside from the fact that reaction (1.4) has a lower cross section at comparable temperatures, the fact remains that it also does not eliminate the problems associated with neutrons. It might be noted that although the energy release (17.6 MeV) of reaction (1.3) is much smaller than the 200 MeV released in a fission reaction, the energy per unit mass in fusion is much larger than its counterpart in fission.

1.2 ENERGY BALANCE AND REACTOR CONDITIONS

Having decided on the most likely fuel cycle, we may now ask what are the minimum conditions that must be met in order

to have a self-sustaining fusion reactor. The "Lawson" (1957) criterion provides the answer, and in its simplest form it states that the energy release in fusion per unit volume is equal to the ion kinetic energy in that volume, or

$$P_{12}\tau = [n_1 n_2 \langle\sigma v\rangle w_{12}]\tau = 3/2(n_1 + n_2)kT \qquad (1.7)$$

where n_1 and n_2 are the particle densities of the reacting ions (*e.g.*, D-T in reaction (1.3)], and w_{12} is the energy release in this reaction. The ion kinetic energy is expressed in terms of temperature T on the right-hand side of the above equation where k is the Boltzmann Constant. The quantity $n_1 n_2 \langle\sigma v\rangle$ is the reaction rate, obtained by appropriate averaging over the velocity distribution of the interacting ions, and τ is the "confinement" time. For the D-T reaction at kT of 10 keV, and with $n_1 = n_2 = n$, the above criterion yields $n\tau \simeq 10^{14}$ sec/cm^3 which simply says that at these temperatures the product of the ion density and confinement time must exceed 10^{14} for a break-even energy balance in a fusion reactor. In arriving at the above figures we have ignored energy losses from the plasma, but even so these figures are fairly accurate. We will examine and derive this criterion in detail later.

At these temperatures, which correspond to about 100 million degrees Kelvin, the fuel gas will be fully ionized, *i.e.*, the electrons will have been completely stripped from the neutral atoms leaving a collection of positively charged particles (ions) and negatively charged particles (electrons) moving about at tremendous speeds or simply a "hot plasma." In their motion the particles of this hot plasma tend to scatter off one another more frequently than they "fuse." In fact the scattering cross section is about a thousand times larger than the reaction cross section, and the particles travel a distance comparable to the earth's circumference before they undergo a fusion reaction. This serves to illustrate the importance of confinement time, which for particle density of 10^{14} cm^{-3} is about one second.

We may conclude this section by stating that the underlying principle of a fusion reactor is the containing of a hot, dense deuterium (or other fuel) gas away from physical walls for a sufficiently long time to allow fusion reactions to take place.

Note the emphasis on keeping the plasma away from physical walls; the fear here is not so much the possibility of melting the container walls as it is cooling the plasma as a result of collisions with these walls, thus destroying the main requisite of a fusion reaction. Examples of operating conditions for steady-state and pulse fusion reactors are shown in Table 1.

Table 1

Parameters for Steady-State and Pulsed Fusion Reactors

	Steady-State		*Pulsed*
Fuel Cycle	D-T	D-He^3	D-T
Fuel Density (cm^{-3})	3×10^{14}	2×10^{14}	2.5×10^{16}
Temperature (keV)	25	300	10
Required time of confinement (sec)	0.5	1.0	0.025
Power Density (w/cm^3)	25	8	18,000
Pressure-ions only (atm)	12	96	400
Min confining mag field (kG)	25	60	140

The reader may wonder whether the conditions cited above have been met in the laboratory or more simply whether "scientific" feasibility of controlled fusion has been demonstrated. The answer is no at the present but hopefully this will be achieved in the not too distant future. The rest of this chapter will be devoted to a discussion of the various approaches to fusion and the obstacles which lie in the road to scientific feasibility. This discussion will effectively be a state-of-the-art summary on the status of fusion research and the time scales often cited for achieving the above objective. It must be emphasized at this point that although fusion power feasibility may permit or deny the possibility of useful power from controlled fusion, it cannot guarantee it. Only engineering or "technological" progress can do that. It is exactly for this reason that substantial research and development activity in this field has begun and is expected to grow in the future.

1.3 APPROACHES TO FUSION

Until a few years ago it was believed that the approach to controlled fusion was exclusively through *magnetic* confinement. Some recent developments concerning laser-fusion or fusion by relativistic electron beams have led to the concept of *inertial* confinement. Both approaches of inertial confinement are comparatively less understood although in the case of laser-fusion, accelerated research in recent years has generated added enthusiasm for the potentialities of this approach. With this in mind we classify the approaches to fusion as follows:

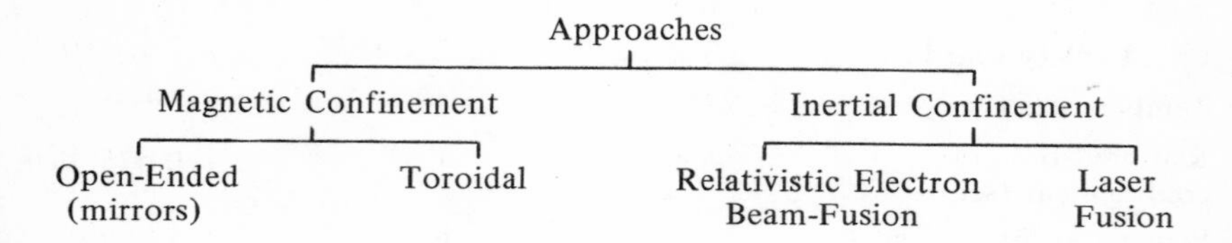

We shall defer any detailed discussion on the inertial approach until Chapter 11; here we will touch upon it only briefly. Rather we will focus our attention on magnetic confinement which has comprised the bulk of research in this field thus far.

1.4 MAGNETIC CONFINEMENT

We recall that the principle of a fusion reactor is the confinement of a very hot plasma for a sufficiently long time to allow the ions to undergo a fusion reaction. We also note that to simply place this plasma in a container is futile because the plasma will strike the container walls and cool off, thus destroying the very condition required for sustained fusion reaction. Moreover, by striking the walls the plasma will knock off high Z (atomic number) material, which further enhances the cooling through radiation loss (bremsstrahlung is proportional to Z^2).

The answer to the confinement problem, as we have learned from elementary physics, lies in placing the plasma in a magnetic field since charged particles can be tied to the field lines much like "beads on a string." They cannot leave the string but they certainly can slide along it. This is in fact the essence of magnetic confinement. Two major confinement schemes are the open-ended or *mirror* machines and the closed-ended or *toroidal* devices.

Magnetic Mirrors

In its simplest form a magnetic mirror machine consists of an open cylinder around which current-carrying conductors are wrapped. If more coils per unit length are wrapped at the ends than at the middle, a field geometry emerges which confines particles between the ends where the magnetic field is strongest. These ends are referred to as mirrors since they reflect charged particles much the same way as optical mirrors reflect light. The field lines of a simple mirror are illustrated in Figure 1.

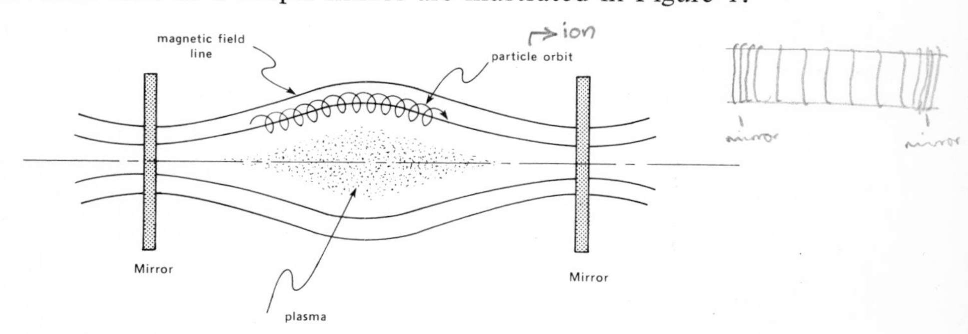

Figure 1. A simple mirror geometry.

The basis for charged particle confinement in magnetic mirrors is the adiabaticity (or constancy) of the magnetic moment of the particle defined as

$$\mu = \frac{W_\perp}{B} \tag{1.8}$$

where $W_\perp = \frac{1}{2}mv_\perp^2$ is the kinetic energy of the particle perpendicular to the field line, and B is the strength of the magnetic field. In the absence of (or until) interactions with other particles the total energy of the particle is also constant, *i.e.*,

$$W = W_\perp + W_\parallel = \text{constant} \tag{1.9}$$

where $W_{||} = \frac{1}{2}mv_{||}^2$ is the energy parallel to the magnetic field. Since, from (1.8), $W_\perp = \mu B$ and μ is constant it follows from (1.9) that

$$W_{||} = W - \mu B \tag{1.10}$$

We note that as B becomes large at the mirrors, μB becomes large and equal to W itself thus leaving no parallel energy, *i.e.*, the particle gets reflected. The same thing happens at the other mirror with the result that the plasma becomes confined–but not altogether. Those particles whose orbits are along the field lines at or near the center line of the device have a low magnetic moment and μB for them at the mirrors is not enough to cancel W. The result is that these particles escape confinement unless the field at the mirrors is made very large (effectively infinite to prevent leakage) which might be technologically impossible and/or economically prohibitive. The mirror machine is therefore inherently leaky, and because of this very characteristic many people seriously doubt its candidacy for a fusion reactor.

Nevertheless some ingenious ideas have recently been proposed to "save" the mirror; they deal primarily with the recovery of the enrgy lost by the escaping particles. Direct conversion of the energy of these particles to electricity appears to be both feasible and efficient, but much more work (as we will note in Chapter 9) needs to be done before this method can make the mirror system practical.

As for particle escape across the magnetic field, it may be reasonably inhibited if the field is made sufficiently strong so that the particle radius of gyration is small. This can be readily seen from the familiar expression for the radius of gyration of a charged particle in a magnetic field, *i.e.*,

$$\rho = \frac{v_\perp}{\omega_c} = \frac{\text{velocity} \perp \text{to field}}{\text{cyclotron frequency}} \tag{1.11}$$

and since

$$\omega_c = \frac{eB}{mc} \tag{1.12}$$

where e is the particle charge, and m is its mass, then it follows that

$$\rho \sim \frac{1}{B} \tag{1.13}$$

In this case (and for a stable plasma) particles can reach the container wall only through "classical" diffusion, *i.e.*, only collisions can give rise to a step-by-step diffusion (random walk) across the field lines. If this diffusion time is comparable or longer than the confinement time of the Lawson criterion, then the device in question will surely be a serious contender for a reactor. So far, unfortunately, no device has exhibited classical diffusion for reasons which we will examine shortly; instead, many have exhibited what is known as "Bohm" diffusion which occurs on a much faster time scale. In order to appreciate the difference in the two time scales it may be desirable to briefly sketch the dependence of the two diffusion coefficients on plasma parameters.

The classical diffusion coefficient for a charged particle across a uniform magnetic field can be written as (Glasstone and Lovberg, 1960)

$$D_{\perp} = \frac{\rho^2}{\tau} = \rho^2 \upsilon \tag{1.14}$$

where υ is the collision frequency, τ is the collision time, and ρ is the gyroradius given by equation (1.11). Since υ can be expressed in terms of the collision mean free path, λ, as (assuming $v_{\|} = 0$)

$$\upsilon = \frac{v_{\perp}}{\lambda} \tag{1.15}$$

then by substitution for υ and ρ in equation (1.14) we get

$$D_{\perp} = \frac{v_{\perp}^3}{\lambda \Omega^2} \sim \frac{v_{\perp}^3}{\lambda B^2} \tag{1.16}$$

The collision mean free path is inversely proportional to the collision cross section which for charged particles is the familiar Coulomb cross section σ_c. This cross section is in turn inversely proportional to the square of the energy of the relative motion so that $\lambda \sim T^2$ where T is the characteristic temperature for the particle velocity distribution. If we insert this information in equation (1.16) and again replace $v_{\perp}^3$ by $T^{3/2}$ we obtain

$$D_{\perp} \sim \frac{T^{-1/2}}{B^2} \tag{1.17}$$

which simply states that higher plasma temperatures and stronger fields are especially desirable for classical diffusion. Thus if magnetically confined plasma obeys the above classical diffusion, then indeed there should be no serious confinement problem if by some means we could raise B and/or T at will. The plasmas of interest have not followed classical diffusion.

It has not because of collective oscillations known as instabilities, the most serious of which we will examine shortly. These instabilities can give rise to turbulence which in turn can lead to enhanced diffusion across field lines characterized by "Bohm" diffusion. This diffusion has been observed in many experiments and the dependence of its coefficient on the important plasma parameters cannot be derived in the simple and straightforward manner used in connection with classical diffusion. Nevertheless it has been shown that

$$D_B = \frac{kT}{16e}\,\frac{1}{B} \sim \frac{T}{B} \tag{1.18}$$

which indicates that Bohm diffusion becomes larger (*i.e.*, worse) at higher temperatures and is less inhibited by larger magnetic fields than its classical counterpart because of its dependence on the first power of B. In fact it is reasonable to define a "confinement quality" parameter for fusion devices by dividing equation (1.17) by (1.18) or

$$D_\perp = \frac{D_B}{A} \tag{1.19}$$

so that when A=1 plasma diffusion goes à la Bohm. For fusion reactors we find that we must have $A \geqslant 100$, and we shall see later what value of this parameter some of the current promising devices have.

Toroidal Devices

It occurred to many people in the early days of fusion research that mirror end losses can easily be eliminated by simply bringing the two ends of the straight cylinder on themselves, forming a "torus" or a donut shaped device as shown in Figure 2.

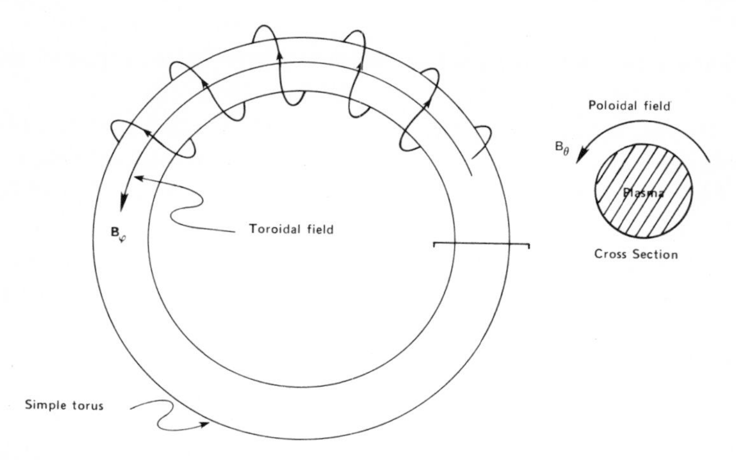

Figure 2. Toroidal system with and without magnetic well.

The confining magnetic field is then produced by also winding current-carrying coils around the torus as shown in the same figure. It might be noted that in the simple torus the plasma always sees a magnetic field curvature which is everywhere "concave" towards it. This feature is shared by mirror plasmas in the center of the machine (see Figure 1) and plays an important role in the subsequent discussion.

1.5 MAGNETOHYDRODYNAMIC (MHD) INSTABILITIES

It was pointed out earlier that plasma confinement necessary for fusion conditions has not been achieved because of instabilities. Individually, charged particles can be confined indefinitely by magnetic fields; but when many of them (as we must have in order to achieve desired power densities) are put together in the presence of electric and magnetic fields generated by their own charges and currents, they tend to move in unison in the form of collective oscillations. Under certain conditions these oscillations can grow to such large amplitudes as to destroy confinement in times much shorter than the necessary confinement time. In terms of the parameter A defined earlier, these processes give rise to values of A much smaller than 100, or equivalently they lead to particle escape across field lines in times much shorter than classical time. These processes are referred to as instabilities and they have constituted the major obstacle to controlled fusion thus far.

The most serious of them are the MHD instabilities which are effectively low frequency, long wave length oscillations that could lead to gross disassembly of the plasma. The "flute"-type MHD instability is driven by "bad" field curvature which is the concave curvature the plasma sees in toroids and in the center of mirrors as pointed out earlier. The relative size of the diffusion which the MHD instability can produce can be seen in Figure 3 where the MHD diffusion coefficient is shown relative to the others

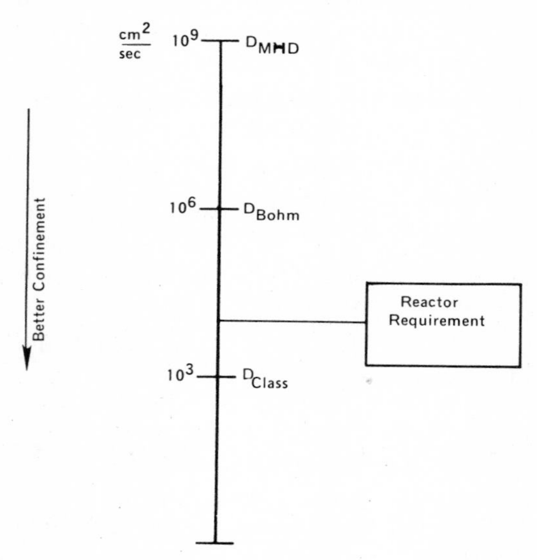

Figure 3. Various diffusion coefficients.

discussed. It has been called an MHD instability because it is analogous to the Rayleigh-Taylor instability in hydrodynamics where it has been known for many years that when the boundary of a heavy fluid supported by a lighter fluid is perturbed the system is unstable, *i.e.*, the perturbation grows in amplitude. The reason for that is that the system tends to seek a state of lowest potential energy and, in so doing, the "free energy" it possesses goes into the oscillations. This phenomenon is very similar to the familiar stability-instability concepts in mechanical systems illustrated in Figure 4.

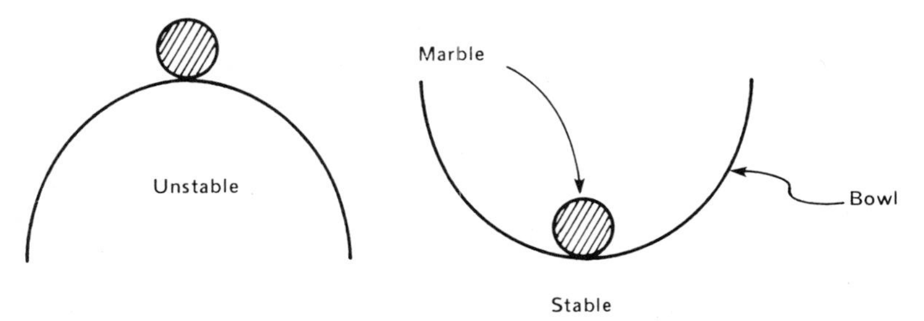

Figure 4. Elementary definition of stability and instability.

In the case of mirror machines the heavy fluid is the plasma; it is supported against a gravitational force (which in this case is the centrifugal force experienced by the particles in their motion along curved field lines) by (vacuum) magnetic field which can be thought of as a very light fluid with a unit mass of $B^2/8\pi$. Since the same bad curvature is seen by plasmas in simple toruses, it is clear that these devices are also susceptible to the "flute" MHD instability (Rosenbluth and Longmire, 1957).

The remedy to this dangerous instability was found in the early sixties when Ioffee and his co-workers in the Soviet Union realized that reversing the field curvature in the plasma region would stabilize the system against this mode. In the hydrodynamic counterpart this is equivalent to placing the lighter fluid on top of the heavy fluid which we know to be stable. By placing current-carrying conductors (Ioffee bars) along the mirror machine and letting the current flow in them alternately in opposite directions, the resulting magnetic field (including the simple mirror field) can be so tailored as to provide "good" or "convex" curvature over most of the plasma. The new field geometry has the shape of a magnetic "well" with the field strength being minimum at the center of the device. Along the axis of symmetry the field must be kept finite so as not to destroy the adiabaticity of the particle magnetic moment which, we recall, is essential for mirror confinement. By contrast to the simple mirror the plasma in a "minimum B" geometry sees a magnetic field whose strength increases outwardly in every direction with little or no chance of destroying its confining ability through MHD instabilities. The stabilizing and straight mirror fields are illustrated in Figure 5.

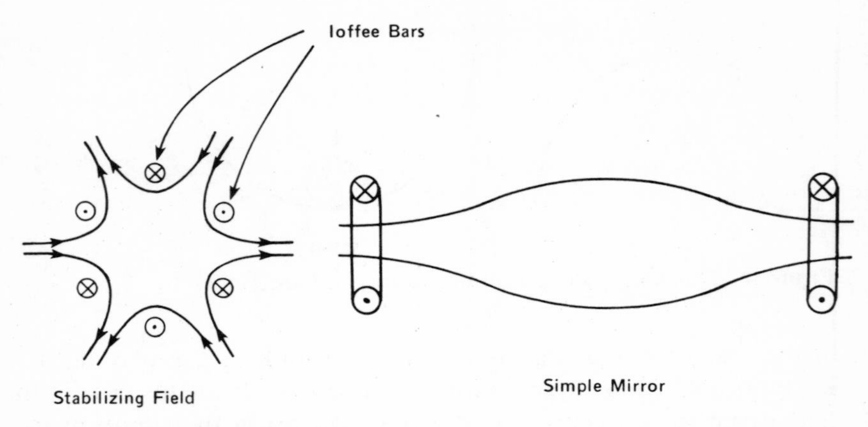

Figure 5. Mirror with Ioffee bars.

The 2X-II device at the Lawrence Livermore Laboratory is a mirror device which has shown marked improvement toward reactor conditions through the use of magnetic wells. This is illustrated in Figure 6 which shows a plot of plasma density versus time at different stages of the experiment. The lowest curve which

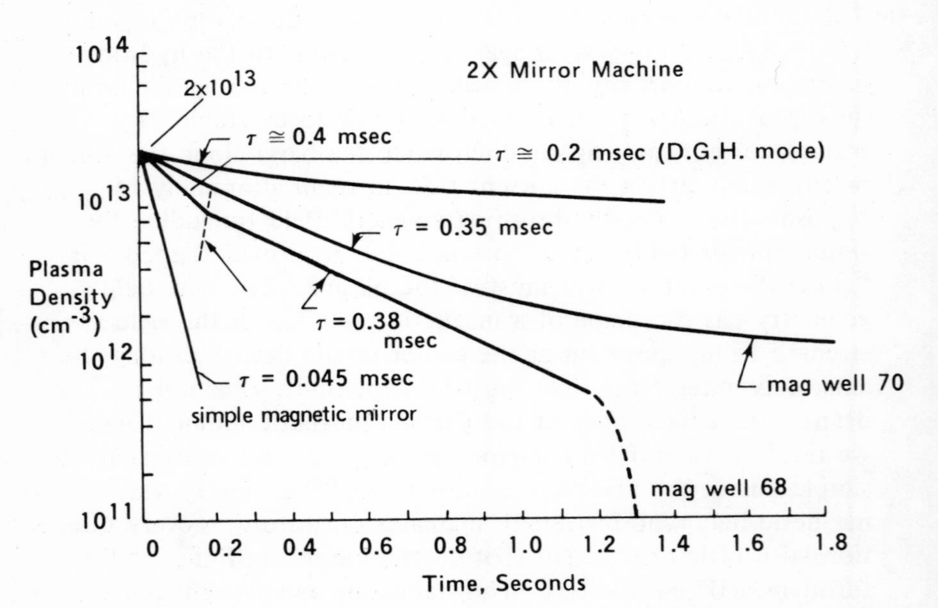

Figure 6. Plasma decay in 2X-II experiment at Livermore (Post, 1970).

corresponds to a simple mirror shows that the plasma decayed in about 45 microseconds. The remaining curves depict the improvement in confinement time due to magnetic wells and suppression of other types of instabilities yet to be discussed. The top curve represents plasma density vs. time for the case of classical diffusion.

The stabilizing effect of magnetic wells on some current fusion experiments and their status relative to reactor conditions as expressed by the Lawson criterion is shown in Table 2. We note from this data that the confinement time for the Russian Tokamak T-3 corresponds to $A \simeq 80$ or simply 80 Bohm times indicating a reasonably quiescent plasma. The plasma in this device is almost good enough for reactor; it is approximately a factor of 10 down in every category.

Table 2

Recent Fusion Experiments Data

	Tokamak T-3	*Mirror 2X-II*	*Theta Pinch Syla IV*	*Lawson Criterion*
Plasma density (n/cm^3)	2×10^{13}	5×10^{13}	5×10^{16}	5×10^{14}
Maximum plasma temp. (T_i keV)	0.6	8	3.2	10
Confinement time (msec)	25	0.4	0.01	200

1.6 MICROINSTABILITIES

The reason current fusion devices have not reached reactor conditions (in spite of universal adoption of magnetic wells) is that they suffer from other (than MHD) types of instabilities. This latter group is particularly sensitive to the detailed microscopic behavior of the plasma, and as a result they are generally referred to as microinstabilities. In contrast to the MHD (macroinstabilities) modes, these instabilities are not as disastrous, but they can lead to local turbulence and enhanced (Bohm) diffusion across field lines. They are effectively independent of the curvature of the magnetic field; rather they are driven by "anisotropy" in particle velocity distribution as well as spatial inhomogeneities in density and/or temperature.

For mirror machines it is essentially mandatory that the velocity distribution of the particles be anisotropic and that most of the particle energy be perpendicular to the magnetic field. This minimizes end losses. Such anisotropy is a source of "free energy" which drives instabilities in mirror devices at frequencies of the order of the ion cyclotron frequency [see equation (1.12)]. The Dory-Guest-Harris (1965) mode shown in Figure 6 as well as the "double-humped" instability of Hall, Heckrotte, and Kammash (1965), have been identified as some of the early impediments in 2X-II experiments. The latter can be corrected by skillfully eliminating the "cold" ion species knocked out of the vacuum chamber by the plasma. The first can be corrected by spreading the energy distribution of the injected ions (rather than injecting them monoenergetically). As a result the improvement in confinement from 1968 to 1970 emerged as shown in the figure.

In order to set up magnetic wells in toroidal devices one must superimpose on the toroidal field a *poloidal* field as demonstrated in Figure 2. This can be accomplished through external windings as in the Stellarator or by currents induced in the plasma as in Tokamak T-3. It may also be accomplished by placing current-carrying internal rings as in Levitrons and other so-called "multipole" devices. In Tokamak the current in the plasma serves yet another important purpose–that of heating the plasma through "ohmic" heating. Though quite successful thus far, this method of heating has limitations in that the plasma resistivity (main element in this process) decreases with increasing temperature. Moreover, ohmic heating tends to preferentially heat the plasma electrons; thus at higher temperatures radiation losses (by bremsstrahlung) can also increase. Hence this heating approach quickly reaches a point of diminishing returns and other means of heating the plasma to reactor conditions must be found. This is, in fact, one of the major areas of intense research, and it belongs to the "fusion technology problems" which must be solved before a fusion reactor can become a reality.

The combined (toroidal and poloidal) field in Tokamak (and other toruses) gives rise to a magnetic well–to counter MHD modes as well as to provide "shear" to combat the microinstability associated with the density gradient. This instability is usually referred to as the "universal" mode since laboratory plasmas universally possess density gradients. It is this instability which many believe to be the source of the anomalous (Bohm) diffusion seen in many fusion experiments, and whose absence in Tokamak is attributed to just the "right" amount of shear provided by the current in the plasma. These results have been duplicated in Princeton's Symmetric Tokamak (ST). They were further verified in another experimental device called the "spherator" which was designed to test independently the effects of magnetic wells and shear. Illustrations of some recent results from the spherator are shown in Figure 7.

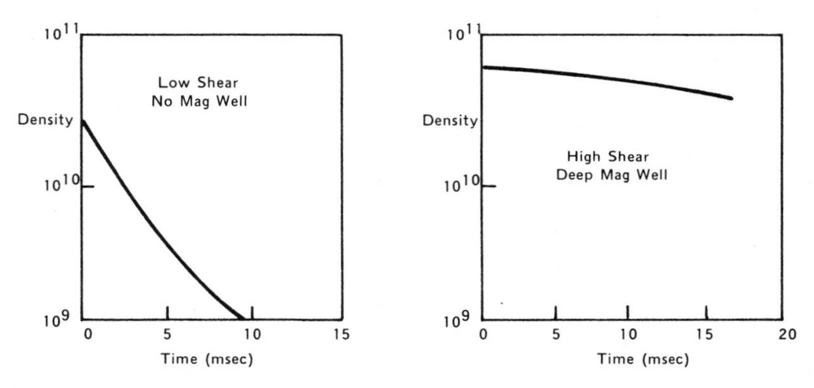

Figure 7. Stabilizing effects of shear and magnetic well.

1.7 WHENCE FUSION POWER FEASIBILITY

It is clear from the recent experimental results of 2X-II, T-3, model ST and others that fusion power feasibility may not be far from realization. The scaling laws from current experiments to reactor conditions will be tested in numerous experiments, soon to be constructed. Among them is PLT or Princeton Large Torus. If it and other devices (*e.g.*, the Tokamak Fusion Test Reactor at Princeton) perform as expected, there is no reason to believe that scientific feasibility will not be demonstrated by the early part of the next decade. This optimism is based on the much better understanding we now have of plasma physics than we had a decade or so ago.

A successful fusion power feasibility demonstration in the early eighties will not, of course, guarantee the possibility of power from controlled fusion. Only technological progress on a large set of engineering problems will. We refer to the many engineering and technological problems which must be solved before a fusion reactor can become a viable commercial energy source. Among these technological problems to be solved are:

- **A.** neutronics and blanket design.
- **B.** plasma fueling, heating and ignition,
- **C.** surface phenomena and radiation damage,
- **D.** engineering design of high magnetic field systems,
- **E.** energy storage and conversion,
- **F.** plasma impurity control, and
- **G.** economic and environmental aspects.

Research efforts in these areas of controlled fusion have already begun and are expected to accelerate in the near future. Although formidable, none of the technological problems are considered insurmountable and a demonstration plant is expected toward the end of the century. If these predictions materialize (and it is believed they will) a commercial plant should not be very far behind, *i.e.*, around the turn of the century.

The remaining chapters will be devoted to elementary but fairly extensive consideration of the major technological problems. An effort will always be made to present the material from basic principles so that at the end the reader will at least have some appreciation for the magnitude of these problems.

REFERENCES

1. Artsimovich, L. A. "Tokamak Devices," *Nuclear Fusion, 12,* 215 (1972).
2. Clarke, J. F. "Energy Containment and Scaling in Tokamaks," ORNL-TM-4585 (May 1974).
3. Coppi, B. and J. Rem. "The Tokamak Approach in Fusion Research," *Scientific American* (July 1972).
4. "Fusion Power–An Assessment of Ultimate Potential," Wash. 1239 (February 1973).
5. Galbraith, D. L. and T. Kammash. "Thermal Efficiency of Mirror Reactors with Direct Conversion," *Nuclear Fusion, 11,* 575 (1971).

6. Glasstone, S. and R. H. Lovberg. *Controlled Thermonuclear Reactions.* (New York: Van Nostrand, 1960).
7. Gough, W. C. "Fusion Energy and the Future," in *The Chemistry of Fusion Technology,* D. M. Gruen, Ed. (New York: Plenum Publishing Corp., 1972).
8. Gough, W. C. and B. J. Eastlund."The Prospects of Fusion Power," *Scientific American.* (February 1971).
9. Krall, N. A. and A. W. Trivelpiece. *Principles of Plasma Physics.* (New York: McGraw-Hill, 1973).
10. Lawson, J. D. *Proc. Phys. Soc. London B70,* 1 (1957).
11. Lidsky, L. M. "Fusion Power," *Technology Rev.* (January 1972).
12. Post, R. F. "Controlled Fusion Research and High Temperature Plasmas," *Annual Review of Nuclear Science, 20,* 509 (1970).
13. Post, R. F. "Controlled Fusion Research, An Application of the Physics of High Temperature Plasmas," *Rev. Mod. Phys., 28,* 338 (1956).
14. Quinn, W. E., E. M. Little, F. L. Ribe, and G. A. Sawyer. *Proc. of Culham Conference,* Vol. I, IAEA Vienna (1971).
15. Ribe, F. L. "Fusion Reactor Systems," Los Alamos Scientific Laboratory Report LA-UR 74-758 (1974).
16. Ribe, F. L., T. A. Oliphant, and W. E. Quinn. Los Alamos Scientific Laboratory Report LA-3294-MS (1965).
17. Rose, David J. "Controlled Nuclear Fusion Status and Outlook," *Science,* (May 21, 1971).
18. Rose, D. J. "Engineering Feasibility of Controlled Fusion," *Nuclear Fusion, 9,* 183 (1969).
19. Rose, D. J. and M. Clark. *Plasmas and Controlled Fusion.* (New York: MIT Press, Wiley, 1961).
20. Rosenbluth, M. N. and C. Longmire. *Annals of Physics, 1,* 20 (1957).

CHAPTER 2

BASIC PROCESSES AND BALANCES IN FUSION REACTORS

We recall from Chapter 1 that the first step in the road to fusion power is the demonstration of scientific feasibility of controlled fusion. This, effectively, amounts to experimental verification of the Lawson criterion. We will now derive this and other balance relations, examine their implications and, in so doing, discuss some of the fundamental plasma processes relevant to fusion reactors.

2.1 ENERGY BALANCE AND IGNITION TEMPERATURE

For a power-producing reactor we would expect the energy output to substantially exceed the energy input; but the minimum operating condition is obtained when these two energies are equal. The energy input must be sufficient to heat the plasma and compensate for the radiation losses, while the energy output includes that produced by the fusion reactions. For a hot plasma in a magnetic field the radiation losses are principally bremsstrahlung and cyclotron radiation; the latter, of course, does not exist in the case of inertial confinement. In the interest of completeness we will include both processes in the balance relation and see later under what conditions one can ignore the cyclotron radiation. If we designate the power dissipation from the plasma by radiation by P_{rad} which we note to be

$$P_{rad} = P_b + P_c = P_{brem} + P_{cyc} \tag{2.1}$$

and let the confinement time be τ then the energy loss by radiation from the plasma is τP_{rad}. If we now let P_R be the power realized from fusion reactions then the energy production becomes τP_R. For a plasma with an equal number of electrons and ions and at the same temperature, the energy input required to heat and maintain the plasma at temperature T during time τ is given by

$$E_{in} = 3nkT + \tau P_{rad} \tag{2.2}$$

where we have used $n_e = n_i = n$, $T_e = T_i = T$, and k is the Boltzmann constant. If we equate the energy input to the plasma with the energy output by fusion reactions, we get the Lawson criterion in one of its simpler forms, *i.e.*,

$$3nkT + \tau P_{rad} = \tau P_R \tag{2.3}$$

If on the other hand we assume that the total energy $(\tau P_R + E_{in})$ can be converted with an efficiency ϵ (to be discussed in Chapter 16) then the condition for the reactor to have a net power gain is $\epsilon(\tau P_R + E_{in}) > E_{in}$, *i.e.*,

$$\epsilon \frac{3nkT + \tau P_{rad} + \tau P_R}{3nkT + \tau P_{rad}} > 1 \tag{2.4}$$

or, in a more simplified form,

$$\epsilon \left[1 + \frac{P_R/3n^2kT}{\left(\frac{1}{n\tau} + \frac{P_{rad}}{3n^2kT}\right)}\right] > 1 \tag{2.5}$$

If we use the often cited efficiency of $\epsilon = 1/3$ then we can write

$$R = \frac{P_R/3n^2kT}{\frac{1}{n\tau} + \frac{P_{rad}}{3n^2kT}} > 2 \tag{2.6}$$

which effectively places conditions on the quantities ($n\tau$) and the temperature T. The above relation is generally referred to as the Lawson criterion although for many of the discussions to follow, it would be adequate to use the form given by equation (2.3) which we now write as

$$\tau P_R - \tau(P_b+P_c) = 3nkT \tag{2.7}$$

We proceed now to obtain expressions for the various terms in this equation.

2.2 BREMSSTRAHLUNG POWER

When an electron moves in the field of an ion it experiences an acceleration and as a result emits continuous x-radiation, referred to as bremsstrahlung. The energy radiated per unit time classically is given by

$$P_b = \frac{dE}{dt} = \frac{2}{3}\frac{e(eZ)a^2}{c^3} \tag{2.8}$$

where e is the electronic charge, eZ is the ion charge, and a is the acceleration. Since the lightest ion is much heavier than the electron, we may view the above interaction as occurring between a stationary ion and an electron moving with a velocity v at an impact parameter b as shown in Figure 8.

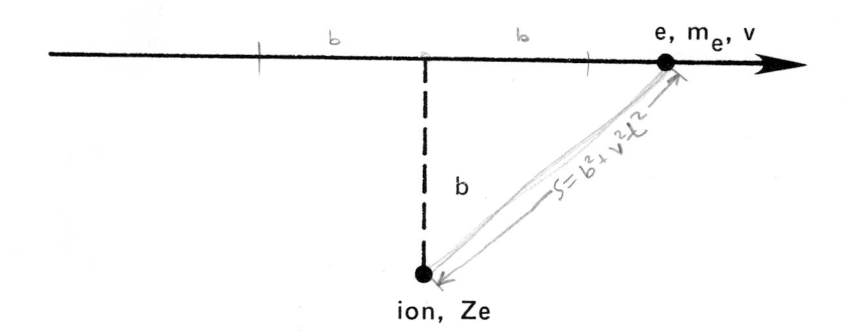

Figure 8. Coulomb collision between light and heavy charged particles.

We will assume that the electron path remains straight during the interaction for which the force is the familiar coulomb force e^2Z/b^2, and the corresponding acceleration is $a = e^2Z/m_eb^2$. This value of the force is the maximum since it corresponds to the closest distance "b," and if we assume that it persists during the time of interaction we can calculate the latter by equating the two impulses or the areas under the two curves shown in Figure 9.

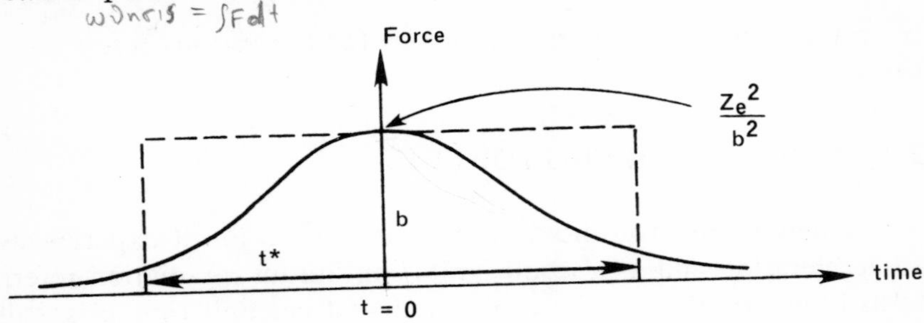

Figure 9. Variation of the coulomb force with time.

If we let t* be the effective interaction time, then

$$\frac{Ze^2}{b^2}t^* = \int_{-\infty}^{+\infty}\frac{Ze^2\ b\ dt}{(b^2+v^2t^2)^{3/2}} = \frac{2Ze^2}{bv}\int_{0}^{\infty}\frac{dx}{(1+x^2)^{3/2}}$$

$$= 2Ze^2/bv \qquad (2.9)$$

or

$$t^* = \frac{2b}{v}$$

When we substitute this result in equation (2.8) we get the energy radiated by an electron as a result of interaction with one ion or

$$E_{rad} = \frac{2}{3}\frac{e^2}{c^3}\frac{Z^2e^4}{m_e^{\ 2}b^4}\frac{2b}{v} = \frac{4}{3}\frac{Z^2e^6}{m_e^{\ 2}c^3b^3v} \qquad (2.10)$$

If in addition this result is multiplied by the electron flux n_ev and the ion density n_i, the result is the energy radiated per unit impact area for all electron-ion collisions in a unit volume of b.

The total bremsstrahlung power per unit volume will then be obtained by multiplying by the element of area $2\pi b\,db$ and integrating over all impact parameters. The result is

$$P_b = \frac{8\pi e^6 Z^2 n_e n_i}{3m_e^2 c^3} \int_{b_{min}}^{\infty} \frac{db}{b^2} \tag{2.11}$$

and if we choose b_{min} to be the de Broglie wave length of the electron, $\hbar/m_e v$, it further becomes

$$P_b = \frac{8}{3} \frac{\pi e^6 Z^2 n_e n_i}{m_e^2 c^3} \frac{m_e v}{\hbar} \tag{2.12}$$

For a Maxwellian velocity distribution of the electrons it is possible to replace v in the above equation by the thermal velocity, or

$$v = \sqrt{\frac{3kT_e}{m_e}}$$

so that the total radiated power can be readily approximated by

$$P_b \simeq \frac{8}{\sqrt{3}} \cdot \frac{\pi e^6 Z^2 n_i n_e}{\hbar c^3 \; m_e^{3/2}} \sqrt{kT_e} \tag{2.13}$$

A factor called the "Gaunt" factor is usually incorporated in the above expression to correct for quantum effects. At high temperatures this factor assumes the value $2\sqrt{3}/\pi$; so if we include it, allowing for more than one ion specie, and substituting the appropriate numerical values we get

$$P_b \simeq 5.35 \times 10^{-31} \; n_e \sum_i (n_i Z^2) T_e^{1/2} \; \frac{\text{watts}}{\text{cm}^3} \tag{2.14}$$

where the electron temperature is in kilovolts. For a (D-D) fuel cycle the above expression becomes

$$P_b(\text{D D}) \simeq 5.35 \times 10^{-31} \; n_D^2 \; T_e^{1/2} \; \frac{\text{watts}}{\text{cm}^3} \tag{2.15}$$

while for a 50-50 D-T mixture it becomes

$$P_b(\text{DT}) \simeq 2.14 \times 10^{-30} \; n_D n_T \; T_e^{1/2} \; \frac{\text{watts}}{\text{cm}^3} \tag{2.16}$$

2.3 CYCLOTRON (SYNCHROTRON) RADIATION

A rigorous evaluation of cyclotron radiation from a hot plasma is quite complicated and must take into account such factors as plasma transparency, absorption and reflection. We will first use the classical formula to estimate this power loss and establish its dependence on plasma parameters; then we will state the more exact relations which take into account the various factors mentioned above. Because of their heavy mass the radiation by the ions will be negligible and thus ignored as in the case of bremsstrahlung. For an electron in a constant magnetic field B the acceleration it experiences is simply the centripital acceleration v^2/R where R is the radius of gyration. Substituting this into equation (2.8) we obtain

$$P_c = \frac{2}{3} \frac{e^2}{c^3} \frac{v^4}{R^2}$$

but $R = v/\omega_{ce}$, where $\omega_{ce} = eB/m_e c$, is the cyclotron frequency which leads to

$$P_c = \frac{2}{3} \frac{e^4 B^2}{m_e^2 c^5} v^2 \qquad (2.17)$$

Once again for a Maxwellian distribution we can substitute the temperature and obtain for the cyclotron power per unit volume

$$P_c \simeq \frac{4}{3} \frac{e^4 B^2}{m_e^3 c^5} n_e k T_e$$

If we now assume that the plasma pressure is exactly balanced by the magnetic field pressure, *i.e.*,

$$\frac{B^2}{8\pi} = (n_i k T_i + n_e k T_e) \qquad (2.18)$$

and further assume that $n_i = n_e = n$, and $T_i = T_e$, we obtain

$$P_c \simeq \frac{64\pi e^4 k^2}{3 m_e^3 c^5} n_e^2 T_e^2 \qquad (2.19)$$

which may be written for T_e in kilovolts as

$$P_c \simeq 5 \times 10^{-32}\ n^2 T_e^2\ \frac{\text{watts}}{\text{cm}^3} \tag{2.20}$$

Comparison of this result with that of bremsstrahlung, equation (2.14), reveals that for low temperatures the latter dominates (and hence P_c may be neglected), whereas at high temperatures cyclotron radiation could substantially exceed bremsstrahlung because of its dependence on the square of the temperature.

When reflection from the vacuum wall is ignored and plasma transparency is taken into account, the energy loss rate due to synchrontron radiation by an electron becomes

$$\frac{dT_e}{dt} = -0.26B^2 T_e \left(1 + \frac{T_e}{204} + ...\right) K_\ell \tag{2.21}$$

(Rose, 1969) where the transparency coefficient K_ℓ depends on the electron temperature T_e and on the quantity $\mathcal{L}$ given by

$$\mathcal{L} = L\omega_p^2/c\omega_{ce} = CL\beta B/\frac{k(T_e+T_i)}{e} \tag{2.22}$$

In this expression L is the plasma size for an equivalent slab, perhaps equal to the radius for cylindrical plasma; $\omega_p = (4\pi ne^2/m)^{1/2}$ is the electron plasma frequency, ω_{ce} is the electron cyclotron frequency and $\beta = nkT/B^2/8\pi$ is the ratio of the plasma pressure to the confining magnetic field pressure. For 15 keV $< T_e <$ 50 keV, $10^3 < \mathcal{L} < 10^5$, the transparency coefficient assumes the value

$$K_\ell = 2.1 \times 10^{-3}\ T_e^{7/4}/\mathcal{L}^{1/2} \tag{2.23}$$

which when put into equation (2.21) shows the approximate dependence on T_e^4 which strongly inhibits T_e (or actually dT_e/dt) from rising. The power loss as given by equation (2.21) is further multiplied by a constant smaller than one when wall reflectivity is taken into consideration; but the interesting and useful fact is that synchrotron radiation is considerably reduced when $\beta \rightarrow 1$, and for this case the power radiated per unit area becomes

$$P_c = 9.7 \times 10^{-32}\ n^{3/2}\ T_e^{\,2}\ \frac{\text{watts}}{m^2} \tag{2.24}$$

The loss represented by this equation is truly negligible in all applications although attaining $\beta \approx 1$ in fusion systems is not a simple matter.

2.4 POWER FROM FUSION REACTIONS

We turn now to the calculation of energy production by fusion reactions assuming two ion species A and B. The power density is given by

$$\text{Power density} = \text{Reaction rate} \times \text{Energy/Reaction} \tag{2.25}$$

where the reaction rate may be written as

$$r = \int_{\vec{v}_1}\int_{\vec{v}_2} d^3v_1\, d^3v_2\, f^A(\vec{v}_1)\, f^B(\vec{v}_2)\, |\vec{v}_2 - \vec{v}_1|\, \sigma(|\vec{v}_2 - \vec{v}_1|) \tag{2.26}$$

Here f^A and f^B are the velocity distribution functions for ions A and B, respectively, and $\sigma(|\vec{v}_{rel}|)$ is the fusion reaction cross section. If we assume that the ions have Maxwellian distribution and that the two species are at thermal equilibrium with one another at temperature T, then we can write

$$r = n_A n_B \left(\frac{m_A}{2\pi kT}\right)^{3/2} \left(\frac{m_B}{2\pi kT}\right)^{3/2} \iint d^3v_1\, d^3v_2\, |\vec{v}_2 - \vec{v}_1| \times$$

$$\sigma(|\vec{v}_{rel}|)\, e^{-\frac{1}{2kT}(m_A v_1^{\,2} + m_B v_2^{\,2})} = n_A n_B \langle \sigma v \rangle \tag{2.27}$$

where n_A and n_B are the particle densities of A and B and m_A and m_B are the respective masses. In order to evaluate the above expression it is convenient to transform to center of mass coordinates; *i.e.*, we let

$$V = \frac{m_A v_1 + m_B v_2}{M = m_1 + m_2} = \text{velocity of center of mass}$$

$$\mu = \frac{m_A + m_B}{M} = \text{reduced mass}$$

and note that $d^3 v_1 \, d^3 v_2 = d^3 V \, d^3 v$ where $\vec{v} = \vec{v}_2 - \vec{v}_1$. If we substitute into equation (2.27) we get

$$r = n_A n_B \left(\frac{M}{2\pi kT}\right)^{3/2} \left(\frac{\mu}{2\pi kT}\right)^{3/2} \iint d^3 V d^3 v \; v\sigma(v) e^{-\frac{MV^2}{2kT} - \frac{\mu v^2}{2kT}}$$

and upon carrying out the $\vec{V}$ and angular integrations in $\vec{v}$ we further obtain

$$r = n_A n_B \left(\frac{\mu}{2\pi kT}\right)^{3/2} 4\pi \int v^3 \, dv \, \sigma(v) e^{-\frac{\mu v^2}{2kT}} \qquad (2.28)$$

The reaction cross section is the familiar "Gamow" cross section given by

$$\sigma(E_{rel}) = \frac{A}{E_{rel}} e^{-\frac{B}{\sqrt{E_{rel}}}} , \quad E_{rel} = \tfrac{1}{2}\mu v^2 \qquad (2.29)$$

where the constants A and B depend on the reaction. With this cross section the reaction rate becomes

$$r = n_A n_B \left(\frac{8kT}{\pi\mu}\right)^{1/2} \frac{A}{kT} \int \frac{dE}{kT} e^{-\left(\frac{E}{kT} + \frac{B}{\sqrt{E}}\right)} \qquad (2.30)$$

where E is understood to be the relative energy of motion. The above integral is of the form

$$\int_d^\infty dx \, e^{-x - a/\sqrt{x}} \simeq 2\sqrt{\frac{\pi}{3}} \left(\frac{a}{3}\right)^{1/3} e^{-3\left(\frac{a}{2}\right)^{2/3}}$$

whose approximate value is shown on the right-hand side. Making use of this approximation we can finally write

$$r = 4n_A n_B \sqrt{\frac{2}{3\mu}} \left(\frac{1}{2}\right)^{1/3} AB^{1/3} \left(\frac{1}{kT}\right)^{2/3} e^{-3\left(\frac{B^2}{4kT}\right)^{1/3}} \tag{2.31}$$

and if we now put in the appropriate constants, A and B for the D-D and D-T reactions, we get

$$r_{DD} = 2.3 \times 10^{-14}\, n_D^2 \left(\frac{1}{kT}\right)^{2/3} e^{-18.8/(kT)^{1/3}} \quad cm^2/sec$$

$$r_{DT} = 3.7 \times 10^{-12}\, n_D n_T \left(\frac{1}{kT}\right)^{2/3} e^{-20/(kT)^{1/3}} \quad cm^2/sec \tag{2.32}$$

In order to calculate the power density we must multiply the above expressions by the corresponding energy release per reaction. For the D-D reaction an average of 8.3 MeV energy is deposited in the reacting system, and at 10 keV temperature equation (2.32) gives $r \approx 8.6 \times 10^{-19}\ cm^3/sec$, so that the power density at that temperature becomes

$$P_{DD}(10\ keV) \approx 3 \times 10^{-31}\ n_D^2\ \frac{watts}{cm^3} \tag{2.33}$$

and at 100 keV we obtain

$$P_{DD}(100\ keV) \approx 10^{-29}\ n_D^2\ \frac{watts}{cm^3}$$

For the D-T reaction the energy remaining in the system is that of the alpha particle, *i.e.*, 3.5 MeV or 5.6×10^{-6} ergs so that the power density becomes

$$P_{DT}(10\ keV) \approx 6 \times 10^{-29}\ n_D n_T\ \frac{watts}{cm^3}$$

$$P_{DT}(100\ keV) \approx 4.5 \times 10^{-28}\ n_D n_T\ \frac{watts}{cm^3} \tag{2.34}$$

We now return to equation (2.6) and observe that both P_R [as given in equations (2.33) and (2.34)], and P_b [as given in equation (2.15)] are functions of n^2 so that R becomes a function of $n\tau$ and T only. We can plot R vs. T for various values of $n\tau$ and as we see below in Figure 10 only certain values of $n\tau$ satisfy the criterion.

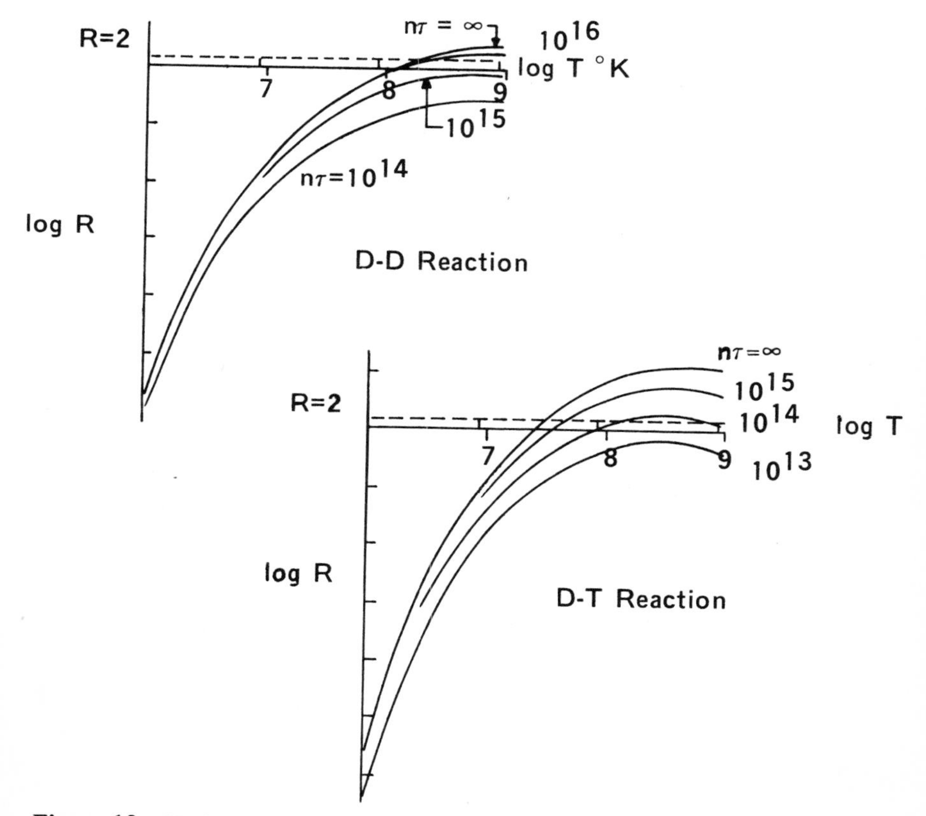

Figure 10. Variation of R with T for various rates of $n\tau$ for D-D and D-T reactions (Lawson, 1957).

It should be noted that in the above plots the cyclotron radiation was ignored since we have assumed that the bulk of the radiation loss is in the form of bremsstrahlung only.

An ideal ignition temperature—or simply the minimum operating temperature for a self-sustaining thermonuclear reactor—can be calculated by simply equating the radiation loss to the energy deposited in the system by the fusion reaction. If again we ignore the cyclotron radiation, we find that the ideal ignition temperatures for D-T and D-D are 4 and 36 keV, respectively.

It appears from the above considerations that since the rate for the fusion reactions increases monotonically (up to about 100 keV) with temperature, the higher the temperature the better. This is not so because any containment scheme for a fusion reactor will be pressure limited, *i.e.*, there is some maximum pressure above which, for one reason or another, it will be impossible to operate. From equation (2.18) we observe that the plasma kinetic pressure is balanced by magnetic field pressure. If this is the limiting pressure then we can write

$$n_{max} = \frac{P_{max}}{kT} \tag{2.35}$$

When the above is substituted in the expression for the reaction rate we obtain

$$r \sim \frac{\langle \sigma v \rangle}{T^2} \tag{2.36}$$

It is the quantity on the right-hand side, the effective reaction parameter, which is truly indicative of the fusion power level as a function of temperature. It is plotted in Figure 11 where the ignition temperature is also shown. The portion of the curve to the left of the ignition point represents the region where bremsstrahlung power exceeds fusion power.

2.5 PARTICLE BALANCE AND BURN-UP FRACTION

We have already seen that regardless of the confinement scheme, particles tend to escape confinement through collisions or other mechanisms. Thus the confinement time τ is an important parameter for fusion reactors. If for no other reason than to "fuel" the reactor, it is necessary to examine the particle balance in a fusion reactor and see what fraction of the particles undergo fusion reactions. If the number of ions injected per

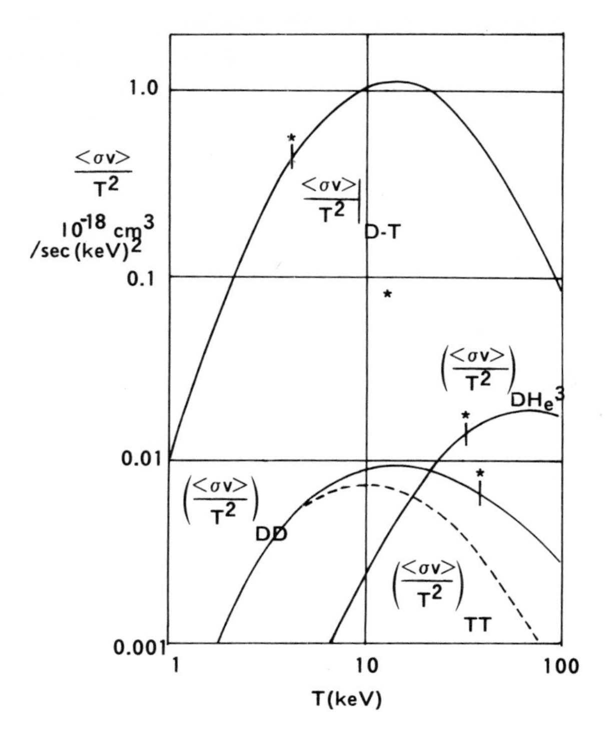

Figure 11. Reaction rate as a function of temperature (Mills, 1967).
* = ignition point = I-P. I-P for TT > 100 keV.

unit time (source) is S_i then the balance equation for the ions can be written as

$$\frac{dn_i}{dt} = S_i - \frac{n_i^2 \langle \sigma v \rangle}{2} - \frac{n_i}{\tau_i} \tag{2.37}$$

where τ_i is the ion lifetime against geometric escape, and the factor of 2 in the reaction term is there because two ions are lost per fusion reaction. For steady-state operation, *i.e.*, $dn_i/dt = 0$, and letting $L_i = n_i/\tau_i$, equation (2.37) becomes

$$0 = S_i - \frac{n_i^2 \langle \sigma v \rangle}{2} - L_i$$

If we now introduce the burn-up fraction f_b:

$$f_b = \frac{S_i - L_i}{S_i}$$

we see that it can be expressed as

$$f_b = \left[1 + \frac{2}{n\tau\langle\sigma v\rangle}\right]^{-1} \tag{2.38}$$

where $\langle\sigma v\rangle$ is that given by equation (2.27), and the subscript i has been dropped since it is not needed. The fractional burn-up is clearly a function of temperature through the term $\langle\sigma v\rangle$ and that is shown in Figure 12 where $n\tau$ from equation (2.7), the simplified Lawson criterion, is also shown for comparison.

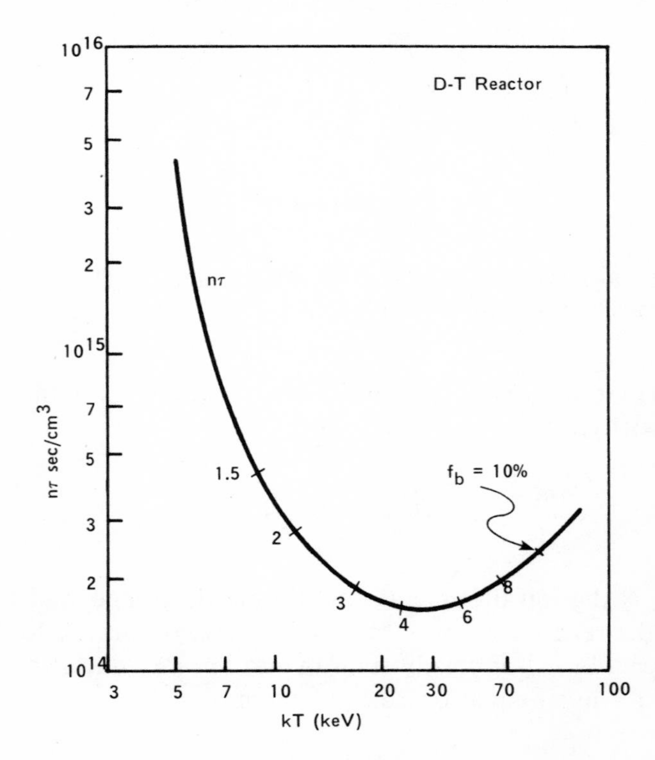

Figure 12. Burn-up fraction as a function of temperature for a D-T reactor.

We can readily see from this figure that in a steady state D-T reactor at a plasma temperature of 10 keV less than 2% of the ions burn, and that only about 8% undergo fusion at T $\approx$ 70 keV.

2.6 A MORE DETAILED PLASMA ENERGY BALANCE

In the discussion of the various balances presented earlier no attempt was made to identify the sources of energy received by the plasma. In this section we will reexamine the plasma energy balance and explicitly present the various energy inputs. We shall follow Rose (1969) and utilize his equations, but in subsequent chapters we will have an opportunity to examine in detail the derivation of these equations and the assumptions leading to the present approximate forms.

We begin with the familiar energy flow diagram shown in Figure 13 where the various species involved are designated with their appropriate parameters, *e.g.*, particle density, energy (or temperature), pressure, etc.

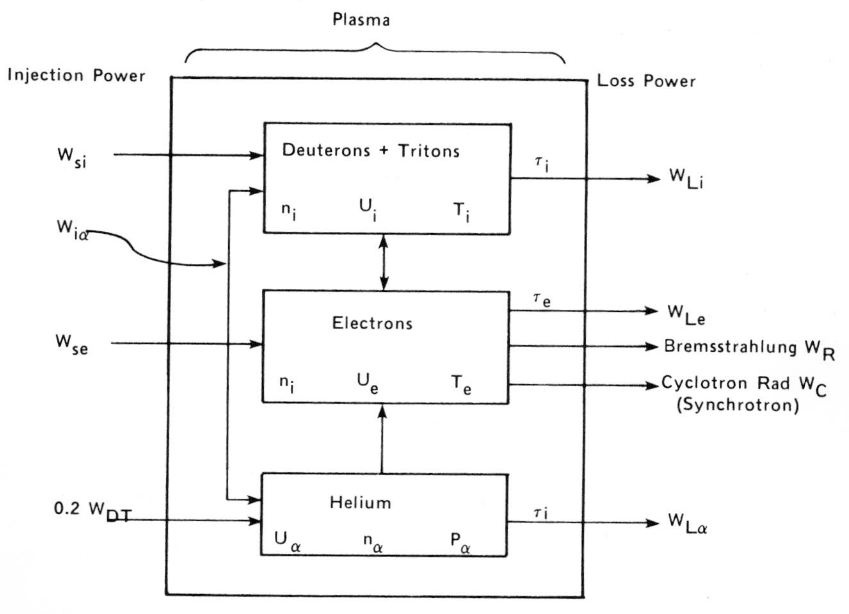

Figure 13. Energy flow diagram for D-T system.

We note that the ions and electrons are injected with power W_{si} and W_{se} per unit volume. This may be actual injection energy or energy received internally, and in the case of complete plasma heating of injected fuel $W_{si} = W_{se} = 0$. The ions can be heated by the α-particles and they can lose or gain energy from the electrons. Their geometric confinement time is τ_i and when they escape they carry with them W_{Li}. For the electrons of the plasma, they also can be heated by the α-particle, and in addition to energy loss W_{Le} due to confinement escape, they also lose energy through bremsstrahlung, W_x, and cyclotron radiation W_c. The alpha particles are assumed to have the same confinement time as the ions, *i.e.*, τ_i, and their power loss due to escape is $W_{L\alpha}$. Clearly, their energy comes from the D-T fusion reaction and it is about one-fifth of the total energy release, *i.e.*, 3.5 MeV.

The total energy content of the ions can be determined from the balance of terms shown in Figure 13. The equation is

$$\frac{d}{dt}\left(\frac{3}{2}n_iT_i\right) = 0 = W_{i\alpha} + W_{si} - W_{Li} - W_{ei} \qquad (2.39)$$

where each of these terms has already been defined. The last term on the right-hand side, which represents the rate of energy loss by the ions as result of collisions with the electrons, can be given by

$$\frac{dT_i}{dt} = \frac{T_i}{\tau_s}\left(1 - \frac{T_e}{T_i}\right) \qquad (2.40)$$

where τ_s is the thermalization time. This time is a function of the electron density $n_e = n_i = n$ and the electron temperature. If the numerical value is used, equation (2.40) becomes

$$\frac{dT_i}{dt} = -\frac{8 \times 10^{-19} n_e}{T_e^{3/2}} T_i\left(1 - \frac{T_e}{T_i}\right) \qquad (2.41)$$

where the temperatures are in keV. Multiplying both sides by 3/2 n_i we can write

$$W_{ei} = 1.2 \times 10^{-18} \frac{n^2 T_i}{T_e^{3/2}}\left(1 - \frac{T_e}{T_i}\right)\frac{\text{keV}}{\text{sec-m}^3} \qquad (2.41)$$

The rate of energy delivery to the ions from the α-particles is equal to the rate of production of α-particles by fusion reactions times that portion of their energy which goes to the ions. If we call this portion $U_{\alpha i}$, then

$$W_{i\alpha} = \tfrac{1}{2}(\text{fusion reaction rate}) \times U_{\alpha i} \tag{2.42}$$

The factor of ½ is included to allow for the fact that in a D-T reactor, D-D reactions also take place but do not produce α-particles. The reaction rate is simply $n_i^2/2 <\sigma v>$, so that equation (2.42) now becomes

$$W_{i\alpha} = \tfrac{1}{4} n_i^2 <\sigma v> U_{\alpha i} \frac{\text{keV}}{\text{sec m}^3} \tag{2.43}$$

The other positive term in the balance equation is the injected power W_{si}. If, once again, the source is S_i, and the energy per injected ion is V_i, then

$$W_{si} = S_i V_i = V_i \left[\frac{n}{\tau_i} + \frac{n^2 <\sigma v>}{2} \right] \frac{\text{keV}}{\text{sec m}^3} \tag{2.44}$$

where equation (2.37) has been used for S_i for a steady-state system. It is perhaps useful to rewrite the above equation in terms of the burn-up fraction f_b. From equation (2.38) we note that

$$S_i = \frac{n/\tau}{1 - f_b}$$

so that the injection power now becomes

$$W_{si} = V_i \left[\frac{n}{\tau(1-f_b)} \right] \tag{2.45}$$

The remaining loss term W_{Li} is simply the number of ions lost per unit time times the energy carried off per ion. If we ignore the reappearance of any part of T_i in the fusion products, then the number of ions lost by geometric escape can be set equal to the source; so that utilizing equation (2.37) we obtain

$$W_{Li} = \frac{3}{2} T_i S_i = \frac{3}{2} T_i \left[\frac{n}{\tau_i} + \frac{n^2 <\sigma v>}{2} \right] \frac{keV}{sec\ m^3} \tag{2.46}$$

Combining all these effects, the ion energy balance equation can be put in the form

$$<\sigma v> \left\{ U_{\alpha i} - 2V_i - 3T_i \right\} - \left\{ 6T_i - 4V_i \right\} / n\tau_i$$

$$- \frac{4.8 \times 10^{-18}}{T_e^{3/2}} T_i \left\{ 1 - \frac{T_e}{T_i} \right\} = 0 \tag{2.47}$$

We turn now to the electrons energy content and write

$$\frac{d}{dt} \left(\frac{3}{2} n_e T_e \right) = W_{\alpha e} + W_{se} + W_{ei} - W_{Le} - W_x - W_c \tag{2.48}$$

where by comparison to the ion balance equation we have the additional loss terms of bremsstrahlung W_x and cyclotron radiation W_c. The first four terms on the right-hand side are of the same form as the corresponding ion terms except now they contain electron parameters. The power received by the electron from the α-particles is again

$$W_{\alpha e} = \frac{n^2}{2} <\sigma v> U_{\alpha e} \frac{keV}{sec\ m^3} \tag{2.49}$$

where $U_{\alpha e}$ is the portion of the α-particle energy that goes to the electrons. As we shall see in subsequent chapters, at low temperatures the electrons receive most of the α-particle energy and as they get hotter the portion received by the ions increases. The electron injected energy is $S_e V_e$ and in steady state, utilizing equation (2.37), it becomes

$$W_{se} = \left[\frac{n}{\tau_e} + \frac{n^2 <\sigma v>}{2} \right] V_e \frac{keV}{sec\ m^3} \tag{2.50}$$

The energy received as result of collisions with ions, W_{ei}, and that lost due to geometric escape, W_{Le}, now becomes

$$W_{ei} = 1.2 \times 10^{-18} \frac{n^2 T_i}{T_e^{3/2}} \left(1 - \frac{T_e}{T_i} \right) \frac{keV}{sec\ m^3} \tag{2.51}$$

$$W_{Le} = \frac{3}{2} T_e \left[\frac{n}{\tau_e} + \frac{n^2 <\sigma v>}{2} \right] \frac{\text{keV}}{\text{sec m}^3} \tag{2.52}$$

Radiative loss rate due to bremsstrahlung per electron can be expressed as

$$\frac{dT_e}{dt} = -2 \times 10^{-21} \, n_i T_e^{1/2} \, \frac{\text{keV}}{\text{sec}} \tag{2.53}$$

which corresponds to equation (2.15) derived earlier. We can see this by multiplying the above equation by 3/2 n_e and applying the conversion factor 1keV/sec = 1.6×10^{-16} watts, *i.e.*,

$$W_X = 4.8 \times 10^{-37} \, n_i n_e T_e^{1/2} \text{ watts/m}^3$$

$$= 4.8 \times 10^{-37} \, n^2 T_e^{1/2} \tag{2.54}$$

If, as suggested by Rose, we include the adjustable parameter C_1 to account for the effect of changing bremsstrahlung, then

$$W_X = C_1 \frac{3}{2} n_e \frac{dT_e}{dt} = C_1 (3 \times 10^{-21}) n^2 T_e^{1/2} \frac{\text{keV}}{\text{sec m}^3} \tag{2.55}$$

Analogously, if we include a parameter C_2 to account for reflectivity from the wall, then using equation (2.21) the cyclotron radiation loss becomes

$$W_c = C_2 \frac{3}{2} n_e \left\{ 0.26 \, B^2 T_e \left(1 + \frac{T_e}{204} + \ldots\right) K_\ell \right\}$$

We now introduce the dimensional size parameter D, as

$$D = \beta^{3/2} (LB)^{1/2} \tag{2.56}$$

which if substituted in equations (2.22) and (2.23), and then in the above expression for W_c, yields

$$W_c = 6 \times 10^{-28} C_2 n^2 T_e^{11/4} (T_e + T_i)^{3/2} \left(1 + \frac{T_e}{204}\right) D \frac{\text{keV}}{\text{sec m}^3} \tag{2.57}$$

Combining these results and multiplying by $4/n^2$ the electron energy balance becomes

$$<\sigma v> \left\{ U_{\alpha e} + 2V_e - 3T_e \right\} - \left\{ 6T_e - 4V_e \right\} / n\tau_e$$

$$+ \frac{4.8 \times 10^{-18}}{T_e^{3/2}} T_i \left\{ 1 - \frac{T_e}{T_i} \right\} - 1.2 \times 10^{-20} C_1 T_e^{1/2}$$

$$- 2.4 \times 10^{-27} C_2 T_e^{11/4} (T_e + T_i)^{3/2} (1 + \frac{T_e}{204})/D = 0 \quad (2.58)$$

The above equation and equation (2.47) determine the equilibrium electron and ion temperatures T_e and T_i if the following quantities are specified *a priori*:

ion and electron injection energies V_i, V_e
burn-up fraction f_b
ratio of ion to electron confinement times $\tau_i/\tau_e = j$
bremsstrahlung parameter C_1
cyclotron radiation reflection adjustment C_2
size parameter D

Plots of the equilibrium temperatures vs burn-up fraction are shown in Figure 14 and 15 for various radiation conditions, and for plasma heating due to alpha particles only, *i.e.*, $V_e = V_i = 0$.

The solid curves in Figure 14 represent the variation of the equilibrium temperatures with f_b in the absence of cyclotron radiation, while the dotted ones are for much radiation. If bremsstrahlung is also absent then T_e and T_i will intersect at the temperature which corresponds to equipartition of the alpha energy or $U_{\alpha e} = U_{\alpha i}$. That temperature is approximately 33 keV and is indicated by the isolated point on both figures. At low fractional burn-up there is little energy to heat much fuel and consequently both T_e and T_i are low, but $T_e > T_i$ since $U_{\alpha e} > U_{\alpha i}$ in this region. At high f_b, say 0.1 or more, both species are heated considerably, and since in this region the electrons are sufficiently hot, more of the a-particle energy goes to the ions, *i.e.*, $U_{\alpha i} > U_{\alpha e}$ and therefore T_i exceeds T_e. The effect of easy cyclotron radiation escape on the equilibrium temperatures is also vividly demonstrated in Figure 14. At low

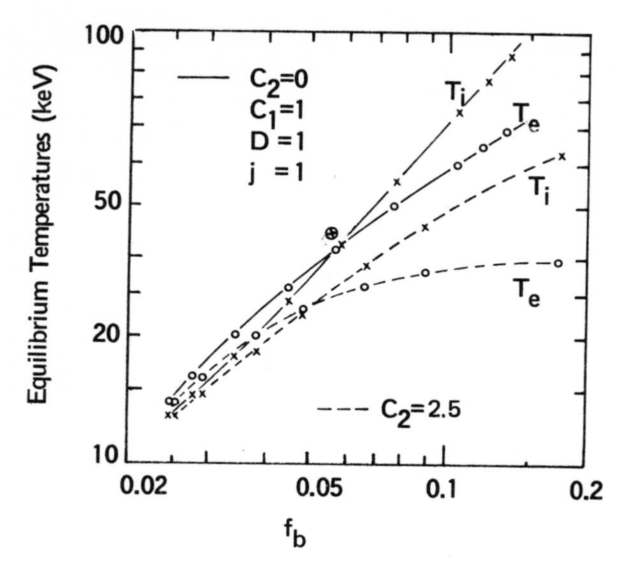

Figure 14. Ion and electron temperature as a function of burn-up for a self-heated plasma (Rose, 1969).

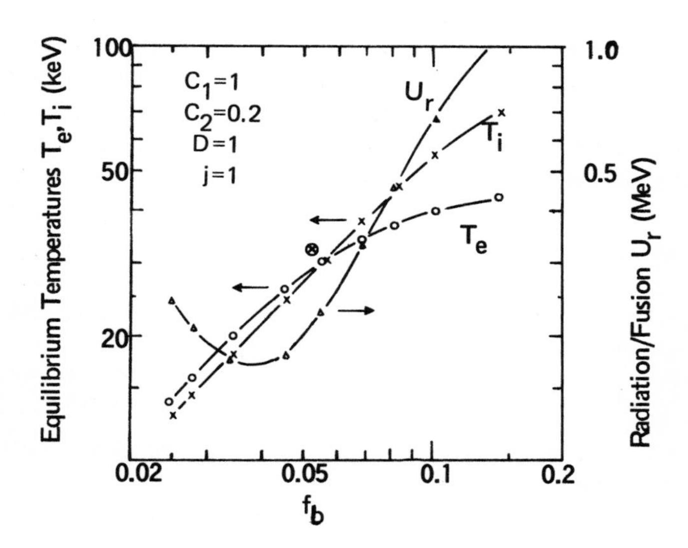

Figure 15. Ion and electron temperature and radiation in a self-heated plasma (Rose, 1969).

T_e, the effect is small since W_c [see equation (2.57)] is also small there. But at high f_b, T_e is kept down as seen from the values of T_e on the solid and dotted curves corresponding to f_b, say, of 0.1. The electron temperature drops from 49 keV at C_2=0 to 28.7 keV at C_2=2.5, which represents an optically thin plasma with low beta and poorly reflecting walls. This results in colder ions because the direct ion-electron energy transfer decreases with decreasing T_e, and also $U_{\alpha i}$ is lowered as T_e decreases. This is demonstrated in the figure where at $f_b = 0.1$, T_i goes from 64 to 39.5 keV, a larger drop than was suffered by the electrons.

An intermediate cyclotron radiation case is shown in Figure 15 where a plot of U_r, total radiation vs. f_b, is illustrated. This quantity which affects the heat load on the vacuum wall is minimum at $f_b = 0.04$ and increases on both sides of this point. At low burn-up and hence low temperature U_r consists mostly of the bremsstrahlung component U_x. As T_e decreases further the bremsstrahlung per fusion decreases, but U_x rises because the fusion rate decreases even faster. On the other hand, at higher $f_b (>0.04)$, U_x decreases and becomes a negligible part of U_r in comparison with U_c. In this region U_r is strongly dependent on temperature, and hence on fractional burn-up. It must be noted that this region is especially applicable to open-ended systems where the electron temperature is generally quite high. In closed systems (current systems) the electron temperature is small by comparison to its counterpart in open-ended systems, although it substantially exceeds that of the ions.

REFERENCES

1. Fermi, E. *Nuclear Physics,* Rev. ed. (Chicago: The University of Chicago Press, 1950).
2. Gamow, G. and C. L. Critchfield. *Theory of Atomic Nucleus and Nuclear Energy–Sources.* (Oxford, Eng.: Oxford University Press, 1949).
3. Glasstone, S. and R. H. Lovberg. *Controlled Thermonuclear Reactions.* (New York: Van Nostrand, 1960).
4. Lawson, J. D. *Proc. Phys. Soc. London B70,* 1 (1957).
5. Mills, R. J. *Nuclear Fusion, 7,* 223 (1967).
6. Post, R. F. "Controlled Fusion Research: An Application of the Physics of High Temperature Plasmas," *Rev. Mod. Phys., 28,* 338 (1956).
7. Rose, D. J. *Nuclear Fusion, 9,* 183 (1969).

CHAPTER 3

SOME ASPECTS OF THE NEUTRONICS IN FUSION REACTORS

We have seen that the major portion of the fusion energy in a D-T cycle is carried away by the neutron. In this regard energy extraction from a fusion reactor is similar to that of a fission reactor in that the neutron energy must somehow be extracted from the "blanket" through a thermal cycle. Moreover, the blanket in the fusion reactor serves the very important purpose of breeding tritium which, we recall, does not occur in nature and must be produced to provide "fuel." For these reasons, and also for estimating neutron irradiation effects within the various structural components, detailed information regarding spatial variation of neutron flux and energy is indeed essential. In their penetration of the various materials which constitute the vacuum wall and the blanket (see Figure 16) the neutrons might suffer elastic scattering, inelastic scattering, and absorption leading to various neutron-induced nuclear reactions and atomic displacements. A true assessment of neutron flux and energy as a function of position requires the use of transport theory and numerical codes. For our purpose here it will be sufficient to estimate these quantities using "diffusion" theory, where neutron scattering is assumed isotropic and the various cross sections are taken to be constant and independent of space and energy.

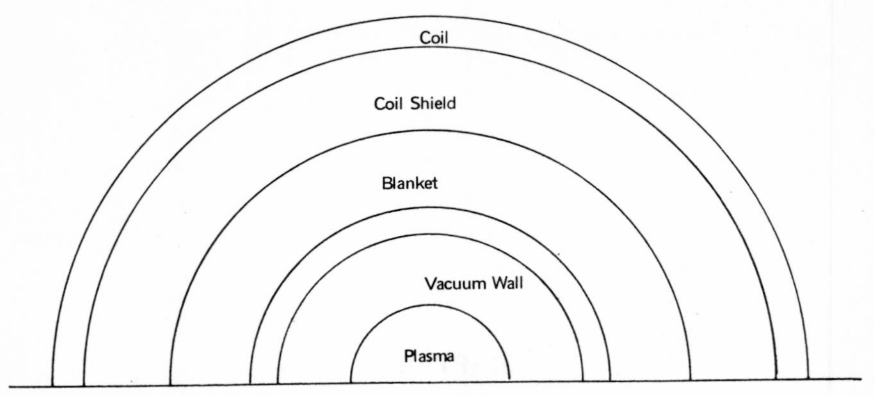

Figure 16. Cross section of fusion reactor.

3.1 NEUTRON DIFFUSION

Assuming the cylindrical geometry shown in Figure 17 we will consider the plasma as a line source emitting S neutrons per second per unit length in an infinite hollow cylinder whose inner

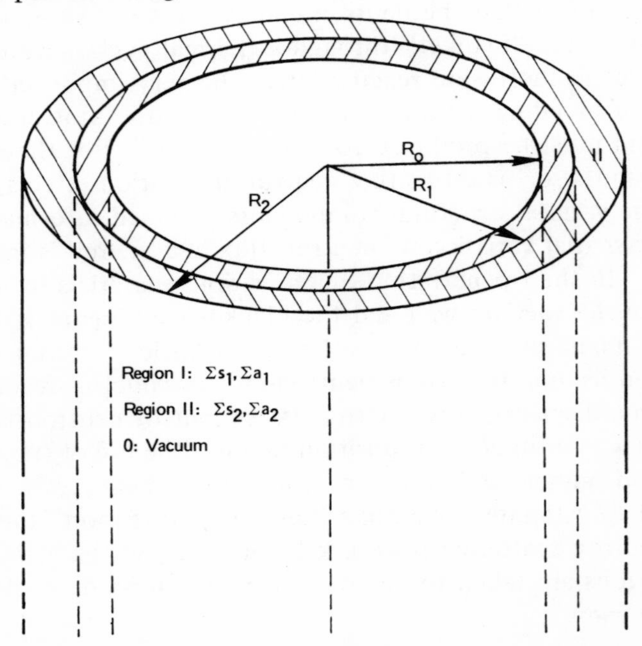

Figure 17. Neutron diffusion in cylindrical geometry.

radius is R_0 and outer radius R_2, consisting of two media designated by I and II. The first cylindrical medium with outer radius R_1 represents the vacuum wall while II represents the blanket; each possesses different cross sections. Since we would very much like the neutrons produced in the plasma to end up in the blanket where we hope to extract their energy, we must choose the material of the vacuum wall to have a very low neutron absorption cross section. The opposite must, of course, be true in the case of the blanket.

The familiar form of the diffusion equation is

$$D\nabla^2\phi - \Sigma_a\phi + S = \frac{\partial n}{\partial t} \tag{3.1}$$

where D is the diffusion coefficient, Σ_a is the macroscopic absorption cross section, ϕ = nv is the flux, n is the neutron number density, and S is the source. For steady state, and introducing the source as a boundary condition, the above equation becomes

$$\nabla^2\phi - \frac{\phi}{L^2} = 0 \tag{3.2}$$

where the diffusion length L has been used in accordance with the definition

$$\frac{1}{L^2} = \frac{\Sigma_a}{D} \tag{3.3}$$

Since we have assumed that the neutron source (*i.e.*, the plasma) is a line source, then we can write

$$\text{Source} = S\,\delta\,(r) \tag{3.4}$$

where S is the source strength in neutrons per unit length per sec and $\delta(r)$ is the Dirac delta whose properties are $\delta(r) = 1$, $r = 0$; $\delta(r) = 0$, $r \neq 0$. Because of cylindrical symmetry, equation (3.2) reduces to

$$\left(\frac{d^2}{dr^2} + \frac{1}{r}\frac{d}{dr} - \frac{1}{L^2}\right)\phi(r) = 0 \tag{3.5}$$

whose solution is

$$\phi(r) = A\, I_0(\frac{r}{L}) + BK_0(\frac{r}{L}) \tag{3.6}$$

where I_0 and K_0 are the modified Bessel functions of zero order. Applying this result to the regions of interest we get

$$\text{Region I:} \quad \phi_1(r) = A_1 I_0(\frac{r}{L_1}) + B_1 K_0(\frac{r}{L_1})$$

$$\text{Region II:} \quad \phi_2(r) = A_2\, I_0(\frac{r}{L_2}) + B_2 K_0\, (\frac{r}{L_2}) \tag{3.7}$$

where the constants of integration will be obtained from the following boundary conditions:

$$\text{A.} \quad \phi_2(R_2) = 0$$

$$\text{B.} \quad \phi_1(R_1) = \phi_2(R_1)$$

$$\text{C.} \quad -D_1 \frac{d\phi_1}{dr}\bigg|_{R_1} = -D_2 \frac{d\phi_2}{dr}\bigg|_{R_1}$$

$$\text{D.} \quad J_0\bigg|_{R_0} = -D_1 \frac{d\phi_1}{dr}\bigg|_{R_0} = \frac{S}{4\pi R_0} \tag{3.8}$$

The first two conditions represent continuity of flux across boundaries, and the last two represent continuity of the neutron current (or derivative of flux) across these boundaries. If we now define the constants α, β, and γ as

$$\alpha = \frac{K_1(x_{01})I_0(x_{11}) + I_1(x_{01})K_0(x_{11})}{I_1(x_{01})} \tag{3.9}$$

$$\beta = \frac{K_0(x_{12})I_0(x_{22}) - K_0(x_{22})I_0(x_{12})}{I_0(x_{22})} \tag{3.10}$$

$$\gamma = \frac{K_1(x_{11})I_1(x_{01}) - K_1(x_{01})I_1(x_{11})}{I_1(x_{01})} \tag{3.11}$$

where the subscript 1 on the Bessel functions indicates their order, and their arguments are given by

$$x_{ij} = \frac{R_i}{L_j}, \qquad \begin{matrix} i = 0,1,2 \\ j = 1,2 \end{matrix} \tag{3.12}$$

with

$$\frac{SL_1}{2\pi R_0 D_1} = S_0 \frac{L_i}{D_1} = S_0{}' \tag{3.13}$$

the constants of integration become

$$A_1 = \frac{SL_1}{2\pi R_0 D_1} \frac{1}{I_1(R_0/L_1)} + B_1 \frac{K_1(R_0/L_1)}{I_1(R_0/L_1)} \tag{3.14}$$

$$B_1 = \frac{S_0{}'}{a} \frac{I_0(x_{11})}{I_1(x_{01})} + \frac{\beta}{a} B_2 \tag{3.15}$$

$$A_2 = -B_2 \frac{K_0(R_2/L_2)}{I_0(R_2/L_2)} \tag{3.16}$$

$$B_2 = \frac{S_0}{I_1(x_{01})} \frac{\gamma I_0(x_{11}) + a I_1(x_{11})}{(D_2/L_2)\eta a - (D_1/L_1)\beta\gamma} \tag{3.17}$$

The quantity η is given by

$$\eta = \frac{K_0(x_{22})I_1(x_{12}) + K_1(x_{12})I_0(x_{22})}{I_0(x_{22})} \tag{3.18}$$

and if we note that

$$\gamma I_0(x_{11}) + a I_1(x_{11}) = K_1(x_{11})I_0(x_{11})K_0(x_{11})I_1(x_{11})$$

$$= \frac{1}{x_{11}} = \frac{L_1}{R_1} \tag{3.19}$$

then equation (3.17) may be written as

$$B_2 = S_0 \frac{L_1}{R_1} \frac{1}{I_1(x_{01})} \frac{1}{(D_2/L_2)\eta\alpha - (D_1/L_1)\beta\gamma} \tag{3.20}$$

For a diffusion material of high mass number the diffusion coefficient may be written as

$$D = \frac{\Sigma_s}{3\Sigma_t^2} \tag{3.21}$$

where Σ_t is the total macroscopic cross section, *i.e.*, the sum of the absorption and scattering cross sections. Substituting the above equation into equation (3.3) we obtain for the diffusion length

$$\frac{1}{L^2} = \frac{3\Sigma_a\Sigma_t}{\Sigma_s} \tag{3.22}$$

If we now assume that medium I (the vacuum wall) is a nearly pure scatterer, *i.e.*, $\Sigma_{a1} \simeq 0$, then $1/L_1 \rightarrow 0$; if medium II (the blanket) is taken as an almost pure absorber, *i.e.*, $\Sigma_{s2} \simeq 0$, then $D_2 \simeq 0$, and the solutions given by equation (3.7) will greatly simplify. These assumptions are quite drastic and make the validity of diffusion theory even more questionable. We will use them, however, in the interest of simplicity and numerical estimates. For medium I we can then write

$$\begin{aligned} I_0(x) &\xrightarrow[x\rightarrow 0]{} 1 + (x^2/4) \\ I_1(x) &\xrightarrow[x\rightarrow 0]{} x/2 \\ K_0(x) &\xrightarrow[x\rightarrow 0]{} -\ln x \\ K_1(x) &\xrightarrow[x\rightarrow 0]{} 1/x \end{aligned} \tag{3.23}$$

From equation (3.9) we also obtain

$$a \xrightarrow[x \to 0]{} \frac{2}{(x_{01})^2} - \ln x_{11} \simeq \frac{2}{(x_{01})^2} = \frac{2}{(R_0/L_1)^2} \qquad (3.24)$$

In the above equation both x_{01} and x_{11} are small, but only the first term is kept since it is the dominant one. For medium II, the argument of the Bessel functions tends to infinity because it is a highly absorbing material. Thus using the asymptotic expansions we obtain for β in equation (3.10) and η in equation (3.18),

$$\beta = \eta \simeq \frac{e^{-x_{12}}}{\sqrt{2\pi x_{12}}} \xrightarrow[x_{12} \to \infty]{} 0 \qquad (3.25)$$

while for γ, equation (3.11), which depends on the properties of medium I, we get

$$\gamma \simeq -\frac{x_{11}^2 - x_{01}^2}{x_{11} x_{01}^2} \qquad (3.26)$$

With these values, the constants of integration A_1 and B_1 become

$$B_1 \simeq \frac{S}{2\pi D_1} \quad ; \qquad A_1 = 0 \qquad (3.27)$$

which leads to the following expression for the neutron flux in the vacuum wall

$$\phi_1 = -B_1 \ln \frac{r}{L_1} = -\frac{S}{2\pi D_1} \ln \frac{r}{L_1} \qquad (3.28)$$

If we follow a similar procedure in connection with the quantities pertaining to the blanket, medium II, we will find that $A_2 = 0$, and

$$\phi_2 \simeq B_2 \sqrt{\frac{L_2}{2\pi r}} \; e^{-r/L_2} \qquad (3.29)$$

These two equations allow us in principle to estimate the nuclear heating in the blanket. As a simple illustration let us consider the neutron "heat load" incident on the vacuum wall.

This heating rate is computed from the incident neutron current which can be calculated directly from equation (3.28) and the fourth boundary condition in equation (3.8), or equivalently from the source current since the region between the source and the wall is pure vacuum. The current is

$$J_o = \frac{S}{2\pi R_o} \tag{3.30}$$

and if we use the dimensions used in a typical CTR design shown in Figure 18, *i.e.*, R_o = 200 cm, then the current becomes

$$J_o = \frac{S}{400\pi} \quad \frac{\text{neuts}}{\text{cm}^2\text{-sec}} \tag{3.31}$$

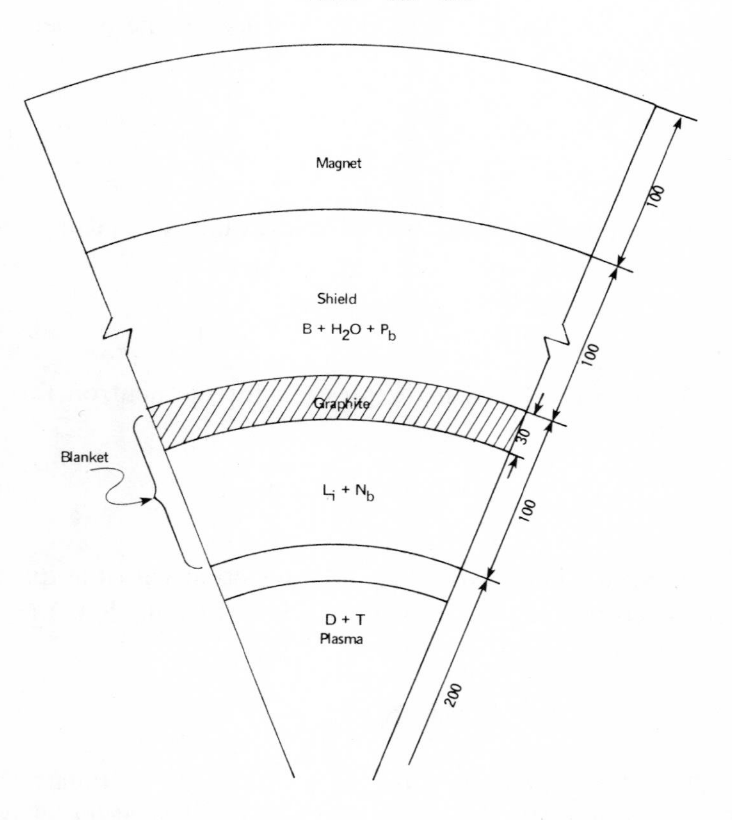

Figure 18. CTR design.

The energy of a neutron emitted in a D-T reaction is 14.5 MeV so that the energy current may be written as

$$J_E = \frac{S}{400\pi} \times 14.5 \times 10^3 \; \frac{\text{keV}}{\text{cm}^2\text{-sec}}$$

In order to calculate S let us consider a plasma column of 100 cm^2 area at a temperature of 10 keV. This area corresponds to a plasma radius of about 5 cm which is indeed very small compared to 200, and hence the assumption of a line source is not unreasonable. The number of neutrons emitted per cm per second by this plasma is

$$S = \tfrac{1}{4}\, n_i^{\,2} \langle\sigma v\rangle_{D\text{-}T} \times 100$$

At a temperature of 10 keV $\langle\sigma v\rangle_{D\text{-}T} \simeq 10^{-16}$ cm^3/sec and if the plasma density $n_i \simeq 10^{15}$/cm^3, then

$$S = \frac{10^{16}}{4} \; \frac{\text{neuts}}{\text{cm-sec}}$$

Substituting in the expression for the energy current and converting to watts/m^2 we find

$$J_E = \frac{10^{16}}{16 \times 10^2\,\pi} \times 14.5 \times 10^3 \; \frac{\text{keV}}{\text{cm}^2\text{-sec}}$$

$$= \frac{14.5}{16\pi} \times 10^{17} \; \frac{\text{keV}}{\text{cm}^2\text{-sec}}$$

$$= \frac{14.5}{16\pi} \times 10^{17} \times 10^{-16} \; \frac{\text{watts}}{\text{cm}^2}$$

$$= \frac{14.5}{16\pi} \times 10^5 \; \frac{\text{watts}}{\text{m}^2}$$

$$= 0.03 \text{ Megawatts/m}^2 = \frac{\text{MW}}{\text{m}^2}$$

A heating rate of 10 MW/m^2 incident on the vacuum wall has been suggested as an acceptable neutron thermal flux on the basis of an economic evaluation of fusion power (Steiner, 1970). If the above neutron heating is representative, then it is clear that it is well within the acceptable limit.

3.2 TRITIUM BREEDING AND DOUBLING TIME

We recall that fusion reactors which utilize the D-T fusion reaction must have the means to breed tritium. Lithium appears to be the only fertile element with any likelihood of producing the required tritium from interactions with neutrons. Natural lithium is composed of two stable isotopes Li^6 (7.4%) and Li^7 (92.6%), and both undergo neutron interaction resulting in tritium. These reactions are

$$Li^6\,(n,\alpha)T + 4.78\ \text{MeV}$$

and

$$Li^7\,(n,n'\alpha)T - 2.47\ \text{MeV}$$

where the second is an endothermic reaction which yields a slow neutron in addition to the alpha particle and tritium. The first reaction has a thermal neutron cross section of about 950 barns (barn = 10^{-24} cm^2), and the values of the cross section for neutron energies between 0.01 MeV and 11 MeV are shown in Figure 19.

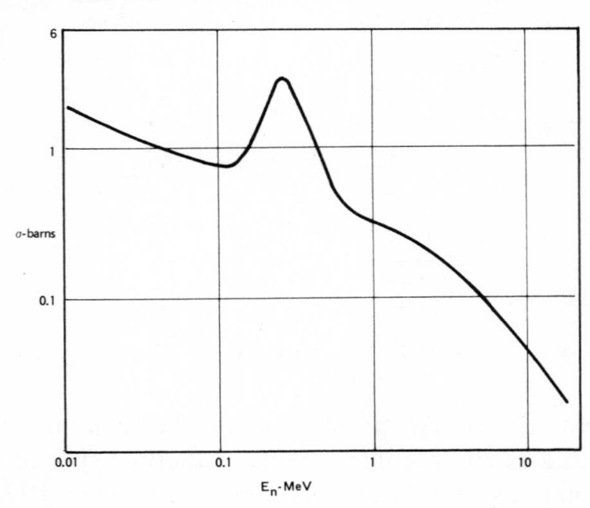

Figure 19. Li^6 (n,T) $H_e{}^4$ cross section vs neutron energy.

By contrast the Li^7 neutron reaction has a reasonable cross section for neutron energy above 4 MeV as illustrated in Figure 20. Therefore, it appears that fusion neutrons will have to be substantially moderated for their interaction with Li^6, while slowing down is not as crucial for their interaction with Li^7.

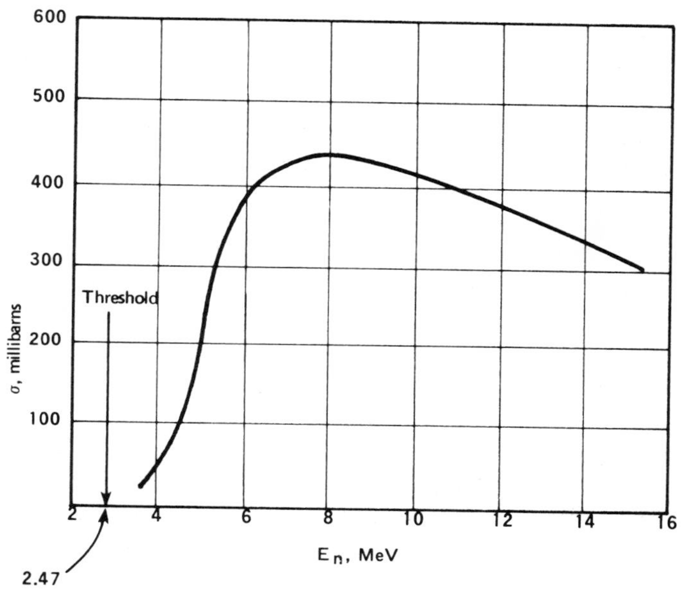

Figure 20. Li^7 (n,n$'\alpha$)T cross section vs neutron energy.

In order to calculate the breeding ratios in the blanket we first must examine the various interactions which the fusion neutron undergoes in the vacuum wall. Assuming that the wall is mainly niobium (Nb), the following reactions will take place for high energy neutrons (14 MeV). The probability of these interactions is indicated by the corresponding cross sections.

Nb Wall at Neutron Energy of 14 MeV

Reaction	Cross Section
${}_{41}Nb^{93}(n,n')_{41}Nb^{93m}$	?
${}_{41}Nb^{93}(n,2n)_{41}Nb^{92}$	1300 mb (max)

Reaction	Cross Section
${}_{41}Nb^{93}(n,2n){}_{41}Nb^{92m}$	400 mb
${}_{41}Nb^{93}(n,p){}_{41}Zr^{93}$	~40 mb
${}_{41}Nb^{93}[(n,d)+(n,p+n)]{}_{40}Zr^{92}$	336 mb
${}_{41}Nb^{93}(n,\alpha){}_{39}Y^{90}$	~10 mb

A proper evaluation of the breeding ratio requires a detailed analysis of the neutron distribution in both space and energy. Although quite simplified, the calculation we will carry out here is reasonably realistic and gives results which compare favorably with more exact calculations.

We shall assume a one-dimensional spatial variation and a two-group energy distribution for the neutrons in the reactor. In order to check with the more exact calculations of Steiner (1970), we shall use a similar blanket design which utilizes five regions, each of which is assumed to be homogeneous:

Region 1 - vacuum wall, 0.5 cm thick consisting of pure niobium;
Region 2 - vacuum wall coolant plus structure, 3.0 cm thick and consisting of 6% niobium plus 94% natural lithium;
Region 3 - second wall, 0.5 cm thick of pure niobium;
Region 4 - main coolant plus structure, 60.0 cm thick of 6% niobium plus 94% natural lithium;
Region 5 - moderator-reflector, 30.0 cm thick of graphite.

As an added simplification, we shall assume that all the neutrons which reach the graphite (Region 5) will be thermalized and reflected back into Region 4. The two energy regions to be considered are: (A) fast, covering the range from 14 MeV down to 4 MeV; and (B) thermal, covering energies below 4 MeV. The distinction between these two energy regions is due to the fact that above 4 MeV a neutron can react with the Li^7 isotope in the reaction

$$_0n^1 + {}_3Li^7 \longrightarrow {}_2He^4 + {}_1T^3 + {}_0n^1 - 2.47 \text{ MeV}$$

giving rise to a triton and a slower neutron. The significant rise in the cross section of the above reaction at 4 MeV is shown in Figure 20, while below about 3 MeV the absorption cross section drops to a negligibly small value. Figure 19 shows that the major contribution to neutron cross sections in the thermal energy region comes from the Li^6 via the reaction

$$_0n^1 + {}_3Li^6 \longrightarrow {}_2He^4 + {}_1T^3 + 4.78 \text{ MeV}$$

Thus, we shall make the drastic assumption that:

A. all neutrons resulting from the ${}_3Li^7(n, Tn')\alpha$ reaction fall immediately in the thermal energy range, and

B. every neutron which enters the thermal region will interact only through the $Li^6(n,T)\alpha$ reaction.

By insisting that all "thermal" neutrons interact only with Li^6 to give a triton, we do not need the cross section for this reaction, namely $\sigma = 950b$. However, we do need the cross sections for the various possible interactions in the high energy region. From Figure 20 we note that the cross section for $Li^7(n,Tn')\alpha$ reaction goes from 0.3b at 14 MeV to about 0.4b between 6 and 12 MeV dropping to 0.2b at 5 MeV and to virtually zero at 4 MeV (see also Chernilin and Yankov, 1970). Thus, a reasonable average value might be about 0.35b for the high energy region. The cross section for the (n,2n) reaction in niobium is, from the above reference, about 0.45b at 14 MeV, but drops rapidly to 0.07b at 10 MeV, and to zero shortly below that energy. The niobium absorption cross section, however, seems to be reasonably constant, at about 2.0b between 3 and 15 MeV. We shall therefore assume that for niobium, in the high energy region,

$$\sigma_{abs} = 2.0b$$

$$\sigma_{n,2n} = 0.4b$$

Steiner gives, among his results, 0.183 absorptions and 0.073 multiplications per incident neutron due to the structural material (Nb). The ratio of these two values should be just twice the ratio of the cross sections since each (n,2n) reaction gives a multiplication factor of two. Thus,

$$\frac{2\sigma_{n,2n}}{\sigma_{abs}} = \frac{0.073}{0.183} = 0.399$$

which is very closely satisfied by the above assumed values.

We shall assume that each neutron resulting from the (n,2n) reaction in Nb is born into the high energy region, so that the effective absorption cross section for niobium is

$$(\sigma_{abs})_{eff} = \sigma_{abs} - \sigma_{n,2n} = 1.6 \text{ barns}$$

With this value, we can now compute the macroscopic cross section of niobium, which has a density $\rho = 8.4$ gm/cm^3 and an atomic weight A = 92.91 gm/mole. It is

$$\Sigma_{Nb} = \frac{8.4}{92.91}(6.02 \times 10^{23})(1.6 \times 10^{-24}) = 0.08708 \text{ cm}^{-1}$$

The macroscopic tritium production cross section for natural lithium, which has a density $\rho = 0.534$ gm/cm^3, an atomic weight A = 6.94 gm/mole, and contains 92.6% Li^7 and 7.4% Li^6, is

$$\Sigma_{Li} = \frac{0.534}{6.94}(6.02 \times 10^{23})(0.35 \times 10^{-24})(0.926) = 0.01501 \text{ cm}^{-1}$$

In the pure niobium regions, namely Regions 1 and 3, the only macroscopically observable effect is a diminution of the neutron flux through absorption. In Regions 2 and 4, containing 94% lithium and 6% niobium, the macroscopic cross section for removal of neutrons from the high energy beam is

$$\Sigma_R = 0.94(0.01501) + 0.06(0.08708) = 0.01934 \text{ cm}^{-1}$$

Thus, for each neutron removed from the fast beam,

$$\frac{0.94(0.01501)}{0.01934} = 0.7298,$$

tritons are produced through the $Li^7(n,Tn')\alpha$ reaction and an equal number of slow neutrons are produced, which in turn give rise to one triton per slow neutron via $Li^6(n,T)\alpha$ reaction.

If we de note by ϕ_{01} the incident neutron flux on the vacuum wall, then the flux at the interface between Regions 1 and 2, namely ϕ_{12}, is given by

$$\phi_{12} = \phi_{01} \left\{ e^{-(0.08708)(0.5)} \right\} = 0.9574\ \phi_{01}$$

The flux of fast neutrons at the interface between Regions 2 and 3, ϕ_{23}, is then

$$\phi_{23} = \phi_{12} \left\{ e^{-(0.01934)(3.0)} \right\} = 0.9436\ \phi_{12} = 0.9034\ \phi_{01}$$

Of those neutrons absorbed in Region 2, a fraction, namely 0.7298, results in tritons through interaction with Li^7, and the slow neutrons resulting from these reactions give rise to an equal number of tritons as result of interaction with Li^6. Thus, we can write

$$T_2^{\ 6} = T_2^{\ 7} = 0.7298 \left\{ 0.9574\ \phi_{01} - 0.9034\ \phi_{01} \right\}$$

$$= 0.03939\ \phi_{01}$$

At the interface between Regions 3 and 4, the fast flux is

$$\phi_{34} = \phi_{23} \left\{ e^{-(0.08708)(0.5)} \right\} = 0.9574\ \phi_{23} = 0.8649\ \phi_{01}$$

and the flux entering the graphite moderator is then

$$\phi_{45} = \phi_{34} \left\{ e^{-(0.01934)(60.0)} \right\} = 0.3133\ \phi_{34} = 0.2710\ \phi_{01}$$

The tritium production in Region 4 via Li^6 and Li^7 interactions can, therefore, be written as

$$T_4^{\ 6} = T_4^{\ 7} = 0.7298 \left\{ 0.8649\ \phi_{01} - 0.2710\ \phi_{01} \right\}$$

$$= 0.4334\ \phi_{01}$$

There is an additional source of tritium, that resulting from the fast flux ϕ_{45} when it is thermalized and reflected back into Region 4. If we call this contribution $T_5{}^6$, then

$$T_5{}^6 = \phi_{45} = 0.2710\ \phi_{01}$$

and the total tritium production in Li^6 becomes

$$T^6 = T_2{}^6 + T_4{}^6 + T_5{}^6 = 0.7438\ \phi_{01}$$

The production from Li^7 is

$$T^7 = T_2{}^7 + T_4{}^7 = 0.4728\ \phi_{01}$$

so that the total tritium production rate is

$$T = T^6 + T^7 = 1.217\ \phi_{01}$$

Table 3 shows a comparison of these results (highly simplified) with those of Steiner's more exact calculations. Since we have neglected scattering from lithium and niobium, which should increase the transfer of neutrons from the fast group to the thermal group, hence increasing T^6 while reducing absorption and T^7, the present calculation should underestimate T^6 while overestimating T^7. Moreover, since the (n,2n) cross section for niobium drops very quickly as the neutron energy decreases, we have overestimated the neutron flux slightly by assuming that the neutrons produced by this reaction can themselves enter into (n,2n) reactions with niobium. On the other hand, some of the neutrons produced by the $Li^7(n,Tn')\alpha$ reactions will probably still be fast enough to initiate (n,2n) or $Li^7(n,Tn')\alpha$ reactions, so that in this respect we have underestimated the fast flux.

Table 3

Tritium Breeding

	Present Calculation	*Steiner's*
T^6	$0.7438\ \phi_{01}$	$0.90\ \phi_{01}$
T^7	$0.4728\ \phi_{01}$	$0.43\ \phi_{01}$
T	$1.217\ \phi_{01}$	$1.33\ \phi_{01}$

We turn now to the calculation of the tritium doubling time which is a very important element in a tritium-based power economy. If we assume that the tritium inventory in the plasma is negligible compared to that in the blanket, and call the latter I, then we can write for the time rate of change of this inventory

$$\frac{dI}{dt} = \frac{dI^+}{dt} - \frac{dI^-}{dt} \tag{3.32}$$

where the (+) and (-) indicate generation and consumption, respectively. Since the breeding ratio T is simply

$$T = \frac{dI^+}{dt} \Big/ \frac{dI^-}{dt} \tag{3.33}$$

and if we use the definition

$$\frac{di^-}{dt} = \frac{1}{I}\frac{dI^-}{dt} \tag{3.34}$$

then equation (3.32) can be written as

$$\frac{dI}{I} = \frac{di^-}{dt}(T-1)dt \tag{3.35}$$

Integrating this equation with the specific rate of tritium consumption (di^-/dt) treated as a constant, we obtain the doubling time t_2 as

$$t_2 = \frac{0.693}{(di^-/dt)(T-1)} \tag{3.36}$$

If we now use the definition of the specific rate of consumption, namely, the product of the specific power and the fusion-energy equivalent, and utilize the estimates:

Specific power of a fusion reactor	$\simeq$	500 MW (Th) per kg of tritium
Fusion-energy equivalent	$=$	1.54×10^{-4} kg of tritium consumed per MW (Th) per day

then equation (3.36) becomes

$$t_2 = \frac{2.47 \times 10^{-2}}{T-1} \text{ years} \quad (3.37)$$

On the basis of this equation, a fusion reactor operating with a breeding ratio of 1.3 obtains a doubling time of approximately 0.08 years. This extremely short doubling time would be very desirable in the early stages of fusion reactor deployment when the availability of tritium would be quite limited.

3.3 NEUTRON RADIATION DAMAGE IN REACTOR MATERIALS

In addition to the favorable contributions they make in power production and tritium breeding, fusion neutrons can cause serious radiation damage in the materials which make up the fusion reactor. These high energy neutrons collide with the atoms of the materials they traverse and knock them out of their lattices. The damage produced relates to the slowing down of the recoil or "knock-on" atoms. As they slow down to rest, these primary recoil atoms collide with other atoms of the target material often ejecting them from their normal lattice sites to slow down in their turn. In this process a cascade of displaced atoms is produced along with an equal number of vacant lattice sites. Because atomic collisions are inelastic, excitation and ionization of the electrons in the medium take place; but since these have only small kinetic energies, they are unable to displace atoms. In metals, at least, the electronic excitations are without influence on the damage, except insofar as they represent an energy loss from the displacement cascade. Depending on the target temperature and upon their concentration, the point defects produced in the displacement cascade may annihilate one another, agglomerate into clusters or extended defects of various kinds, or interact with impurity atoms or with previously existing dislocations. The surviving defects and their interactions govern the changes observed in the physical properties of neutron-irradiated materials. This bewildering variety of changes in the physical properties of materials is not very well understood and is yet controversial.

In this section we will attempt to examine in an elementary way the physics of atomic displacement and assess the incremental damage by estimating the number of displaced atoms by 14 MeV neutrons. Since the same phenomenon occurs in fission reactors a comparison will be made between the effects of a fusion neutron and a hypothetical fission neutron of 1.41 MeV energy. In the following calculations we will ignore the cascading effect and assume that the damage is due only to the primary knock-ons. We follow closely the work of Myers (1969) and begin by assuming that the scattering of the (1.41 MeV) thermonuclear neutrons by the target atoms is elastic and isotropic in the center of mass system. Then from the conservation relations we find that the neutron energy after collision, E, is given by

$$\frac{E}{E_n^\circ} = \frac{1}{2}\left[(1+\alpha) + (1-\alpha)\cos\theta\right] \tag{3.38}$$

where θ is the scattering angle in the C. M. System, E_n° is the incident neutron energy (14.1 MeV in our case), and

$$\alpha = \left(\frac{1-A}{1+A}\right)^2 \tag{3.39}$$

with A being the mass number of the target nucleus. The energy transfer to the knock-on is simply $\Delta E = 1-E$, or from equation (3.38)

$$\Delta E = \frac{E_n^\circ}{2}(1-\alpha)(1-\cos\theta) \tag{3.40}$$

We note that the maximum energy transfer occurs at $\theta = \pi$ with a value

$$(\Delta E)_{max} = E_n^\circ (1-\alpha) = \frac{4A}{(1+A)^2} E_n^\circ \tag{3.41}$$

while for $\theta = 0$, *i.e.*, glancing collision, the energy transfer is zero. If we substitute equation (3.41) into (3.40) we can rewrite the latter as

$$\Delta E = \frac{4A}{(1+A)^2} E_n^\circ \sin^2\frac{\theta}{2} = (\Delta E)_{max}\sin^2\frac{\theta}{2} \tag{3.42}$$

Following the development of Seitz and Koehler (1956), we shall consider the screened coulomb scattering of the knock-on in its lattice of stationary A atoms. The closest distance of approach in the collision of two charged particles Ze and ze is obtained by equating the potential and kinetic energies, *i.e.*,

$$\frac{Zze^2}{b} = \frac{1}{2} M_o V^2 \tag{3.43}$$

or

$$b = \frac{2Zze^2}{M_o V^2}$$

where M_o is the reduced mass and V is the relative velocity. In the case of the knock-on interaction with a similar stationary atom in the lattice, $M_o = M/2$ where M is the mass of A, and V is the velocity of the knock-on, so equation (3.43) becomes

$$b = \frac{4Z^2 e^2}{MV^2} \tag{3.44}$$

We introduce the screening parameter ζ which we define as

$$\zeta = \frac{b}{a} \tag{3.45}$$

where a is the screening radius of the coulomb field of the nucleus by the orbital electrons; it has the value

$$a = \frac{a_h}{\sqrt{2}\, Z^{1/3}} \tag{3.46}$$

with a_h being the Bohr radius of the hydrogen atom. For $\zeta \gg 1$ the screening is strong, and it is weak for $\zeta \ll 1$. Since classical collision theory is being employed, it is desirable to also introduce an approach parameter which reflects on the validity of this theory. It can be written as

$$K = \frac{b}{\lambda} \tag{3.47}$$

where λ is the de Broglie wave length of the incident particle. For $K \gg 1$ the classical theory of scattering is valid, as may be seen from the direct application of the uncertainty principle:

$$\Delta b \Delta p \geqslant \hbar$$

where Δb is uncertainty in the position of the target atom at impact parameter b. If we accept an uncertainty in the momentum equal to itself, *i.e.*, $\Delta p = p$, then

$$(\Delta b)_{min} = \frac{\hbar}{p} \tag{3.48}$$

For a stationary target the change in momentum is simply the momentum transfer which is equal to the impulse. From equation (2.9) we recall that the effective collision time is $2b/V$, so that

$$p = \frac{Z^2 e^2}{b^2} \frac{2b}{V} = \frac{2Z^2 e^2}{bV}$$

Putting this in equation (3.48), and realizing that $(\Delta b)_{min}$ is now the de Broglie wave length, we get

$$\lambda = \frac{\hbar V}{2Z^2 e^2} b$$

or

$$\frac{\lambda}{b} = \frac{\hbar V}{2Z^2 e^2}$$

For a classical collision theory to be valid we would expect λ/b to be very small or

$$\frac{b}{\lambda} = K \gg 1$$

which is the result given by equation (3.47).

The energy received by the lattice atom from the knock-on is simply that given by equation (3.42) which we will now rewrite using T as the energy, *i.e.*,

$$T = T_m \sin^2 \frac{\theta}{2} \tag{3.49}$$

with

$$T_m = \frac{4A}{(1+A)^2} E_n^\circ \tag{3.50}$$

being the maximum recoil energy. It is also of interest to determine the minimum energy that can be transmitted.

We note first that in the case of weak screening the classical scattering equations are appropriate down to values of the angle θ as small as ζ [see equation (3.45)], while for strong screening the minimum value of θ is ζ/K which is very small. For the purposes of calculations the values of these parameters for like atoms are

$$K = 2Z^2 \left(\frac{R_H M}{T_m m_e}\right)^{1/2} \tag{3.51}$$

$$\zeta = 4\sqrt{2}\,Z^{7/3} \frac{R_H}{T_m} \tag{3.52}$$

$$\zeta/K = 2\sqrt{2}\,Z^{1/3} \left(\frac{R_H m_e}{T_m M}\right)^{1/2} \tag{3.53}$$

where again M is the mass of the atom whose mass number is A, m_e is the electron mass, and R_H is the Rydberg constant for hydrogen equal to 13.54 eV. The numerical values are shown in Table 4 for incident fusion neutrons in a selection of appropriate reactor materials.

In obtaining these values the average, $T_m{}^* = T_m/2$, rather than the maximum energy transfer was used in equations (3.51-3.53). We observe from this table that K is increasingly larger than one, that ζ/K is indeed a small angle, and that there is a progression to strong screening as the knock-on energy decreases. Since our purpose is to determine the minimum energy transfer, we return to equation (3.49) and note that it becomes for small angles

$$T \simeq T_m \frac{\theta^2}{4} \tag{3.54}$$

which in turn becomes for weak screening

$$T_w = \frac{T_m}{4} \zeta^2 \sim 8Z^{14/3} \frac{R_H^2}{T_m} \tag{3.55}$$

Table 4

Screening Parameters

Material	*Z*	*A*	T_m* (keV)		*K*		ζ		ζ/K x 10^{-5}	
			fu	*fi*	*fu*	*fi*	*fu*	*fi*	*fu*	*fi*
Be	4	9	2540	254	9.5	30	7.6×10^{-4}	7.6×10^{-3}	8	26
C	6	12	2000	200	28	86	2.5×10^{-3}	2.5×10^{-2}	9	29
Al	13	27	970	97	280	-	0.030	0.30	11	35
Ti	22	48	565	56.5	-	-	0.18	1.8	13	41
Ni	28	59	465	46.5	-	-	0.39	3.9	14	45
Nb	41	93	296	29.6	-	-	1.5	15	16	51
Ta	73	181	155	15.5	-	1.3×10^{5}	11	10	19	61

fu = fusion, fi = fission

and for strong screening

$$T_s \simeq \frac{T_m}{4}\left(\frac{\zeta}{K}\right)^2 \simeq 2Z^{2/3} R_H \frac{m_e}{M} \qquad (3.56)$$

The numerical values of T_W and T_S are shown in Table 5 where the unity values in the T_W column are assumed to apply in the transition regime of $\zeta \sim 1$.

Table 5

Minimum Energy Transfer T_a

	T_W (eV)			T_s (eVx10^3)
	fu	*fi*	*0.1 fi*	
Be	0.38	3.8	38	4.3
C	3.2	32	1	4.1
Al	242	1	-	3.0
Ti	1	1	-	2.4
Ni	1			1.5
Nb	1			1.9
Ta				1.4

The differential scattering cross section for the collisions we have been discussing can be written as

$$d\sigma = 2\pi x dx = b^2 \pi \frac{V^2 M_o^2}{2M_2} \frac{dQ}{Q^2} \tag{3.57}$$

(Evans, 1955) where x is the impact parameter, dQ is the energy transfer to the target particle whose mass is M_2 and velocity V, M_o is the reduced mass and b is the closest distance of approach defined earlier. For like atoms, $M_o = M_2/2$, and the above expression becomes

$$d\sigma = \frac{\pi b^2}{4} (\tfrac{1}{2} M_2 V^2) \frac{dQ}{Q^2}$$

Now if we realize that $\frac{1}{2} M_2 V^2$ is the maximum energy that can be transferred, *i.e.*, T_m, then replacing Q by T we obtain

$$d\sigma = \frac{\pi b^2}{4} T_m \frac{dT}{T^2} \tag{3.58}$$

The total cross section for a collision in which the energy transfer is between T_m and T is obtained by integrating or

$$\sigma = \frac{\pi b^2}{4} \left(\frac{T_m}{T} - 1 \right) \tag{3.59}$$

Of special interest here is not only the minimum energies which can be transferred, *i.e.*, T_W and T_S, but also the minimum energy E_d needed to displace an atom to an interstitial position. This energy has been taken by Seitz to be 25 eV on grounds that it takes 6 eV to sublime an atom from the surface of a solid. The displaced atom will leave a vacancy in its lattice site, and the combined interstitial and vacancy is a "Frenkel" defect. We shall use 25 eV for all the materials considered here. For $T = E_d$ equation (3.59) gives the total cross section for this process or

$$\sigma_d = \frac{\pi b^2}{4} \frac{T_m}{E_d} \tag{3.60}$$

noting that for such displacements $T_m/E_d \gg 1$. Similarly for T_a we can also write

$$\sigma_a = \frac{\pi b^2}{4} \frac{T_m}{T_a} \tag{3.61}$$

which is applicable to all other coulomb collisions. In the above expression T_a can be T_w or T_s depending on the circumstances, but as we have already seen most of the collisions in the materials of interest tend to be of the strong screening type.

The average energy transmitted to the struck atoms in a displacement type of collision is given by

$$\bar{T}_d = \int_{E_d}^{T_m} T d\sigma_d / \sigma_d$$

which upon using equations (3.58) and (3.60) it further becomes

$$\bar{T}_d = E_d \ln \frac{T_m}{E_d} \tag{3.62}$$

In a similar manner we can also write

$$\bar{T}_a = T_a \ln \frac{T_m}{T_a} \tag{3.63}$$

Clearly, the energy loss per unit path by the knock-on in coulomb encounters is

$$\left.\frac{dE}{dx}\right|_c = n_o \int_{T_a}^{T_m} T d\sigma(T)$$

where n_o is the density of target atoms. With equation (3.58) the above integral can be readily carried out to give

$$\left.\frac{dE}{dx}\right|_c = 2\pi n_o \frac{Z^4 e^4}{MV^2} \ln \frac{T_m}{T_a} \tag{3.64}$$

where use was made of equation (3.44) for the closest distance of approach, and noting that E and T_m are equivalent for like atoms. The corresponding expression for energy per unit path lost in producing displacements is

$$\left.\frac{dE}{dx}\right|_d = 2\pi n_o \frac{Z^4 e^4}{MV^2} \ln \frac{T_m}{E_d} \qquad (3.65)$$

and therefore the fraction of the energy dissipated in collisions which give rise to displacements is simply the ratio of equation (3.65) to (3.64) or

$$R_d = \frac{\ln T_m/E_d}{\ln T_m/T_a} = \frac{\ln E/E_d}{\ln E/T_a} \qquad (3.66)$$

In addition to the elastic scattering—some of which may result in atomic displacements— the knock-on loses energy in ionizing the medium. We shall presume that above a transition energy, E_t, ionization will be the predominant process in the slowing down of the knock-on. Seitz and Koehler find that the interaction which is most important to the loss of energy by the knock-on due to ionization of the conduction electrons in the target materials is that due to the coulomb field of weakly screened charges. They find that the energy loss per unit path for this process is given by

$$\left.\frac{dE}{dx}\right|_e = 2\pi n_o \frac{Z^2 e^4}{m_e V^2} Z \left(\ln \frac{\epsilon}{\epsilon_o} + 1.08\right) \qquad (3.67)$$

for ϵ larger or comparable to ϵ_o, the energy of the Fermi level electrons. In this expression Z is the number of conduction electrons per atom, ϵ is the energy of an electron having the same velocity as the knock-on, *i.e.*,

$$\epsilon = \frac{m_e E}{M} \qquad (3.68)$$

and ϵ_o is given by

$$\epsilon_o = 3^{2/3}\ \pi^{4/3}\ \frac{\hbar}{2m_e}\ \rho^{2/3} \tag{3.69}$$

with ρ being the number of electrons per unit volume. The Fermi energy can be written in this case as

$$\epsilon_o = (3\pi^2\ n_o Z)^{2/3}\ (a_h)^2\ R_H \tag{3.70}$$

and it is of the order of 10 eV. In the regime of $\epsilon \ll \epsilon_o$ the loss rate becomes

$$\frac{dE}{dx} \simeq 2\ n_o\ \sigma\ \epsilon \tag{3.71}$$

with σ being the total scattering cross section. This has been further approximated by

$$\left.\frac{dE}{dx}\right)_e = (12\pi)\ (76.8)\ (n_o z\epsilon a_h)^2\ Z^{2/3} \tag{3.72}$$

It has been suggested (Kinchin and Pease, 1955) that the transition energy, E_t, below which the ionization loss mechanism becomes unimportant is somewhat less than the energy of the electrons at Fermi. It takes the value

$$E_t = \frac{1}{16}\ \frac{M}{m_e}\ \epsilon_o \tag{3.73}$$

which along with ϵ_o is shown in Table 6 for the materials of interest.

With these results we can now obtain the ratio R of the ionization loss rate to the collision rate: for $\epsilon \gtrsim \epsilon_o$.

$$R(E)_u = \left.\frac{dE/dx)_e}{dx/dx)_c}\right|_{\text{upper}} = \frac{zM}{m_e Z^2}\ \frac{\ln \frac{m_e E}{M\epsilon_o} + 1.08}{\ln E/T_a} \tag{3.74}$$

Table 6

Fermi and Transition Energies

Material	ϵ_o (eV)	E_t (keV)
Be	14.5	15
C	8.6	12
Al	11.7	36
Ti	5.4	30
Ni	8.3	56
Nb	5.2	55
Ta	5.3	110

and for the lower range, $\epsilon \ll \epsilon_o$

$$R(E)_L = 227 \frac{m_e Z}{MZ^{10/3}} \left(\frac{E}{R_H}\right)^2 \frac{1}{\ln E/T_a} \tag{3.75}$$

Of special interest, of course, is the portion of the knock-on energy above E_t which is associated with coulomb (non-ionizing) collisions. That fraction is simply,

$$f_c = \frac{dE/dx)_c}{dE/dx)_c + (dE/dx)_e} = \frac{dE/dx)_c}{(dE/dx)_{total}}$$

and using equation (3.74) it becomes

$$f_c = \frac{1}{1 + R(E)} \tag{3.76}$$

The associated collision energy can therefore be written as

$$E_c = \int_{E_t}^{E} \frac{dE}{1+R(E)} + E_t, \qquad E > E_t \tag{3.77}$$

so that that portion of it which is associated with displacements can be readily given by

$$E_{dis} = R_d E_c \tag{3.78}$$

It seems reasonable to expect that the average number of atoms displaced by a knock-on of energy $\bar{E}$ would be

$$\bar{\upsilon} \simeq \frac{\bar{E}}{2E_d} + 1 \tag{3.79}$$

as discussed by Kinchin and Pease. A more refined representation is given by Seitz and Koehler to be

$$\bar{\upsilon} = (0.107 + 0.561 \ln \frac{T_m}{E_d}) \frac{1}{1 - E_d/T_m} \tag{3.80}$$

for $E/E_d \geqslant 4$. In the case of like atoms, $T_m = E$, and E is greater than $\sim 10^3$, so that

$$\bar{\upsilon} \simeq 0.561 \ln \frac{E}{E_d} \simeq 0.561 \frac{\bar{E}}{E_d} \tag{3.81}$$

The total number of displaced atoms is the number of pockets of energy $\bar{E}$ which are available times $\bar{\upsilon}$ or

$$N = \frac{E_{dis}}{\bar{E}} \bar{\upsilon} = 0.561 \frac{R_d E_c}{E_d} \tag{3.82}$$

In order to calculate this quantity for the materials of interest, equation (3.77) must first be computed for E_c so that the result can be inserted in the above equation. This is done numerically, and Table 7 shows the input for these calculations. Equation (3.74) is first used for R(E) down to that value of E, namely E_u, for which $R(E)_u$ vanishes. Equation (3.75) is subsequently used for R(E) from E_u to E_t, and the results for these two steps are shown in Tables 8 and 9, respectively.

We observe that E_{cs_L} is increasingly closer to E_u-E_t from T_i to T_a. This is primarily due to equation (3.72) and indicates that the choice of the value for E_t is not very critical for these

Table 7

Input Data

Material	*Z*	*A=M*	*z*	$\epsilon_o(eV)$	$\epsilon(eV)$ *fu*	$\epsilon(eV)$ *fi*	*E(keV)* *fu*	*E(keV)* *fi*
Be	4	9	2	14.5	15.5	15.5	2540	254
C	6	12	1	8.6	91.0	9.1	2000	200
Al	13	27	3	11.7	19.7	2.0	970	97
Ti	22	48	1	5.4	6.4	0.64	565	56.5
Ni	28	59	1.2	8.3	4.3	0.43	465	46.5
Nb	41	93	1	5.2	1.7	0.17	296	29.6
Ta	73	181	1	5.3	0.47	0.05	155	15.5

Table 8

Superthreshold Energy
E_{cs_u} due to $R(E)_u$

Material	E_u *(keV)*	T_a *(eV)*	$(E\text{-}E_u)$ *(keV)* *fu*	$(E\text{-}E_u)$ *(keV)* *fi*	E_{cs_u} *(keV)* *fu*	E_{cs_u} *(keV)* *fi*
Be	80	$9.6\text{x}10^5\ E^{-1}$	2460	174	7.2	1.3
C	70	$6.3\text{x}10^6\ E^{-1}$	1930	130	8.8	1.8
Al	300	$2.3\text{x}10^8\ E^{-1}$	670	-	5.2	0
Ti	165	1	400	-	42	0
Ni	300	1	165	-	43	0
Nb	-	-	-	-	0	0
Ta	-	-	-	-	0	0

Table 9

Superthreshold Energy
E_{cs_L} due to $R(E)_L$

	Ta (eV)	E_u-E_t (keV)		E_{cs_L} (keV)	
		fu	fi	fu	fi
Be	$9.6 \times 10^5 \ E^{-1}$	65	65	0.2	0.2
C	1	58	58	2.3	2.3
Al	1	264	64	14.9	9.8
Ti	1	135	26	98	25
Ni	1	244	-	160	0
Nb	1	246	-	220	0
Ta	1.43×10^{-3}	45	-	45	0

materials. If we add these effects we can rewrite equation (3.77) as

$$E_c = E_{cs_u} + E_{cs_L} + E_t \tag{3.83}$$

for $E \geqslant E_t$ and

$$E_c = E \tag{3.84}$$

for $E < E_t$ as shown in Table 10.

Table 10

Total Collision Energy (keV)

	Be	C	Al	Te	Ni	Nb	Ta
Fusion	22.4	23.1	56.1	170	259	225	155
Fission	16.5	16.1	45.8	55.0	46.5*	29.6*	15.5*

*Governed by equation (3.84)

The values of R_d as given by equation (3.66) are not very sensitive to the arguments of the logarithmic terms. Again, arbitrarily, E is taken as E_c, and T_a is taken as T_s. The results are shown in Table 11.

Table 11

Values of R_d

	Be	*C*	*Al*	*Ti*	*Ni*	*Nb*	*Ta*
Fusion	0.44	0.44	0.47	0.49	0.49	0.50	0.47
Fission	0.43	0.43	0.45	0.45	0.44	0.43	0.40

When all these results are put into equation (3.82) we get the number of displaced atoms due to the first knock-on. They are shown in Table 12.

Table 12

Number of Displaced Atoms (N)

	Be	*C*	*Al*	*Ti*	*Ni*	*Nb*	*Ta*
Fusion	220	230	590	1850	2850	3350	1650
Fission	160	160	460	550	460	290	140

For the sake of completeness we will include Table 13 showing the total cross section for the materials of interest for 14.1 and 1.41 MeV neutrons. These were used in calculating the displacements and a more exact calculation should of course include the contributions of inelastic scattering which we have totally ignored.

Table 13

Total Scattering Cross Section σ_T (barns)

	Be	*C*	*Al*	*Ti*	*Ni*	*Nb*	*Ta*
Fusion 14.1 MeV	1.5	1.4	1.7	2.3	2.7	4.0	5.2
Fission 1.4 MeV	2.5	2.2	3.0	3.2	3.2	6.0	7.6

We can now use the results of Tables 12 and 13 to estimate the maximum displacement accumulation. This is indicated by D which is given by

$$\frac{D}{\phi t} = 100 N \sigma_T \times 10^{-24} \quad (3.85)$$

where ϕ is the neutron flux in neutrons/cm^2-sec, t is the irradiation time in seconds, N and σ_T as given in the tables, and D is the per cent of atoms displaced. For the same ϕt one can compare the fusion to fission accumulation by calculating the ratio

$$a = \frac{D_{fu}}{D_{fi}} \quad (3.86)$$

These results are shown in Table 14 and Figures 21 and 22.

Table 14

Fusion and Fission Displacement Accumulation

	$\frac{D}{\phi t}(\times 10^{-20})$		
Material	*fu*	*fi*	a
Be	3.3	4.0	0.83
C	3.2	3.5	0.91
Al	10.0	13.8	0.73
Ti	43	17.6	2.4
Ni	77	14.7	5.2
Nb	135	17.4	7.8
Ta	86	10.6	8.1

As pointed out earlier, the extent of radiation damage by neutrons on the physical properties of these materials is still a matter of considerable debate and speculation. For example, some studies have been made on the effect of neutron irradiation

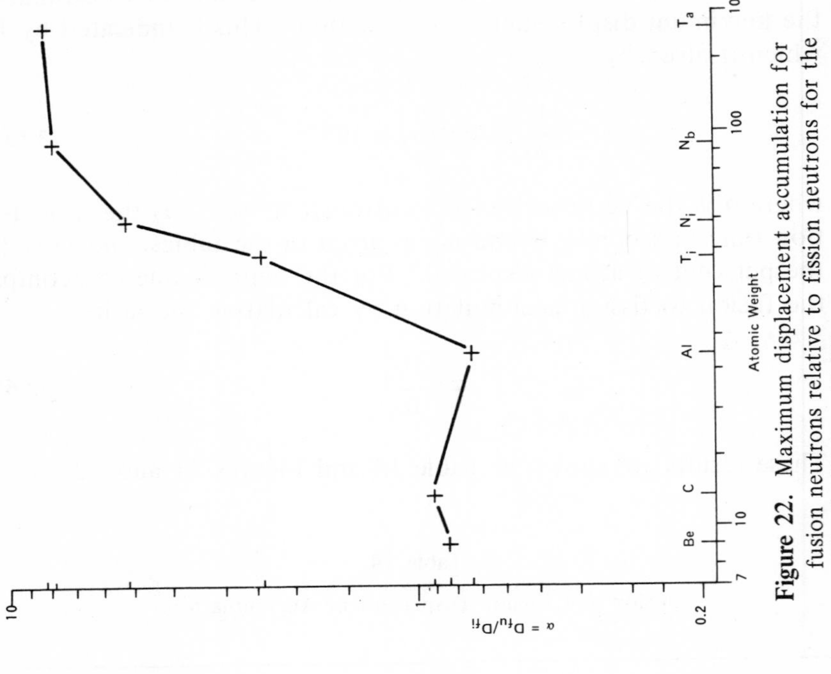

Figure 22. Maximum displacement accumulation for fusion neutrons relative to fission neutrons for the same ϕt (Myers, 1969).

Figure 21. Maximum displacement accumulation for fusion neutrons (Myers, 1969).

on the rupture strength of stainless steel and a zirconium alloy at various temperatures. The effects were nearly negligible as shown in Figures 23 and 24.

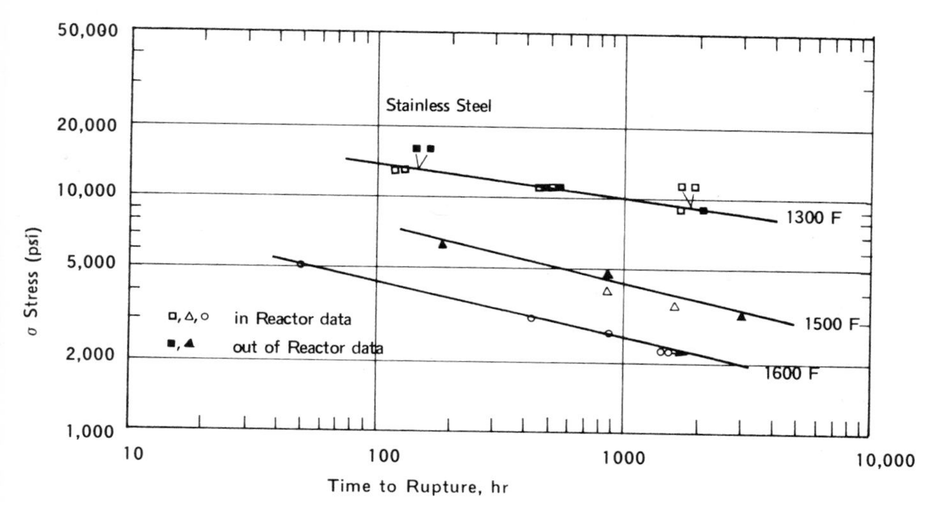

Figure 23. Rupture strength vs time; neutron flux 3 x 10^{13} (>1 MeV) (Myers, 1969).

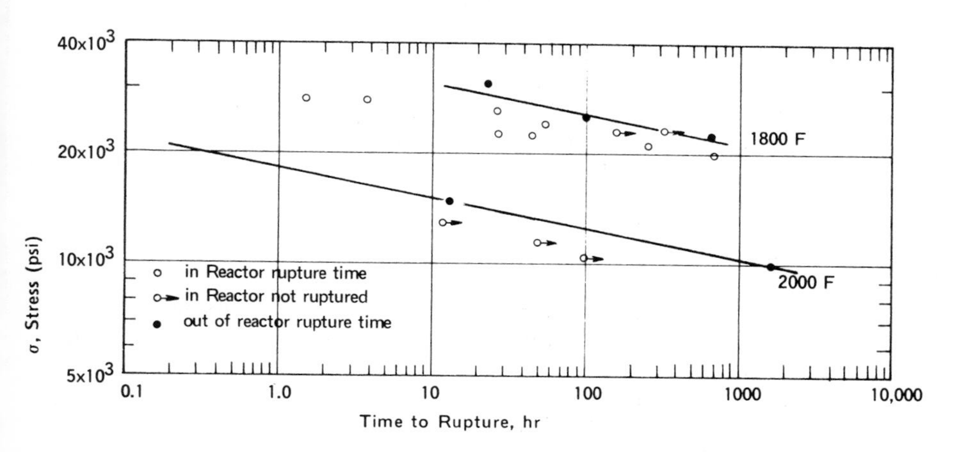

Figure 24. Rupture stress vs time for columbium – 1 per cent zirconium alloy (Myers, 1969).

It may be desirable to estimate the time it takes for a 100 per cent displacement in materials subjected to a neutron flux from a fusion plasma. If such a flux is about 10^{15} neutrons/cm^2-sec, then in N_b the time it takes to attain 100 per cent level is

$$t_{100} = \frac{10^{20}}{(1.35)(10^{15})\,\pi \times 10^{7}} \sim 10^{-3} \text{ years} \qquad (3.92)$$

This is indeed a short time but we must keep in mind that the analysis is approximate, and a more exact examination might reveal that this figure is unduly pessimistic. Much work remains to be done before a true assessment of neutron damage can be made.

REFERENCES

1. Chernilin, Y. F. and G. B. Yakov. "Nuclear Data for Thermonuclear Reactors," translation from Russian of report IAE-1986 Moscow (1970).
2. Crocker, V. S., S. Blow, and C. J. H. Watson. "Nuclear Cross Section Requirements for Fusion Reactors," Culham Laboratory Report CLM-P240 (1970).
3. Evans, R. D. *The Atomic Nucleus.* (New York: McGraw-Hill, 1955).
4. Glasstone, S. and M. C. Edlund. *The Elements of Nuclear Reactor Theory.* (New York: Van Nostrand Co., 1952).
5. Homeyer, W. G. "Thermal and Chemical Aspects of the Thermonuclear Blanket-Problem," Technical Report No. 435, MIT Research Laboratory of Electronics, Cambridge, Mass. (1965).
6. Impink, A. J., Jr. "Neutron Economy in Fusion Reactor Blanket Assemblies," Technical Report No. 434, MIT Research Laboratory of Electronics, Cambridge, Mass. (1965).
7. Kinchin, G. H. and R. S. Pease. *Reports on Progress in Physics, 18,* 1 (1955).
8. Lee, J. D. "Tritium Breeding and Direct Energy Conversion," in *The Chemistry of Fusion Technology,* D. M. Gruen, Ed. (Plenum Publishing Corp., 1972).
9. Myers, B. "Some Observations on 14 MeV Neutron Radiation Effects on Reactor Materials," BNES, Nuclear Fusion Reactor (1969).
10. Seitz, F. *Discussion of The Faraday Society, 5,* 271 (1949).
11. Seitz, F. and J. S. Koehler. *Solid State Physics Advances in Research and Applications,* Vol. 2. (New York: Academic Press, 1956), p. 425.
12. Smith, A. B., J. F. Whalen, and D. T. Guenther. "Fast Neutron Processes in Niobium–Measurement and Evaluation," AP/CTR Technical Memorandum No. 4 (1973).

13. Steiner, D. *Nuclear Applications and Technology, 9,* 83 (1970).
14. Steiner, D. and S. Blow. "Neutronics Calculations on a Fusion Reactor Benchmark Model," Culham Laboratory Report CLM-P345 (1973).

CHAPTER 4

PHYSICS OF NEUTRAL BEAM HEATING

4.1 INTRODUCTION

We recall from the introductory chapter that heating and maintaining the plasma at thermonuclear temperatures are perhaps the most critical aspects of fusion reactors. Several methods of heating have been used to bring the plasma to the temperature discussed in Chapter 1, but there is no indication that these methods will necessarily be successful in bringing the plasma to reactor temperatures. We recall, for example, that ohmic heating, which has been used successfully thus far in Tokamaks, has its limitations in that the plasma resistivity decreases with increasing electron temperature, and the current in the plasma gives rise to instabilities which render this method ineffective past a certain point. The various heating methods that have been proposed or discussed and their possible utilization in open- or closed-ended systems for low or high beta plasmas are shown in Table 15.

It is generally believed that neutral injection will be the ultimate method of heating and maintaining the fusion plasma in magnetically confined systems. Although it is somewhat premature to say this with any assurance at this time, there are indications that this method will be the least beset by problems and in principle it has no serious limitations. Although we will discuss some of the other heating methods in some detail subsequently, we will examine the neutral injection method extensively, especially with regard to the underlying physics and the applicability to both open- and closed-ended systems. The main feature

Table 15

Various Heating Methods (Forsen 1971)

	System		
Heating Method	*Low β Open*	*Low β Closed*	*High β*
Ohmic	?	yes	yes
Turbulent	yes	yes	yes
R.F.	yes	yes	?
Adiabatic Comp.	yes	yes	yes
Shock	?	?	yes
Laser and Relativistic Beam	yes	yes	yes
Plasma Gun	yes	?	?
Neutral Injection	yes	yes	?
Other	?	?	?

of this heating method is the transfer of energy from heavy charged particles to the target plasma through collisions. Energetic ions are passed through a neutralizer (without much change in their energy) and then injected across the magnetic field lines where they become ionized once again and in essence form a source of energetic particles to deliver energy (and possible fuel) to the plasma. Because they are neutralized they are not impeded by the confining magnetic field from entering the target plasma.

Clearly, the physics of the interaction of charged particles with plasma is the same whether these particles are externally injected or internally created as in the case of the alpha particles in a D-T reactor. In some of the examples discussed in Chapter 2, we examined a fusion plasma that was heated internally by the fusion reaction products. Although a more exact and appropriate treatment of this subject must include the interaction of charged particles with plasma in the presence of collective effects, we shall approach the problem in the most elementary way and then assess the validity of this analysis by comparing the results with some of the available more exact treatments.

4.2 PLASMA HEATING BY ENERGETIC HEAVY CHARGED PARTICLES

Heavy energetic charged particles can impart energy to the electrons and ions of a target plasma through collisions. We shall examine this process using Binary Collision Theory where collective effects will be incorporated only through the use of Debye screening in the coulomb cross section. The more rigorous treatments rely on the use of the plasma kinetic equations where collective effects are included in a natural way through the plasma dielectric. We shall see later that the binary collision theory utilized gives results which are within a few per cent of the more exact results in most regions of interest. This is significant in that one can obtain reasonably good estimates without resorting to complicated analysis which requires substantial machine computations.

In the slowing down of an ion (or heavy charged particle) in a plasma, the small angle scattering so predominates that the results are strongly dependent upon the shielding cutoff of the coulomb scattering cross section. Most calculations of slowing down use the "classical" cutoff angle which is strictly valid only in the limit of very small velocities. This was done, for example, in the previous chapter in the analysis of the slowing down of a knock-on in a solid medium. When the conditions for the use of the classical cutoff fail, it would be relatively easy to substitute the "quantum" cutoff angle (see, for example, Bohm, 1951) for the classical; but for many cases, particularly those involving the scattering of ions by other ions, the conditions for the validity of the quantum are not met. It is quite possible for an ion, in slowing down from an energetic state, to pass from a region where the quantum result is valid into a region where neither quantum nor classical values apply, and then into a velocity region where the classical result is generally valid. What we shall present in this analysis is a calculation which incorporates both the classical and quantum cutoff angles, and in some manner provides for the transition between the two. In so doing, we shall identify those regions in velocity space where in the slowing down of a charged particle in plasma the quantum effects play a significant role.

Geometric Considerations of Collision

We begin by considering the geometry of collision between a test particle of mass m and velocity $\vec{v}$, with a field particle of mass M and velocity $\vec{V}$ before collision. After collision, the velocities of the test and field particles are respectively $\vec{v}'$ and $\vec{V}'$. Let us further assume that the angle between $\vec{v}$ and $\vec{V}$ is ψ. We shall call the velocities of the test particle in the center of mass (C.M.) system before and after collision $\vec{W}_1$ and $\vec{W}_1'$, and the velocities of the field particle in this coordinate system $\vec{W}_2$ and $\vec{W}_2'$. The manner in which these velocities are oriented is shown in Figure 25. It should be noted that the two triangles to the right and left of the wavy line are not necessarily coplanar, but may be rotated through an angle ν relative to one another about the dashed line representing $\vec{V}_m'$, the velocity of the center of mass.

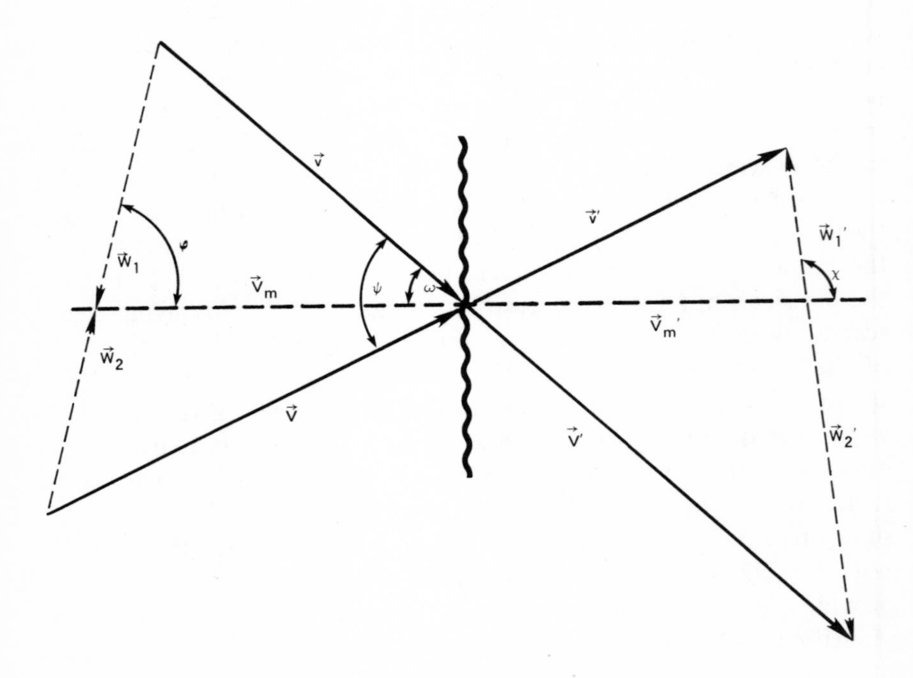

Figure 25. Collision geometry.

We know, of course, that $W_1' = W_1$ and $W_2' = W_2$, *i.e.*, the magnitudes of these velocities, are equal, and that $\vec{W}_1$ and $\vec{W}_2$ must be colinear as must $\vec{W}_1'$ and $\vec{W}_2'$. The relative velocity $\vec{V}_r'$ between the two particles has the magnitude

$$V_r = W_1 + W_2 \tag{4.1}$$

which from Figure 25 is

$$V_r^2 = v^2 + V^2 - 2vV \cos \psi \tag{4.2}$$

From conservation of momentum, the magnitude of the velocity of the center of mass can be written as

$$V_m^2 = \frac{m^2}{(m+M)^2} v^2 + \frac{M^2}{(m+M)^2} V^2 + \frac{2mM}{(m+M)^2} vV \cos \psi \tag{4.3}$$

With the angles φ and ψ defined from Figure 25 [the angle ω shown is useful in obtaining equation (4.3) but drops out of subsequent calculations], we can write the initial and final test particle speeds in terms of V_m and V_r as

$$v^2 = V_m^2 + \frac{M^2}{(m+M)^2} V_r^2 - \frac{2M}{m+M} V_m V_r \cos \varphi \tag{4.4}$$

$$(v')^2 = V_m^2 + \frac{M^2}{(m+M)^2} V_r^2 + \frac{2M}{m+M} V_m V_r \cos \chi \tag{4.5}$$

The scattering cross section is a function of the change in angle between initial and final velocities (the same for either particle) in the C.M. coordinate system. If we call this angle θ, then

$$\cos \theta = \frac{\vec{W}_1 \cdot \vec{W}_1'}{W_1^2} \tag{4.6}$$

Figure 26 depicts $\vec{W}_1$ and $\vec{W}_1'$ in a coordinate system appropriate to the calculation of the above equation. It is readily found that

$$\cos \theta = - [\cos \varphi \cos \chi + \sin \varphi \sin \chi \cos \nu] \tag{4.7}$$

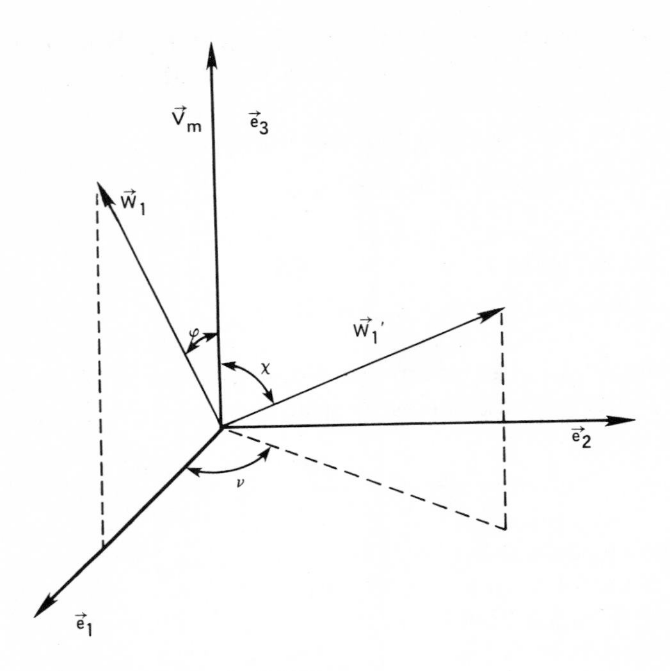

Figure 26. Collision angles in C.M.S.

Average Scattering Cross Section

Since the scattering cross section is also a function of the relative speed, V_r, between the two particles, the number of collisions per unit time experienced by a test particle of velocity, v, which we can represent as $n v \bar{\sigma}_s(v)$, is given by

$$n v \bar{\sigma}_s(v) = n \int_0^\infty dV \int_{-1}^{1} d(\cos\psi) \int_{\Omega'} d\Omega' \; V_r \sigma_s(V_r,\theta) \, F[V,\Omega(\psi)] \quad (4.8)$$

where n is the number of field particles per unit volume, and F is the number density of plasma particles having a velocity whose magnitude is V and whose direction is given by $\Omega(\psi)$. In particular, for a Maxwellian plasma we have

$$F[V,\Omega(\psi)] = 2\pi \left(\frac{M}{2\pi kT}\right)^{3/2} V^2 \, e^{-MV^2/2kT} \quad (4.9)$$

The solid angle into which the test particle is scattered, $d\Omega'$, is just

$$d\Omega' = \sin\chi \, d\chi \, d\nu \tag{4.10}$$

and from equation (4.2)

$$d(\cos\psi) = -\frac{V_r}{vV} dV_r \tag{4.11}$$

then equation (4.8) can be written as

$$\bar{\sigma}_s(v) = \frac{2\pi}{v^2}\left(\frac{M}{2\pi kT}\right)^{3/2} \int_0^\infty Ve^{-\frac{MV^2}{2kT}} \int_0^{2\pi}\int_0^{\pi}\int_{|V-v|}^{V+v} V_r^2 \, \sigma_s(V_r,\theta)\sin\chi \, d\chi \, d\nu \, dV_r \, dV \tag{4.12}$$

For an unshielded coulomb potential, $\sigma_s(V_r,\theta)$ is given by the standard formula

$$\sigma_s(V_r,\theta) = \left(\frac{e_1 e_2}{2\mu V_r^2}\right)^2 \frac{1}{\sin^4(\theta/2)} \tag{4.13}$$

where e_1 is the charge of the test particle, e_2 is the charge of the plasma ion or electron, and μ is the reduced mass. However, for collisions with large impact parameters, for which θ is very small, it is necessary to take into account shielding by other ions and electrons. There are commonly two approaches taken to the scattering cross section for a shielded coulomb potential: for $(e_1 e_2)/(\hbar V_r) \ll 1$, the problem can be solved and it is found that (see Bohm, 1951)

$$\sigma_s(V_r,\theta) = \left(\frac{e_1 e_2}{2\mu V_r^2}\right)^2 \frac{1}{[\sin^2(\theta/2) + \frac{1}{4}(\hbar/\mu\lambda_D V_r)^2]^2} \tag{4.14}$$

where λ_D is the Debye length, given by

$$\lambda_D^2 = \frac{kT}{ne_2^2} \tag{4.15}$$

On the other hand, for $(e_1 e_2)/(\hbar V_r) \gg 1$, the interaction may be described by classical mechanics, and it is usually assumed that

$\sigma_s(V_r,\theta)$ is given by equation (4.13) for $\theta \geqslant (2e_1e_2)/(\mu V_r^2 \lambda_D)$ and is zero for θ less than this value.

A roughly equivalent solution is to use a form similar to equation (4.14) for all values of V_r namely,

$$\sigma_s(V_r,\theta) = \left(\frac{e_1 e_2}{2\mu V_r^2}\right)^2 \frac{1}{[\sin^2(\theta/2) + \frac{1}{4}\theta_o^2(V_r)]^2} \qquad (4.16)$$

where $\theta_o(V_r) = \hbar/(\mu V_r \lambda_D)$ for $V_r \gg (e_1e_2)/\hbar$, and $\theta_o(V_r) = (2e_1e_2)/(\mu V_r^2 \lambda_D)$ for $V_r \ll (e_1e_2)/\hbar$. While the behavior of equation (4.16) is vastly different from that of equation (4.13) with a cutoff at θ_o, in the region $o \leqslant \theta \leqslant \theta_o$, the two forms give similar or identical results, to first order, in many problems requiring integration over θ. For example

$$\int_{\theta_o}^{\pi} \frac{\sin\theta\, d\theta}{\sin^4(\theta/2)} = 2\left[\frac{1}{\sin(\theta_o/2)} - 1\right] \simeq 8/\theta_o^2$$

$$\int_{o}^{\pi} \frac{\sin\theta\, d\theta}{[\sin^2(\theta/2) + \frac{1}{4}\theta_o^2]^2} = \left[\frac{1}{(\theta_o^2/4)} - \frac{1}{1+\theta_o^2/4}\right] \simeq 8/\theta_o^2$$

and

$$\int_{\theta_o}^{\pi} \frac{\sin^2(\theta/2)\sin\theta\, d\theta}{\sin^4(\theta/2)} \simeq 4\ln(2/\theta_o)$$

$$\int_{o}^{\pi} \frac{\sin^2(\theta/2)\sin\theta\, d\theta}{[\sin^2(\theta/2) + \frac{1}{4}\theta_o^2]^2} \simeq 4\ln(2/\theta_o)$$

This identicality of results does not hold for all possible functions of θ which may be used to multiply $\sigma_s(V_r,\theta)$ inside the integral, but the discrepancy is likely to be small unless the multiplying function itself gives great weight to the very small values of θ.

Having chosen the form of the cross section as given by equation (4.16), we are still left with the problem of defining a functional form for $\theta_o(V_r)$ in the region where V_r is of the same order of magnitude as $(e_1e_2)/\hbar$. If the quantity $M/2kT\ (e_1e_2/\hbar)^2$ were either very large or very small we could simply use the appropriate limiting form for θ_o. Unfortunately for many problems of interest, the above quantity is of order unity so that we must somehow fill in the gap between $e_1 e_2/\hbar V_r \ll 1$ and $e_1 e_2/\hbar V_r \gg 1$. It is probably reasonable to assume that $\theta_o(V_r)$ goes fairly smoothly from one form to the other, so that possibly a linear combination of the two would be appropriate. For example, if we let

$$\theta_o(V_r) = \frac{\hbar}{\mu\lambda_D V_r}\left(1 + \frac{e_1 e_2}{\hbar V_r}\right)$$

it would have the desired limiting values. In the interest of simplicity of calculation, however, we shall instead extend the range of assumed validity of each of the limiting forms in toward $V_r = e_1 e_2/\hbar$. Since the two forms become equal when $V_r = 2e_1 e_2/\hbar$, we shall use this as the dividing line between the two regions. Thus

$$\begin{aligned} \theta_o(V_r) &= \frac{2e_1 e_2}{\mu\lambda_D V_r^2} \qquad \text{for } 0 \leqslant V_r \leqslant \frac{2e_1 e_2}{\hbar} \\ &= \frac{\hbar}{\mu\lambda_D V_r} \qquad \text{for } \frac{2e_1 e_2}{\hbar} \leqslant V_r \leqslant \infty \end{aligned} \tag{4.17}$$

In order to facilitate the use of the above equation, we now invert the order of integration over V_r and V in equation (4.12). Then, since $\sin^2(\theta/2) = \frac{1}{2}(1-\cos\theta)$, we get

$$\bar{\sigma}_s(v) = \frac{2\pi}{v^2}\left(\frac{M}{2\pi kT}\right)^{3/2}\left(\frac{e_1 e_2}{\mu}\right)^2 \int_0^\infty \frac{1}{V_r^2} \int_{|V_r - v|}^{V_r + v} \sin\chi \int_0^\pi \int_0^{2\pi} \frac{d\nu\, d\chi\, dV\, dV_r\, V e^{-\frac{MV^2}{2kT}}}{[1 + (\theta_o^2/2) + \cos\varphi\cos\chi + \sin\varphi\sin\chi\cos\nu]^2} \tag{4.18}$$

where the angle φ is a function of V and V_r, as may be seen from equation (4.4). The integration over ν can be readily performed, and if in addition, we define the new variable

$$y \equiv \cos \chi \tag{4.19}$$

and the two parameters

$$a \equiv (1 + \tfrac{1}{2}\,\theta_o^{\,2})^2 - \sin^2 \varphi \tag{4.20}$$

$$b \equiv 2(1 + \tfrac{1}{2}\,\theta_o^{\,2}) \cos \varphi \tag{4.21}$$

then equation (4.18) becomes

$$\bar{\sigma}_s(v) = \frac{4\pi^2}{v^2}\left(\frac{M}{2\pi kT}\right)^{3/2}\left(\frac{e_1 e_2}{\mu}\right)^2 \int_0^\infty \frac{1}{V_r^{\,2}} \int_{|V_r - v|}^{V_r + v} V e^{-\frac{MV^2}{2kT}} \int_{-1}^{1} \frac{(1+\tfrac{1}{2}\theta_o^{\,2}) + y \cos\varphi}{(a + by + y^2)^{3/2}}\, dy\, dV\, dV_r \tag{4.22}$$

After tedious, but straightforward algebraic manipulation the y-integral can be evaluated and the above equation reduces to

$$\bar{\sigma}_s(v) = \frac{8\pi^2}{v^2}\left(\frac{M}{2\pi kT}\right)^{3/2}\left(\frac{e_1 e_2}{\mu}\right)^2 \int_0^\infty \frac{dV_r}{V_r^{\,2}\,\theta_o^{\,2}\,(1 + \frac{\theta_o^{\,2}}{4})} \int_{|V_r - v|}^{v + V_r} V e^{-\frac{MV^2}{2kT}}\, dV \tag{4.23}$$

Since θ_o is a function only of V_r, the V integral can also be readily performed to yield

$$\bar{\sigma}_s(v) = \frac{8\pi}{v^2}\left(\frac{M}{2\pi kT}\right)^{1/2}\left(\frac{e_1 e_2}{\mu}\right)^2 \int_0^\infty \frac{\sinh\left(\frac{MvV_r}{kT}\right)}{V_r^{\,2}\,\theta_o^2\,(1+\frac{\theta_o^2}{4})}\, e^{-\frac{M}{2kT}(v^2 + V_r^{\,2})}\, dV_r \tag{4.24}$$

If we now substitute the values of $\theta_o(V_r)$ from equation (4.17) we obtain

$$\bar{\sigma}_s(v) = \frac{8\pi}{v^2}\left(\frac{M}{2\pi kT}\right)^{1/2}\left(\frac{e_1 e_2}{\mu}\right)^2 \left\{\left(\frac{\mu\lambda_D}{2e_1 e_2}\right)^2\right.$$

$$\int_0^{2e_1e_2/\hbar} \frac{V_r^2 \sinh\left(\frac{MvV_r}{kT}\right)}{1+\left(\frac{e_1 e_2}{\mu\lambda_D}\right)^2 \frac{1}{V_r^2}} e^{-\frac{M}{2kT}(v^2+V_r^2)} dV_r \qquad (4.25)$$

$$\left. + \frac{\mu\lambda_D^2}{\hbar} \int_{2e_1e_2/\hbar}^{\infty} \frac{\sinh\left(\frac{MvV_r}{kT}\right)}{1+\left(\frac{\hbar}{2\mu\lambda_D}\right)^2 \frac{1}{V_r^4}} e^{-\frac{M}{2kT}(v^2+V_r^2)} dV_r \right\}$$

The remaining integral in this expression cannot be performed without some further approximation. In order to evaluate the appropriateness of such approximations we now define the following dimensionless variable

$$x \equiv \frac{M}{2kT} V_r^2 \qquad (4.26)$$

and the dimensionless constants

$$R^2 \equiv \frac{M}{2kT} v^2 \qquad (4.27)$$

$$\alpha \equiv \frac{M}{2kT}\left(\frac{2e_1 e_2}{\hbar}\right)^2 \qquad (4.28)$$

$$\beta \equiv \frac{M}{2kT}\left(\frac{e_1 e_2}{\mu\lambda_D}\right) \qquad (4.29)$$

$$\gamma \equiv \frac{M}{2kT}\left(\frac{\hbar}{2\mu\lambda_D}\right)^2 \qquad (4.30)$$

With these substitutions, equation (4.25) assumes the form

$$\bar{\sigma}_s(v) = \frac{\sqrt{\pi}}{v^2}\left(\frac{M}{2kT}\right)\left(\frac{e_1 e_2}{\mu}\right)^2 \left\{ \frac{1}{\beta^2}\int_0^{\alpha} \frac{x^3 \sinh(2Rx^{1/2})}{\beta^2 + x^2} e^{-(R^2+x)} \frac{dx}{\sqrt{x}} \right.$$

$$\left. + \frac{1}{\gamma}\int_{\alpha}^{\infty} \frac{x \sinh(2Rx^{1/2})}{\gamma + x} e^{-(R^2+x)} \frac{dx}{\sqrt{x}} \right\} \qquad (4.31)$$

From the numerical values listed in Table 16 for plasmas of interest, it is apparent that for reasonable values of n (say $10^{13} \rightarrow 10^{16}$ cm^{-3}), and kT of the order of one keV or higher, both β and γ are much smaller than unity. Thus it is reasonable to neglect

Table 16

Parameter Values

n=cm^{-3}, kT (keV) (Kammash and Galbraith, 1973)

Test Particle	*Target Particle*	α	β	γ
Electron	Electron	$\frac{5.45\times10^{-2}}{kT}$	$\frac{1.73\times10^{-15}\, n^{1/2}}{(kT)^{3/2}}$	$\frac{5.49\times10^{-29}\, n}{(kT)^2}$
	Deuteron	$\frac{2.00\times10^{2}}{kT}$	$\frac{3.18\times10^{-12}\, n^{1/2}}{(kT)^{3/2}}$	$\frac{5.03\times10^{-26}\, n}{(kT)^2}$
Deuteron	Electron	$\frac{5.45\times10^{-2}}{kT}$	$\frac{8.65\times10^{-16}\, n^{1/2}}{(kT)^{3/2}}$	$\frac{1.37\times10^{-29}\, n}{(kT)^2}$
	Deuteron	$\frac{2.00\times10^{2}}{kT}$	$\frac{1.73\times10^{-15}\, n^{1/2}}{(kT)^{3/2}}$	$\frac{1.49\times10^{-32}\, n}{(kT)^2}$
He^3	Electron	$\frac{2.18\times10^{-2}}{kT}$	$\frac{1.73\times10^{-15}\, n^{1/2}}{(kT)^{3/2}}$	$\frac{1.37\times10^{-27}\, n}{(kT)^2}$
	Deuteron	$\frac{8.00\times10^{2}}{kT}$	$\frac{2.72\times10^{-15}\, n^{1/2}}{(kT)^{3/2}}$	$\frac{1.04\times10^{-32}\, n}{(kT)^2}$

β^2 and γ in the denominators of the two integrals in equation (4.31). The equation then simplifies to

$$\overline{\sigma}_s(v) = \frac{\sqrt{\pi}}{v^2}\left(\frac{M}{2kT}\right)\left(\frac{e_1 e_2}{\mu}\right)^2 \left\{ \frac{1}{\beta^2}\int_0^{\alpha} \sqrt{x}\, e^{-(R^2+x)} \sinh(2Rx^{1/2})\, dx \right.$$

$$\left. + \frac{1}{\gamma}\int_{\alpha}^{\infty} x^{-1/2}\, e^{-(R^2+x)} \sinh(2Rx^{1/2})\, dx \right\} \tag{4.32}$$

These two integrals can now be evaluated in terms of the error function

$$\Phi(z) \equiv \frac{2}{\sqrt{\pi}}\int_0^{z} e^{-t^2}\, dt \tag{4.33}$$

and we obtain after some computation

$$\int_0^{\alpha} \sqrt{x}\, e^{-(R^2+x)} \sinh(2Rx^{1/2})\, dx = \frac{1}{2}\left\{(\sqrt{a}-R)\, e^{-(\sqrt{a}+R)^2} + 2Re^{-R^2}\right.$$

$$\left. -(\sqrt{a}+R)\, e^{-(\sqrt{a}-R)^2}\right\} - \frac{\sqrt{\pi}}{4}(1+2R^2) \left\{\Phi(\sqrt{a}+R)\right.$$

$$\left. - \Phi(\sqrt{a}-R) - 2\,\Phi(R)\right\} \tag{4.34}$$

and

$$\int_{\alpha}^{\infty} x^{-1/2}\, e^{-(R^2+x)} \sinh(2Rx^{1/2})\, dx = \frac{\sqrt{\pi}}{2}\left\{\Phi(\sqrt{a}+R) - \Phi(\sqrt{a}-R)\right\} \tag{4.35}$$

Substituting these results into equation (4.32), and replacing a, β, γ, and R by their component parameters, we obtain finally

$$\overline{\sigma}_s(v) \cong \frac{\pi}{2}\frac{\lambda_D^2}{v^2} \left\{ \frac{1}{\sqrt{\pi}}\left(\frac{2kT}{M}\right)^{1/2} \left[\left(\frac{2e_1 e_2}{\hbar} - v\right) e^{-\frac{M}{2kT}\left(\frac{2e_1 e_2}{\hbar} + v\right)^2}\right.\right.$$

$$\left. - \left(\frac{2e_1 e_2}{\hbar} + v\right) e^{-\frac{M}{2kT}\left(\frac{2e_1 e_2}{\hbar} + v\right)^2} + 2v\, e^{-\frac{Mv^2}{2kT}} \right] \tag{4.36}$$

$$+\left(\frac{kT}{M}+v^2\right)\left[2\Phi\left(\sqrt{\frac{Mv^2}{2kT}}\right)-\Phi\left(\sqrt{\frac{M}{2kT}}\left[\frac{2e_1e_2}{\hbar}+v\right]\right)\right.$$

$$\left.+\Phi\left(\sqrt{\frac{M}{2kT}}\left[\frac{2e_1e_2}{\hbar}-v\right]\right)\right]+\left(\frac{2e_1e_2}{\hbar}\right)^2\left[\Phi\left(\sqrt{\frac{M}{2kT}}\left[\frac{2e_1e_2}{\hbar}+v\right]\right)\right.$$

$$\left.\left.-\Phi\left(\sqrt{\frac{M}{2kT}}\left[\frac{2e_1e_2}{\hbar}-v\right]\right)\right]\right\} \tag{4.36}$$

Physically, $\bar{\sigma}_s(v)$ does not have a profound significance since it is the cross section for an interaction which has no restrictions whatsoever. The foregoing calculation does, however, illustrate the method we shall follow in obtaining the energy loss, which is a meaningful physical quantity. In addition by obtaining equation (4.36) we are in a position to compare these results with those obtained by others. If we take the limit in the above equation as

$$\frac{M}{2kT}\left(\frac{2e_1e_2}{\hbar}\right)^2 \rightarrow 0$$

we obtain

$$\lim_{\alpha\to 0}\ \bar{\sigma}_s(v) = 4\pi\left(\frac{e_1e_2\lambda_D}{\hbar v}\right)^2\ \Phi\left(\sqrt{\frac{Mv^2}{2kT}}\right) \tag{4.37}$$

which, except for a slight difference in the initial factor, is the same result obtained for example by Husseiny and Forsen (1970).

Average Energy Loss

To calculate the probable energy loss we return to equation (4.12) and insert the energy loss,$(m/2)(v^2-v'^2)$, into the integrand, thereby obtaining the quantity

$$L=\frac{2\pi}{v^2}\left(\frac{M}{2\pi kT}\right)^{3/2}\int_0^\infty Ve^{-\frac{MV^2}{2kT}}\int_{|V-v|}^{V+v}V_r^2\int_0^\pi\int_0^{2\pi}\frac{m}{2}(v^2-v'^2)\,\sigma_s(V_r,\theta)\sin\chi\,d\chi\,d\nu\,dV_r\,dV \tag{4.38}$$

The average rate of energy loss is then given by

$$\frac{d\bar{E}}{dt} = - n L v \tag{4.39}$$

From equations (4.4) and (4.5) we find that

$$v^2 - v'^2 = - \frac{2M}{m+M} V_m V_r (\cos\varphi + \cos\chi) \tag{4.40}$$

Since V_m is not one of the integration variables, we eliminate it through equations (4.2) and (4.3) to get

$$V_m^2 = \frac{m}{m+M} v^2 + \frac{M}{m+M} V^2 - \frac{mM}{(m+M)^2} V_r^2 \tag{4.41}$$

Rather than use the resulting square root in equation (4.40) we substitute (4.41) into (4.4) obtaining

$$\frac{2M}{m+M} V_m V_r \cos\varphi = \frac{M}{m+M}(V^2 - v^2) - \frac{M(m-M)}{(m+M)^2} V_r^2 \tag{4.42}$$

so that equation (4.40) now becomes

$$v^2 - v'^2 = \left[\frac{M}{m+M}(v^2 - V^2) + \frac{M(m-M)}{(m+M)^2} V_r^2\right]\left[1 + \frac{\cos\chi}{\cos\varphi}\right] \tag{4.43}$$

This equation does not contain any dependence on the angle ν, so that we may proceed directly to the analog of equation (4.22), *i.e.*,

$$L = \frac{2\pi^2}{v^2}\left(\frac{M}{2\pi kT}\right)^{3/2}\left(\frac{e_1 e_2}{\mu}\right)^2 \mu \int_o^\infty \frac{1}{V_r^2} \int_{|V_r - v|}^{V_r + v} V \left[v^2 - V^2 + \frac{m-M}{m+M} V_r^2\right] e^{-\frac{MV^2}{2kT}}$$

$$\times \int_{-1}^{1} \frac{[1 + y/\cos\varphi]\ [1 + (\theta_o^{\ 2})/2 + y\cos\varphi]}{[a + by + y^2]^{3/2}}\, dy\, dV\, dV_r \tag{4.44}$$

where a and b have been defined in equations (4.20) and (4.21), respectively. The evaluation of the y integral in the above equation is straightforward, though tedious, and when it is performed we obtain

$$L = \frac{2\pi^2 \mu}{v^2}\left(\frac{M}{2\pi kT}\right)^{3/2}\left(\frac{e_1 e_2}{\mu}\right)^2 \int_0^\infty \frac{1}{V_r^2}\int_{|V_r - v|}^{V_r + v} [v^2 - V^2 + \frac{m-M}{m+M} V_r^2]\, V e^{-\frac{MV^2}{2kT}}$$

$$\times \left[\ln\left(\frac{1+\frac{\theta_o^2}{4}}{\frac{\theta_o^2}{4}}\right) - \frac{1}{1 + \frac{1}{4}\theta_o^2}\right] dV_r\, dV \qquad (4.45)$$

As before, θ_o is independent of V, so that we may perform the V integral, which is

$$\int_{|V_r - v|}^{V_r + v} \left[\left(v^2 + V_r^2 \frac{m-M}{m+M}\right) V - V^3\right] e^{-\frac{MV^2}{2kT}}\, dV$$

$$= \left(\frac{2kT}{M}\right) e^{-\frac{M}{2kT}(v^2 + V_r^2)} \left[2v V_r \cosh\left(\frac{MvV_r}{kT}\right)\right.$$

$$\left. - \left(\frac{2M}{m+M} V_r^2 + \frac{2kT}{M}\right) \sinh\left(\frac{MvV_r}{kT}\right)\right] \qquad (4.46)$$

When this result is substituted in equation (4.45), the values of θ_o are taken from equation (4.17), and we make use again of the dimensionless quantities defined in equations (4.26-4.30), we get

$$L = \frac{\sqrt{\pi}\mu}{v^2}\left(\frac{e_1 e_2}{\mu}\right)^2 \left\{ \int_0^\alpha x^{-3/2} \left[\ln\left(\frac{\beta^2 + x^2}{\beta^2}\right) - \frac{x^2}{\beta^2 + x^2}\right] e^{-(R^2 + x)}\right.$$

$$\left[2Rx^{1/2} \cosh(2Rx^{1/2}) - \left(1 + \frac{2Mx}{m+M}\right) \sinh(2Rx^{1/2})\right] dx$$

$$+ \int_\alpha^\infty x^{-3/2} \left[\ln\left(\frac{\gamma + x}{\gamma}\right) - \frac{x}{\gamma + x}\right] e^{-(R^2 + x)}$$

$$\left.\left[2Rx^{1/2} \cosh(2Rx^{1/2}) - \left(1 + \frac{2Mx}{m+M}\right) \sinh(2Rx^{1/2})\right] dx\right\} \qquad (4.47)$$

The two integrals in the above equation appear even less capable of being performed than their counterparts in equation (4.31). Therefore let us note that for kT in keV

$$\frac{\gamma}{\alpha} = 2.52 \times 10^{-22} \left(\frac{Me}{\mu}\right)^2 \frac{n}{Z_1^{\,2} kT} \qquad (4.48)$$

so that, unless n approaches 10^{22} cm^{-3}, γ is certainly negligible compared to x in the region where $x \geqslant \alpha$. Thus in the second integral we can write

$$\ln\left(\frac{\gamma+x}{\gamma}\right) - \frac{x}{\gamma+x} \simeq \ln\left(\frac{x}{\gamma}\right) - 1 \simeq \ln\left(\frac{x}{\gamma}\right) \qquad (4.49)$$

Since $1/\gamma$ is likely to be very large, the exponential factor e^{-x} in the integrand of equation (4.47) causes to be negligible those regions of the integration for which ln(x) is of the same order as, or larger than $1/\gamma$. We can therefore assume, finally, that

$$\ln\left(\frac{\gamma+x}{\gamma}\right) - \frac{x}{\gamma+x} \simeq \ln\left(\frac{1}{\gamma}\right) \qquad (4.50)$$

In considering the first integral of equation (4.47), we also have the ratio β^2/α^2 very small unless n approaches 10^{22} cm^{-3}, *i.e.*,

$$\frac{\beta^2}{\alpha^2} = \frac{\gamma}{\alpha} = 2.52 \times 10^{-22} \left(\frac{me}{\mu}\right)^2 \frac{n}{Z_1^{\,2} kT} \qquad (4.51)$$

Thus, over most of the interval $0 \leqslant x \leqslant a$, β^2 will be negligible compared to x^2. Near x=0, this is no longer true, however, since the interval from x=0 to say x=2 is only a very small fraction of the total interval of integration. The effect of making an error in approximating the integrand in this small region should be negligible, so long as neither the true integrand nor the approximation to it becomes excessively large in this region. It can be readily shown that

$$\lim_{x \to 0} \left[\frac{2Rx^{1/2}\cosh(2Rx^{1/2}) - \left(1 + \frac{2Mx}{m+M}\right)\sinh(2Rx^{1/2})}{x^{3/2}}\right] = \frac{8}{3}R^3 - \frac{4M}{m+M}R \qquad (4.52)$$

Making the assumption that

$$\ln\left(\frac{\beta^2 + x^2}{\beta^2}\right) - \frac{x^2}{\beta^2 + x^2} \simeq \ln(1/\beta^2) \tag{4.53}$$

assures that this factor will not give undue emphasis to the region near x=0, although of course the approximate value is much larger than the true value of this factor (which goes to zero) near x=0. Thus, it is not unreasonable to approximate equation (4.47) by

$$L \simeq \frac{\sqrt{\pi\mu}}{v^2}\left(\frac{e_1 e_2}{\mu}\right)^2 \left\{ \ln\left(\frac{1}{\beta^2}\right) \int_0^{\alpha} x^{-3/2} e^{-(R^2+x)} \left[2Rx^{1/2}\cosh(2Rx^{1/2}) - \left(1 + \frac{2Mx}{m+M}\right)\sinh(2Rx^{1/2})\right] dx + \ln\left(\frac{1}{\gamma}\right) \int_{\alpha}^{\infty} x^{-3/2} e^{-(R^2+x)} \left[2Rx^{1/2}\cosh(2Rx^{1/2}) - \left(1 + \frac{2Mx}{m+M}\right)\sinh(2Rx^{1/2})\right] dx \right\} \tag{4.54}$$

A partial integration of the cosh term reduces this to

$$L \simeq \frac{2\sqrt{\pi}}{v^2}\left(\frac{e_1 e_2}{\mu}\right)^2 \left\{ \ln\left(\frac{1}{\beta^2}\right)\left[x^{-1/2} e^{-(R^2+x)} \sinh(2Rx^{1/2})\right]_0^{\alpha} + \frac{m}{M+m} \ln\left(\frac{1}{\beta^2}\right) \int_0^{\alpha} x^{-1/2} e^{-(R^2+x)} \sinh(2Rx^{1/2})\, dx + \ln\left(\frac{1}{\gamma^2}\right)\left[x^{-1/2} e^{-(R^2+x)} \sinh(2Rx^{1/2})\right]_{\alpha}^{\infty} + \frac{m}{m+M} \ln\left(\frac{1}{\gamma}\right) \int_{\alpha}^{\infty} x^{-1/2} e^{-(R^2+x)} \sinh(2Rx^{1/2})\, dx \right\} \tag{4.55}$$

The integral over $\alpha \leqslant x \leqslant \infty$ is given by equation (4.35) while

$$\int_0^{\alpha} x^{-1/2} e^{-(R^2+x)} \sinh(2Rx^{1/2})\, dx = \frac{\sqrt{\pi}}{2}\left\{ 2\Phi(R) + \Phi(\sqrt{\alpha}-R) - \Phi(\sqrt{\alpha}+R) \right\} \tag{4.56}$$

Substituting these results into equation (4.55), we obtain, noting that $\beta^2/\gamma = \alpha$,

$$L \simeq \frac{\pi}{\mu}\left(\frac{e_1 e_2}{v}\right)^2 \Bigg\{\left[\frac{2m}{M+m}\Phi(R) - \frac{4}{\sqrt{\pi}}Re^{-R^2}\right]\ln(1/\beta^2)$$

$$+ \ln(\alpha)\left[\frac{m}{m+M}\left\{\Phi(R+\sqrt{\alpha}) + \Phi(R-\sqrt{\alpha})\right\}\right.$$

$$\left. + \frac{1}{\sqrt{\pi\alpha}}\left\{e^{-(R+\sqrt{\alpha})^2} - e^{-(R-\sqrt{\alpha})^2}\right\}\right]\Bigg\} \qquad (4.57)$$

As α goes to infinity, the quantity in the square brackets which multiplies $\ln(\alpha)$ goes to zero as $e^{-\alpha}$, so that

$$\lim_{\alpha\to\infty} L \simeq \frac{\pi}{\mu}\left(\frac{e_1 e_2}{v}\right)^2 \left[\frac{2m}{m+M}\Phi(R) - \frac{4}{\sqrt{\pi}}Re^{-R^2}\right]\ln(1/\beta^2) \qquad (4.58)$$

On the other hand, when α goes to zero,

$$\lim_{\alpha\to 0}\left[\frac{m}{m+M}\left\{\Phi(R+\sqrt{\alpha}) + \Phi(R-\sqrt{\alpha})\right\} + \frac{1}{\sqrt{\pi\alpha}}\left\{e^{-(R+\sqrt{\alpha})^2} - e^{-(R-\sqrt{\alpha})^2}\right\}\right]$$

$$= \frac{2m}{M+m}\Phi(R) - \frac{4}{\sqrt{\pi}}Re^{-R^2} \qquad (4.59)$$

so that, with a little cancellation we get

$$\lim_{\alpha\to 0} L \approx \frac{\pi}{\mu}\left(\frac{e_1 e_2}{v}\right)^2 \left[\frac{2m}{m+M}\Phi(R) - \frac{4}{\sqrt{\pi}}Re^{-R^2}\right]\ln(1/\gamma) \qquad (4.60)$$

The only difference between the results in these two limiting cases is in the logarithmic term, as could be observed from equation (4.54) in which the two integrands are identical. From Table 16 it appears that whenever the target particle is an electron, equation (4.60) should be valid for electron temperatures greater than 5 keV and may be as low as 1 keV. If the target particle is a deuteron, equation (4.58) should hold for deuteron temperatures of about 10 keV or less.

Equations (4.58) and (4.60) are identical, depending upon the choice of $\theta_o(V_r)$, with the results of Butler and Buckingham (1962) namely their equations 7 and 8. Our interest in the present results, then, is not so much for the limiting forms of equations (4.58) and (4.60) as it is for the manner described in equation (4.57) in which the energy loss goes from proportionality to $\ln(1/\beta^2)$ into proportionality to $\ln(1/\gamma)$ as α goes from infinity to zero.

Some Numerical Results

If we define the functions

$$f_1(R) \equiv \frac{2m}{m+M} \frac{\Phi(R)}{R} - \frac{4}{\sqrt{\pi}} e^{-R^2} \tag{4.61}$$

and

$$f_2(R,\alpha) \equiv \frac{m}{m+M} \left\{ \frac{\Phi(R+\sqrt{\alpha}) + \Phi(R-\sqrt{\alpha})}{R} \right\}$$

$$+ \frac{1}{R\sqrt{\pi\alpha}} \left\{ e^{-(R+\sqrt{\alpha})^2} - e^{-(R-\sqrt{\alpha})^2} \right\} \tag{4.62}$$

then the rate of energy loss can be written as

$$\frac{d\bar{E}}{dt} = -nvL = -\frac{n\pi}{\mu}(e_1 e_2)^2 \left(\frac{M}{2kT}\right)^{1/2} \left\{ f_1(R) \ln(1/\beta^2) \right.$$

$$\left. + f_2(R,\alpha) \ln(\alpha) \right\} \tag{4.63}$$

Recalling that $\alpha\gamma = \beta^2$, we note that

$$\lim_{\alpha \to 0} \left(\frac{d\bar{E}}{dt}\right) - \lim_{\alpha \to \infty} \left(\frac{d\bar{E}}{dt}\right) = -\frac{n\pi}{\mu}(e_1 e_2)^2 \left(\frac{M}{2kT}\right)^{1/2} f_1(R) \ln(\alpha) \tag{4.64}$$

and

$$\left(\frac{d\bar{E}}{dt}\right) - \lim_{\alpha \to \infty} \left(\frac{d\bar{E}}{dt}\right) = -\frac{n\pi}{\mu}(e_1 e_2)^2 \left(\frac{M}{2kT}\right)^{1/2} f_2(R,\alpha) \ln(\alpha) \tag{4.65}$$

and that the correct value of $(d\bar{E}/dt)$ will be somewhere between the two limiting values as long as

$$0 \leqslant \frac{f_2(R,\alpha)}{f_1(R)} \leqslant 1$$

Moreover, we observe that

$$\frac{\lim\limits_{\alpha \to 0} \left(\frac{d\bar{E}}{dt}\right) - \lim\limits_{\alpha \to \infty} \left(\frac{d\bar{E}}{dt}\right)}{\lim\limits_{\alpha \to \infty} \left(\frac{d\bar{E}}{dt}\right)} = \frac{\ln(\alpha)}{\ln(1/\beta^2)} \tag{4.66}$$

and we see that the importance of the correction term depends upon the magnitudes of the two ratios

$$\frac{f_2(R,\alpha)}{f_1(R)} \quad \text{and} \quad \frac{\ln(\alpha)}{\ln(1/\beta^2)}$$

or

$$\frac{\left(\frac{d\bar{E}}{dt}\right) - \lim\limits_{\alpha \to \infty} \left(\frac{d\bar{E}}{dt}\right)}{\lim\limits_{\alpha \to \infty} \left(\frac{d\bar{E}}{dt}\right)} = \frac{f_2(R,\alpha)}{f_1(R)} \; \frac{\ln(\alpha)}{\ln(1/\beta^2)} \tag{4.67}$$

When the ratio f_2/f_1 is zero, then $(d\bar{E}/dt)$ takes the limiting value

$$\lim_{\alpha \to \infty} \left(\frac{d\bar{E}}{dt}\right) = -\frac{n\pi}{\mu} (e_1{}^2 e_2{}^2) \left(\frac{M}{2kT}\right)^{1/2} f_1(R) \ln(1/\beta^2) \tag{4.68}$$

while when this ratio is one, $(d\bar{E}/dt)$ takes the other limiting value

$$\lim_{\alpha \to 0} \left(\frac{d\bar{E}}{dt}\right) = \frac{-n\pi}{\mu} (e_1 e_2)^2 \left(\frac{M}{2kT}\right)^{1/2} f_1(R) \ln(1/\gamma) \tag{4.69}$$

If the ratio f_2/f_1 is between zero and one, then $(d\bar{E}/dt)$ falls between the two limiting values whose separation is indicated by the ratio $\ln(\alpha)/\ln(1/\beta^2)$. However, the ratio f_2/f_1 may also lie outside this interval, particularly when R is close to R_0, the finite value of R (if such a value exists) for which $f_1(R_0)=0$. In such cases, of course, $(d\bar{E}/dt)$ lies outside the limiting values.

In order to provide some measure of the importance of the correction term we examine some sample numerical values of the quantities $\ln(\alpha)/\ln(1/\beta^2)$ and $f_2(R,\alpha)/f_1(R)$. The cases chosen are for a plasma density of 10^{14} cm^{-3}, with the test particle taken to be a deuteron. These calculations have been done for the slowing down due to deuterons, and also due to electrons. Figures 27 through 29 give these results for the deuteron scatterers, while Figures 30 through 32 give them for electron scatterers.

For equal test particle and plasma field particle masses, $f_1(R)$ goes to zero very close to R=1 (or E=kT). The function $f_2(R,\alpha)$ apparently goes to zero at a value of R very close to, but not identical,to the value R_0 for which $f_1(R)$ is zero. This leads to a blowing up of the $f_2(R,\alpha)/f_1(R)$ curve near E=kT, as shown, for example, by the kT=100 keV curve of Figure 29.

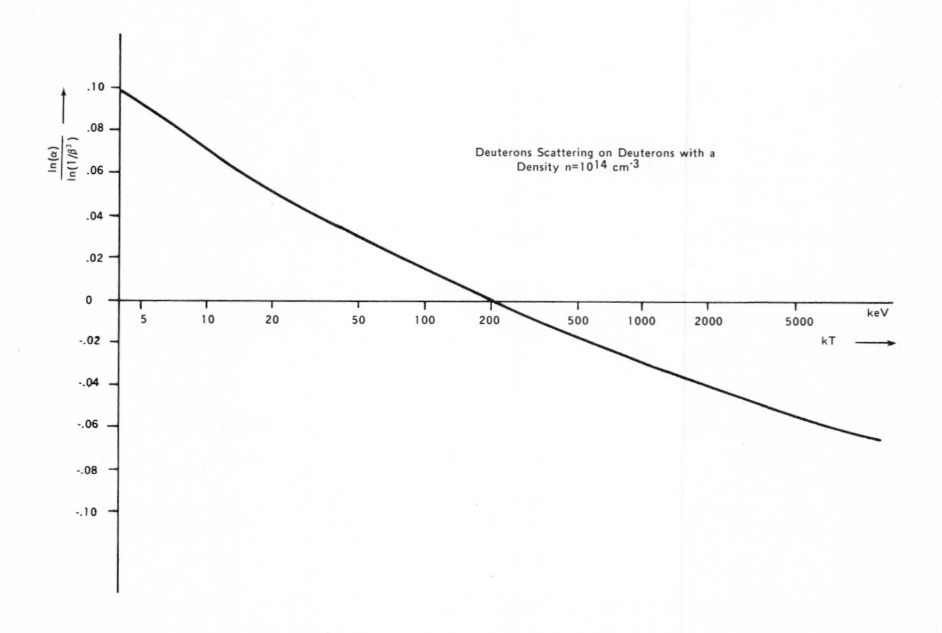

Figure 27. The ratio $\frac{\ln(\alpha)}{\ln(1/\beta^2)}$ vs kT.

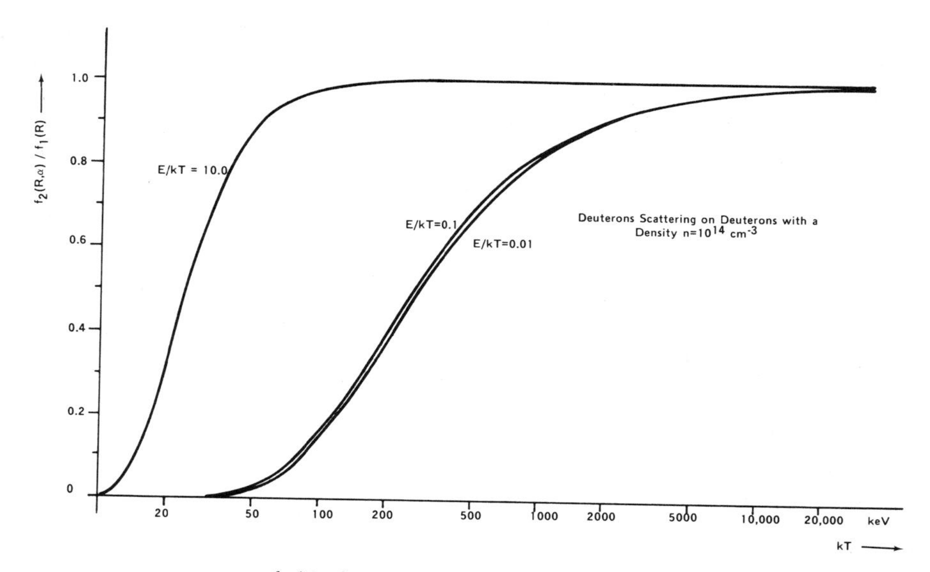

Figure 28. The ratio $\frac{f_2(R,\alpha)}{f_1(R)}$ vs kT.

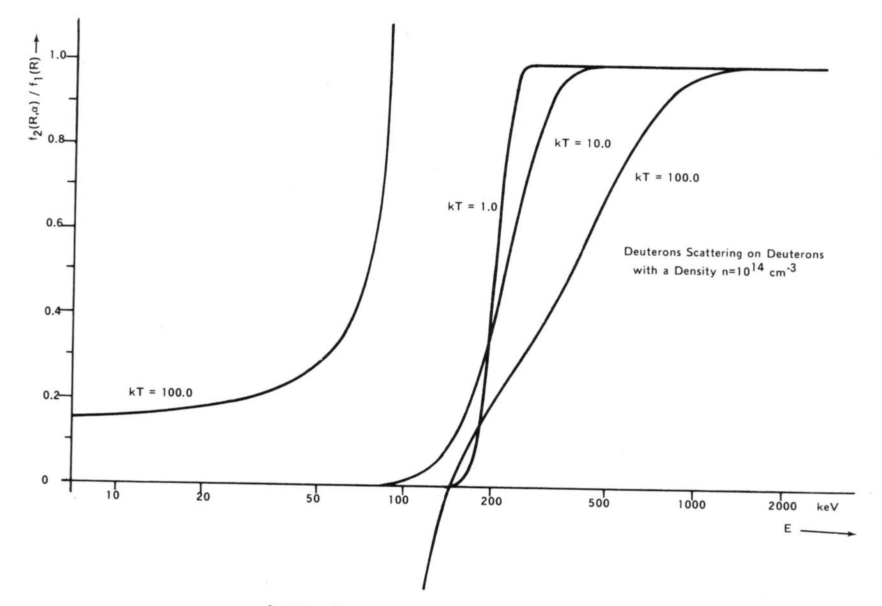

Figure 29. The ratio $\frac{f_2(R,\alpha)}{f_1(R)}$ vs E.

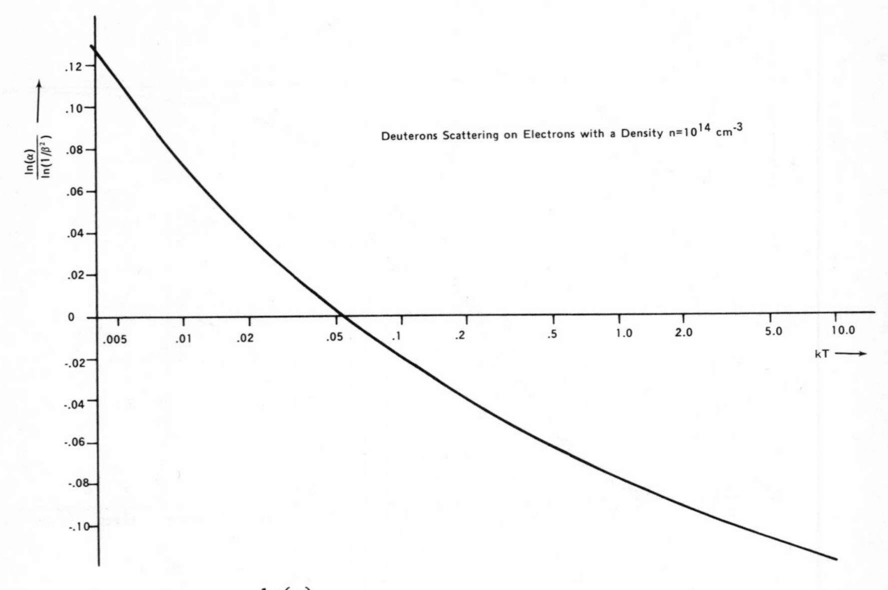

Figure 30. The ratio $\dfrac{\ln(\alpha)}{\ln(1/\beta^2)}$ vs kT.

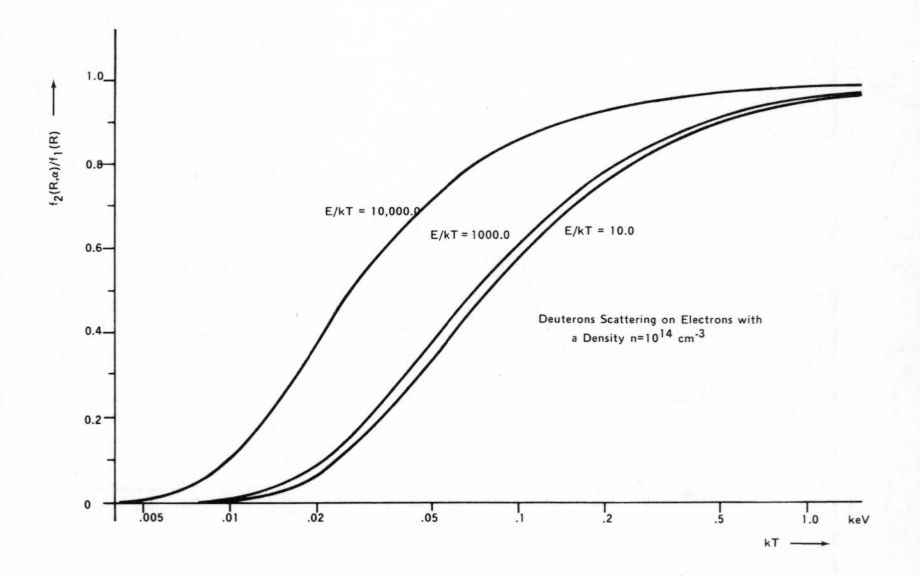

Figure 31. The ratio $\dfrac{f_2(\mathrm{R},\alpha)}{f_1(\mathrm{R})}$ vs kT.

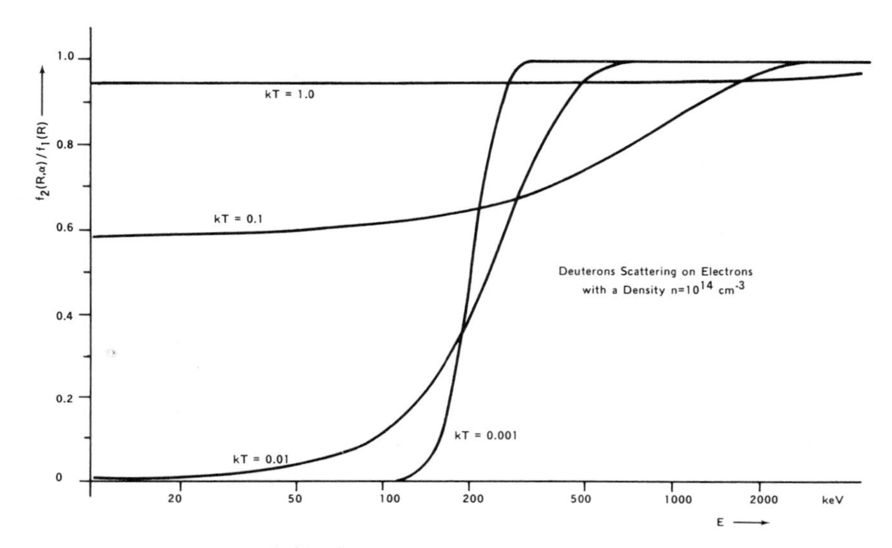

Figure 32. The ratio $\dfrac{f_2(R,\alpha)}{f_1(R)}$ vs E.

For many problems, the separation between the limiting values of $(d\overline{E}/dt)$ is so small as to be negligible, or else the ratio $f_2(R,\alpha)/f_1(R)$ is extremely close to either zero or one, so that $(d\overline{E}/dt)$ can be well described by a proper choice of the limiting expression. However, there are some problems for which neither of these circumstances holds, and the use of either equation (4.68) or (4.69) could lead to significant error. One such case, shown in Figure 33, is that of a deuteron scattering on deuterons, where the plasma deuterons have a temperature of 1.0 keV and a density of 10^{14} per cm^3. If the initial energy of the test particle deuteron is 300 keV or more, it will spend a significant amount of time at each limit, and these limits differ by more than 10 per cent. Similarly if the plasma electrons have a temperature of 1.0 eV, and a density of 10^{14} per cm^3, the cross-over from quantum to classical limits occurs near 200 keV test particle energy, and the separation of these limits is better than 20%. This case is illustrated in Figure 34.

Figures 35 to 37 show a comparison between the results of this analysis and those of Sigmar and Joyce (1971) in which

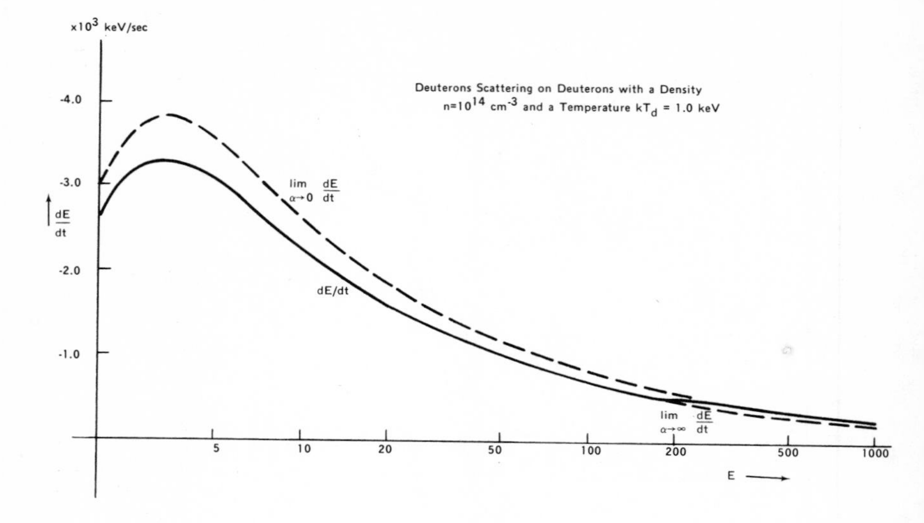

Figure 33. $\frac{dE}{dt}$ vs E.

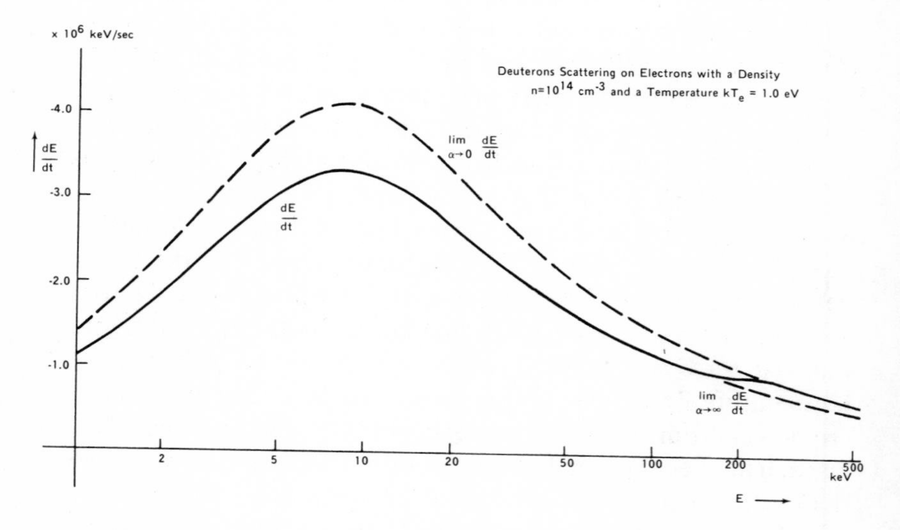

Figure 34. $\frac{dE}{dt}$ vs E.

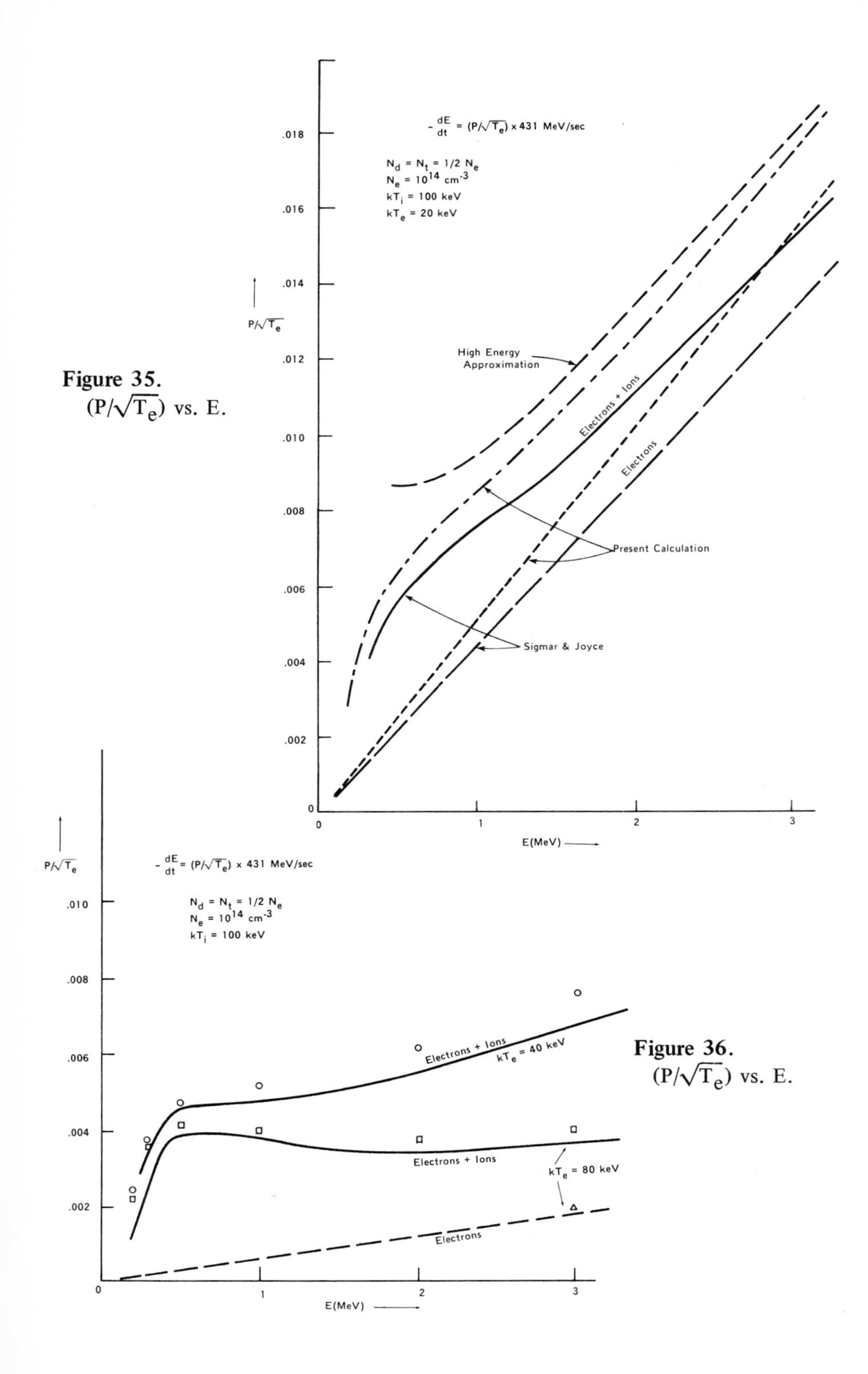

Figure 35. $(P/\sqrt{T_e})$ vs. E.

Figure 36. $(P/\sqrt{T_e})$ vs. E.

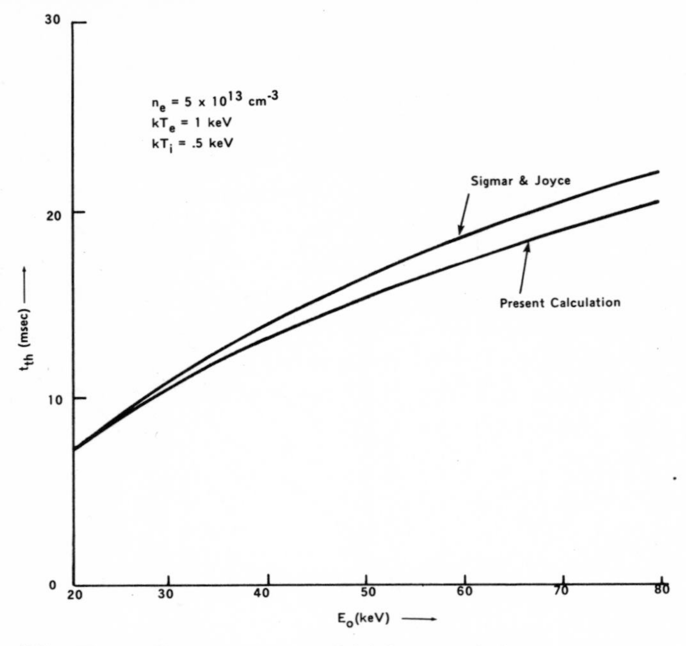

Figure 37. Thermalization time vs initial energy of proton.

plasma collective effects are accounted for more exactly through the use of the Balescu-Lenard equations. For an electron temperature of 20 keV the energy loss rate versus energy for an alpha particle slowing down in a plasma containing an equal number of deuterons and tritons is shown in Figure 35. We note that the present calculation follows the exact one quite well especially in the low energy end, since in this range collective effects are not important and binary collision theory is especially adequate. The "high energy approximation" (Rose, 1969), in which the test particle velocity is taken to be much larger than the thermal velocity of the target particle, breaks down completely in this range since it does not account for the thermalization process. At higher electron temperatures the agreement between the present analysis and the "exact" is even better (see Figure 36) although in all cases the present theory underestimates thermalization times as demonstrated in Figure 37 which shows thermalization time versus injection energy for a proton injected in an electron-proton plasma.

Other Aspects of Charged Particle Collisions

The rate of energy loss obtained above is but one aspect (though a very important one) of the interaction of charged particles with plasma and its relevance to plasma heating in fusion devices. There are others which might be equally important since they shed a special light on the effectiveness and practical utilization of this heating method. To deduce these parameters of interest we return to the energy loss rate equation (4.39), and for simplicity examine it in the classical limit as illustrated by equation (4.58). Specifically we write

$$\frac{d\overline{E}}{dt} = - nvL_{\alpha\to\infty}$$

$$= - \frac{2n\pi}{\mu} \frac{(e_1 e_2)^2}{v} \left[\frac{2m}{m+M} \Phi(R) - \frac{4}{\sqrt{\pi}} R e^{-R^2} \right] \ln \Lambda \tag{4.70}$$

where for the sake of conformity with the literature on this subject we have replaced $\ln(1/\beta^2)$ by $2 \ln \Lambda$, the commonly used logarithmic term. This expression can be further written as

$$\frac{d\overline{E}}{dt} = - \frac{8\sqrt{\pi}n}{M} \frac{(e_1 e_2)^2}{V_{th}} F \left(\frac{v}{V_{th}}\right) \ln \Lambda \tag{4.71}$$

where, as we recall, M is the mass of the target particle and $V_{th} = (2kT/M)^{1/2}$ is its thermal velocity. The function F(x) is related to the error function through

$$F(x) = \frac{1}{x} \int_0^x e^{-t^2} dt - \left(1 + \frac{M}{m}\right) e^{-x^2} \tag{4.72}$$

It is also convenient to express the rate of energy loss in terms of the velocity of the test particle by invoking the equivalence of dE/dt with mv(dv/dt); the result is

$$\frac{d(v^3)}{dt} = - \frac{24\sqrt{\pi}n}{mM} (e_1 e_2)^2 \ln \Lambda \; G(v/V_{th}) \tag{4.73}$$

where G(x) = xF(x). We note that for small arguments, *i.e.*, $x \ll 1$, the function G(x) is given by

$$G(x) \simeq -\frac{M}{m} x + \left(\frac{2}{3} + \frac{M}{m}\right) x^3 + \ldots \tag{4.74}$$

while for large arguments, $x \gg 1$, it assumes the asymptotic form represented by

$$G(x) \simeq \sqrt{\pi}/2 \tag{4.75}$$

It is interesting to note at this point that both G(x) and F(x) vanish when the argument has the value

$$x^2 = \frac{v^2}{V_{th}^{\;2}} = \frac{3}{2}\left(\frac{M}{m}\right)$$

or

$$\frac{1}{2} mv^2 = \frac{3}{2} kT \tag{4.76}$$

This result indicates that there is no energy loss by the test particle when its energy reaches the thermal energy of the field particles. A similar result will be obtained in the next section when we examine the thermal equilibration of two gases of charged particles at different temperatures.

Of special interest to fusion reactors is the relative loss rates to the electrons and ions of the plasma since heating the latter to thermonuclear temperature is the primary objective. If we denote by

$$V_i = \sqrt{\frac{2kT_i}{M_i}}$$

the thermal velocity of the plasma ions whose charge number is Z_i, then for a test particle slowing down in a plasma consisting of electrons and ions at temperatures of T_e and T_i, respectively, the rate of energy loss becomes

$$\frac{d(v^3)}{dt} = -\frac{24\sqrt{\pi}}{m}(e_1 e_2)^2 \ln \Lambda \left\{\frac{Z_i^{\;2}\, n_i}{M_i} G\,(v/V_i) + \frac{n_e}{M_e} G\,(v/V_e)\right\} \tag{4.77}$$

In the extreme case when the incident particle velocity is much larger than the thermal velocities of the target particles (high energy approximation), *i.e.*, $v/V_e \gg 1$; $v/V_i \gg 1$, equation (4.75) can be used to show that

$$\frac{\text{energy transfer rate to electrons}}{\text{energy transfer rate to ions}} \sim \frac{M_i}{M_e}\left(\frac{n_e}{Z_i^2\, n_i}\right) \tag{4.78}$$

which shows that most of the energy is going to the electrons. From the practical point of view it may be more meaningful to ask about that velocity which the incident particle must have in order for the loss rates to the electrons and ions to be equal. If we call that velocity u then we see from equation (4.77) that it satisfies the condition

$$\frac{Z_i^2\, n_i}{M_i}\, G(u/v_i) = \frac{n_e}{M_e}\, G(u/V_e) \tag{4.79}$$

A solution of the above equation is possible when $u/V_i \gg 1$ and $u/V_e \ll 1$, so using equations (4.74) and (4.75) we find

$$u^3 \simeq V_e^3 \left(\frac{M_e}{M_i}\right)\left(\frac{3\sqrt{\pi}}{4}\, Z_i^2\, \frac{n_i}{n_e}\right) \tag{4.80}$$

In the case of equal electron and ion temperatures, *i.e.*, $T_e = T_i$, so that $\frac{1}{2}M_eV_e^2 = \frac{1}{2}M_iV_i^2 = \frac{3}{2}kT$, then the above equation will be replaced by

$$u^3 \simeq V_i^3 \left(\frac{M_i}{M_e}\right)^{1/2} \left(\frac{3\sqrt{\pi}}{4}\, Z_i^2\, \frac{n_i}{n_e}\right); \quad T_e = T_i \tag{4.81}$$

for the "equipartition" velocity u. For a deuterium plasma with $n_e = n_i$; $Z_i = 1$, $M_i/M_e \simeq 3700$ the above relations yield

$$u = V_e/14; \quad T_i \neq T_e$$

$$u = 4V_i \quad ; \quad T_i = T_e$$

Moreover if this same plasma at a temperature of $T_e = T_i = 10$ keV serves as the target of an incident 3.2 MeV proton (*e.g.*, from a D-D reaction), then it is easy to see that

$$\frac{v}{V_i} \simeq 24; \quad \frac{v}{V_e} \simeq 0.4$$

which allows us with the aid of equations (4.74) and (4.75) to show from equation (4.76) that for $n_e = n_i$

$$\frac{\text{energy loss rate to electrons}}{\text{energy loss rate to ions}} = \frac{n_e}{M_e}\left\{\frac{2}{3}\left(\frac{u}{v}\right)^3\right\} \Big/ \frac{Z_i^2\ n_i}{M_i}\frac{\sqrt{\pi}}{2}$$

$$= 170$$

In this case the energy transfer once again favors the electrons since the proton velocity far exceeds the equipartition velocity; in fact, it is six times larger than u.

Another parameter which bears on the effectiveness of this heating method in a magnetically (or for that matter inertially) confined system is the so-called "slowing down time." This time, which represents the time it takes the energy of the incident particle to decrease by a factor of e, takes on added significance when it is compared with the confinement time. Clearly, if the confinement time is much shorter than the slowing down time the heating is ineffective since the particle escapes before it has had a chance to deposit its energy in the plasma. For the sake of illustration we will deduce the slowing down time of a heavy charged particle on electrons only, although the result can be easily extended to the interaction with the ions of the plasma. Our starting point is equation (4.70) from which we recall that m is the mass of the incident particle whose velocity is v and energy is E, and M is the mass of the target particle whose temperature is kT. In the present case $M = M_e$ and $T = T_e$, *i.e.*, those of the plasma electrons, and if once again we assume that v is much smaller than the electron thermal velocity, V_e, then the argument of the error function is small and we can expand it in power series as follows:

$$\Phi(R) = \frac{2}{\sqrt{\pi}}\left[R - \frac{R^3}{3} + \ldots\right] \tag{4.82}$$

Moreover, the reduced mass in this case is $\mu = mM_e/(m+M_e) \simeq M_e$ and upon substitution of these results into equation (4.70) it becomes

$$\overline{\frac{dE}{dt}} = - \frac{2\pi n_e}{M_e E^{1/2}} \sqrt{\frac{m}{2}} (e_1 e_2)^2 \ln \Lambda \left[\frac{8}{3\sqrt{\pi}} \left(\frac{M_e}{m}\right)^{3/2} \left(\frac{E}{T_e}\right)^{3/2} \right]$$

or

$$\overline{\frac{dE}{dt}} = - \frac{8}{3} \frac{\sqrt{2\pi}}{T_e^{3/2}} n_e \frac{M_e^{1/2}}{m} (e_1 e_2)^2 \ln \Lambda \, E$$

or

$$\overline{\frac{dE}{dt}} = - E/\tau_s \tag{4.83}$$

where the slowing down time τ_s is given by

$$\tau_s = \frac{3}{8} \left(\frac{m}{M_e}\right) \sqrt{\frac{M_e}{2\pi}} \frac{T_e^{3/2}}{(e_1 e_2)^2 \ n_e \ln \Lambda} \tag{4.84}$$

with the Boltzmann constant being incorporated in T_e.

An equally important characteristic time is the "thermalization time" which represents the time it takes the test particle to reach an energy equal to the thermal energy of the field particles. If we take the electron and ion temperatures and densities in the plasma to be equal so that $v/V_e \ll 1$ and $v/V_i \gg 1$, then the rate of energy loss to the electrons is exactly that given by equation (4.83) while that to the ions can be immediately obtained from equation (4.70) by noting that $R \gg 1$ and hence $\Phi(R) \to 1$ in this case.

The result is

$$\left(\overline{\frac{dE}{dt}}\right)_{total} = \left(\overline{\frac{dE}{dt}}\right)_e + \left(\overline{\frac{dE}{dt}}\right)_i$$

$$= - \frac{E}{\tau_s} - \frac{4\pi n \ln \Lambda (e_1 e_2)^2}{M_i} \sqrt{\frac{m}{2}} \, E^{-1/2}$$

We find it convenient to normalize the incident particle energy to the electron temperature; thus we let $E' = E/T_e$ and the above equation becomes

$$\frac{dE'}{dt} = -\frac{1}{\tau_s}\left[E' + r^3 E'^{-1/2}\right] \tag{4.85}$$

where

$$r^3 = \frac{3}{4}\frac{\sqrt{\pi}\, m^{3/2}}{M_i M_e^{1/2}} \tag{4.86}$$

with M_i being the mass of the plasma ions. If we call $E'_o = E'(t=0)$ the injection energy, then the solution of equation (4.85) subject to this initial condition is

$$E' = \left[(r^3 + E_o'^{3/2})\, e^{-x} - r^3\right]^{2/3} ; \quad x = \frac{3}{2}\frac{t}{\tau_s} \tag{4.87}$$

Moreover, we can obtain from this result the time t_{th} which corresponds to the case when $E' = E'_{th}$; it is the thermalization time we are after and has the value

$$t_{th} = \frac{2}{3}\tau_s \ln \frac{E_o'^{3/2} + r^3}{E_{th}'^{3/2} + r^3} \tag{4.88}$$

By way of example, if we consider a 50 keV proton slowing down in a hydrogenous plasma whose density is 5×10^{13} cm^{-3} and temperature is 1 keV, we find from the above equation that its thermalization time is about 13 milliseconds which agrees with the value shown in Figure 37. It should be noted however that the thermalization time given above is only approximate and in view of the assumptions made it breaks down near thermalization. A more meaningful and exact quantity will be calculated in the next section.

Nevertheless we can still use the above results to estimate the total amount of energy absorbed by the ions since heating the ions is the main objective of any heating method applied to fusion devices. We recall from equation (4.85) that the second term corresponds to the energy transfer to the ions, and if we integrate it over time we obtain that portion of the incident particle energy which is absorbed by the ions. In other words we write

$$- \Delta E'_i = r^3 \int_0^{x_{th}} dx E'^{-1/2}(x)$$

where x_{th} is the scaled thermalization time defined through $E'(x_{th}) = E'_{th}$. If we now substitute from equation (4.87) for E' and introduce the variable

$$s^3 = (r^3 + E_o'^{3/2}) e^{-x} - r^3$$

we find that we can write for $\Delta E'_i$

$$- \Delta E'_i = 3 r^3 \int_{E_{th}'^{1/2}}^{E_o'^{1/2}} \frac{s\, ds}{(s^3 + r^3)}$$

or

$$- \Delta E'_i = r^3 \left\{ \ln \left[\sqrt{\left(s - \frac{r}{2}\right)^2 + \frac{3}{4} r^2} \Big/ (s+r) \right] + \sqrt{3} \arctan \left[(2s - r)/\sqrt{3}\, r \right] \right\}_{E_{th}'^{1/2}}^{E_o'^{1/2}} \qquad (4.89)$$

We observe that this is a simple monotonic function of the injection energy E'_o, saturating asymptotically as arctan $E_o'^{1/2}$ for large E'_o. The saturation value is

$$- \Delta E'_{i\infty} = \frac{\pi}{\sqrt{3}} r^2 \qquad (4.90)$$

which is determined purely by the value of r which itself is a function of the various masses involved. Equation (4.85) also reveals that the energy of the test particle E'^*, which corresponds to equal transfer to the electrons and ions, is given by

$$E'^* = r^3 E'^{*-1/2}$$

or

$$E'^* = r^2 \qquad (4.91)$$

Assuming that E'_{th} is negligibly small and replaceable by zero, equation (4.89) can be evaluated at E'^{*} and upon comparing the result with equation (4.90) we find

$$\frac{\Delta E'_{i\infty}}{\Delta E'_i \, (E' = E'^{*})} = 2.44 \tag{4.92}$$

The above result suggests injection energies well above the equipartition energy E'^{*}, and the physical explanation of this is as follows. With increasing injection energy the electrons absorb a larger fraction ($\sim E'^{3/2}$) than the ions per unit time. But the time integral over the thermalization process gets a larger contribution from the ions near thermalization such that the total amount of energy absorbed by the ions can increase with injection energy until saturation is reached (see, for example, Figure 40).

When a charged particle slows down in a plasma it is equally as important to know its "slowing down range" as it is to know its slowing down or thermalization time. An injected particle whose range exceeds the characteristic dimensions of the plasma will be of little use as a heating source. The range of a particle whose initial energy is E_o can be written as

$$\lambda = \int_0^{E_o} \frac{v \, dE}{dE/dt} \tag{4.93}$$

If we focus our attention on a case where the incident particle is quite massive and the energy transfer is primarily to the electrons, then using equations (4.82) and (4.70) we can rewrite the above expression as

$$\lambda = \frac{\mu}{2\pi n \, (e_1 e_2)^2 \ln \Lambda} \int_0^{E_o} \frac{3 v^2 \, dE}{8\sqrt{\pi} \, R^3} \tag{4.94}$$

A case of interest is the range of a 3.5 MeV alpha particle in a plasma whose temperature is kT_e. The mass ratio is M_α/M_e = 7344 and since $\ln \Lambda \simeq$ 20, we see that the range is given

$$\lambda_\alpha = 2.16 \times 10^{21} \frac{(kT_e)^{3/2}}{n}, \quad kT_e = \text{keV} \tag{4.95}$$

or

$$\lambda_\alpha = 5.46 \times 10^{16} \frac{T_e^{3/2}}{n} \; ; \; T_e = {}^\circ \text{Kelvin}$$

Because of the assumptions used these results are valid for $T_e \gtrsim 10^8$ °K or $kT_e \gtrsim 10$ keV.

Thermal Equilibrium

Having obtained a formula for $(d\bar{E}/dt)$, it would be reassuring to check and find that a Maxwellian distribution of test particles which obeys this formula will come to thermodynamic equilibrium with the field particles. If the test particles have the distribution

$$F'(v) = 4\pi \left(\frac{m}{2\pi kT'}\right)^{3/2} v^2 e^{-\frac{mv^2}{2kT'}} \tag{4.96}$$

then the average rate of energy exchange between the test particles and the plasma field particles whose temperature is T, is just

$$\left\langle \frac{d\bar{E}}{dt} \right\rangle = \int_0^\infty \frac{d\bar{E}}{dt} F'(v)\, dv \tag{4.97}$$

Since dE/dt is given as a function of the variable R defined by equation (4.27), we note that the test particle distribution, as function of R, can be written as

$$F'(v)\, dv = F'(R)\, dR = \frac{4}{\sqrt{\pi}} A^{3/2} R^2 e^{-AR^2} dR \tag{4.98}$$

where

$$A = \frac{mT}{MT'} \tag{4.99}$$

Thus, from equation (4.63) we can now write

$$\left\langle \frac{d\bar{E}}{dt} \right\rangle = -\frac{4n\sqrt{\pi}}{\mu} e_1^{\ 2} e_2^{\ 2} \left(\frac{M}{2kT}\right)^{1/2} A^{3/2} \left\{ \ln\left(\frac{1}{\beta^2}\right) \int_0^\infty R^2 e^{-AR^2} f_1(R)dR \right.$$

$$\left. + \ln(\alpha) \int_0^\infty R^2 e^{-AR^2} f_2(R,\alpha)\, dR \right\} \tag{4.100}$$

The two integrals over R are readily done, using equations (4.61) and (4.62), to give

$$\int_0^\infty R^2 e^{-AR^2} dR f_1(R) = \frac{1}{A\sqrt{1+A}} \left[\frac{m}{m+M} - \frac{A}{1+A}\right] \tag{4.101}$$

$$\int_0^\infty R^2 e^{-AR^2} f_2(R,\alpha)dR = \frac{1}{A\sqrt{1+A}} \left[\frac{m}{m+M} - \frac{A}{1+A}\right] e^{-\frac{\alpha A}{1+A}} \tag{4.102}$$

And with these results, equation (4.100) becomes

$$\left\langle \frac{d\bar{E}}{dt} \right\rangle = -4n\sqrt{\pi}(e_1 e_2)^2 \left[\frac{mM}{2k^3(mT + MT')^3}\right]^{1/2} \left[kT' - kT\right]$$

$$\left\{ \ln\left(\frac{1}{\beta^2}\right) + e^{-\frac{mkT\alpha}{mkT+MkT'}} \ln(\alpha) \right\} \tag{4.103}$$

which does exhibit the desired result that dE/dt is zero for $T' = T$. Since both equations (4.101) and (4.102) contain the $(T'-T)$ factor, both the approximate limiting forms of (dE/dt), as well as the more exact equation (4.63) exhibit this property of zero net energy exchange when the temperatures are equal.

It might be observed at this point that the above equation can be put in the form

$$\left\langle \frac{dE}{dt} \right\rangle = \frac{T'}{\tau}\left(1 - \frac{T}{T'}\right) \tag{4.104}$$

where τ can be viewed as the thermalization time. We see that this equation is indeed identical to equation (2.40) of Chapter 2 which represented the rate of energy transfer between two species of different temperatures.

Remarks on the Heating Aspect

We are now in a position to reflect on some of the statements made in Chapter 2 regarding alpha particle heating of plasma. From Figure 35 we see that at high alpha particle energy most of the energy transfer is to the electrons. This is expected since at high energy the relative velocity between the alpha particle and the target electron is small and the coulomb cross section is large. As the alpha particle slows down it begins to preferentially heat the ions because its velocity becomes comparable to the ion thermal velocity. This is indicated by the "hump" in the curve at low energies. The "high energy approximation" mentioned earlier does not show this effect because it fails when the test and target particle velocities become comparable—the reason for the upturn in that curve. When Figures 35 and 36 are compared, one readily observes that more of the alpha particle energy is transferred to the ions as the electron temperature is made larger. This means that more efficient heating of the ions can take place if by some means the electrons of the plasma are first heated and kept hot. This fact is especially important in "injection" heating and underlies the reason for its possible utilization as a supplementary heating method in such fusion devices as Tokamak which are first heated ohmically.

The extent to which an alpha particle could heat a mirror plasma is illustrated in Figure 38 which shows the energy degradation of a 3.5 MeV alpha particle as a function of time for various electron temperatures. The straight line labeled τ_{90} denotes the confinement time of the alpha particle based on the assumption that the particle is lost through the mirrors once it undergoes a collision leading to 90-degree deflection. This indicates that large confinement time is required in mirror reactors if substantial alpha particle heating is to take place.

Figure 39 depicts the relative amounts of alpha particle energy going to the electrons and ions of the plasma as a function of the electron temperature. The cross-over occurs at about 32 keV as pointed out in Chapter 2. For a Tokamak plasma

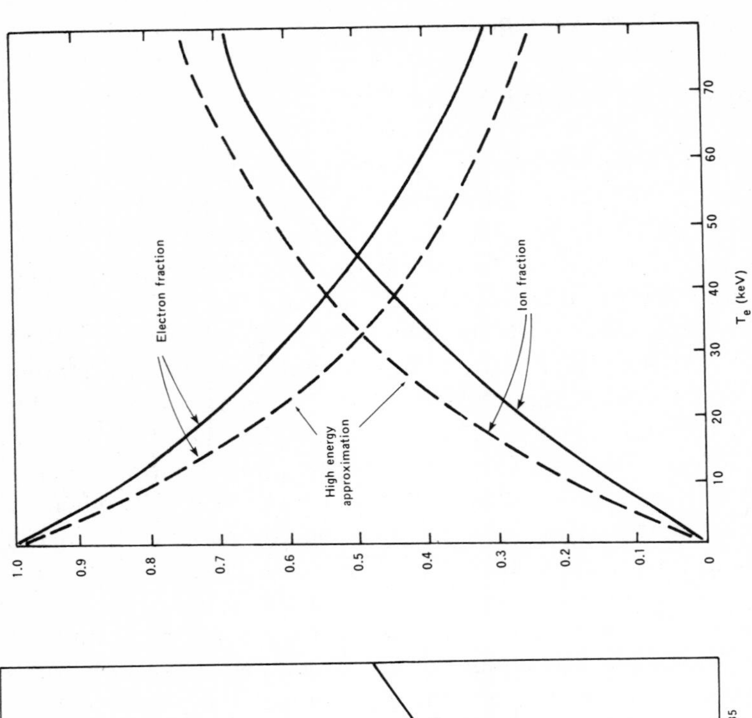

Figure 39. Fractions of alpha particle energy deposited into electrons and ions versus electron temperature.

Figure 38. Alpha particle energy versus time for different electron temperatures.

consisting of protons and electrons with $n = 5 \times 10^{13}$, $T_e = 1$ keV, and $T_i = 0.5$ keV the thermalization time vs injected energy of energetic protons is shown in Figure 37. In this particular example thermalization time is taken to be the time it takes an injected particle to drop in energy to twice the electron temperature, *i.e.*, 2 keV. We see that it ranges from about 8 msec at 20 keV to about 22 msec at 80 keV. Since the confinement time in a device like the T-3 Tokamak is in the range 10-20 msec, it appears that injection heating might be an effective second stage heating for these devices.

In Figure 40 we see the relative fractions of injection energy going into the electrons and ions as well as the total amount of energy going to the ions in the Tokamak plasma mentioned above. We note that the breakeven point between electron and ion absorption occurs at 44 keV, and that the absolute amount of energy going to the ions asymptotes at an injection energy of about 70 keV. This indicates that for the plasma parameters assumed above an optimum injection energy is about 70 keV and that a slight addition of 70 keV protons would increase the ion energy measurably.

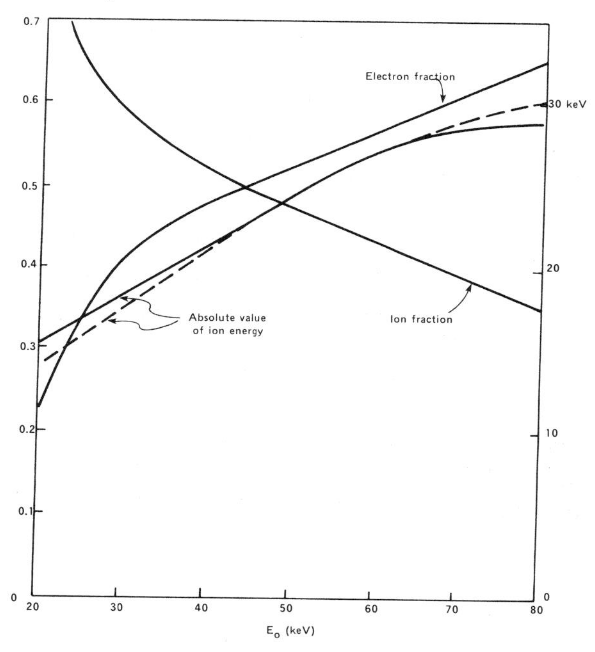

Figure 40. Relative amounts of injected energy deposited into electrons and ions and absolute value of energy deposited into the ions.

Interaction Between Identical Particles

The differential scattering cross section used in the analysis thus far is that derived for distinguishable particles. If the test and target particles are completely identical (including, of course, M=m and $e_2=e_1$), then according to Mott and Massey (1965) we must replace equation (4.13) with the following for unshielded coulomb potential:

$$\sigma_s(V_r,\theta) = C_S P_S + C_A P_A \qquad (4.105)$$

where the functions

$$P_S = \left(\frac{e_1^{\ 2}}{mV_r}\right)^2 \left\{ \sin^{-4}\left(\frac{\theta}{2}\right) + \cos^{-4}\left(\frac{\theta}{2}\right) + \frac{2\cos\left(\frac{e_1^{\ 2}}{\hbar V_r}\ln\left[\tan^2\left(\frac{\theta}{2}\right)\right]\right)}{\sin^2\left(\frac{\theta}{2}\right)\cos^2\left(\frac{\theta}{2}\right)} \right\} \qquad (4.106)$$

and

$$P_A = \frac{e_1^{\ 2}}{mV_r^{\ 2}} \left\{ \sin^{-4}\left(\frac{\theta}{2}\right) + \cos^{-4}\left(\frac{\theta}{2}\right) - \frac{2\cos\left(\frac{e_1^{\ 2}}{\hbar V_r}\ln\left[\tan^2\left(\frac{\theta}{2}\right)\right]\right)}{\sin^2\left(\frac{\theta}{2}\right)\cos^2\left(\frac{\theta}{2}\right)} \right\} \qquad (4.107)$$

are, respectively, the cross sections resulting from a description of the collision by a purely symmetric or purely antisymmetric wave function. The coefficients C_S and C_A depend upon the magnitude of the particle spin and upon the type of statistics (Fermi-Dirac or Bose-Einstein) which apply. For example, when the particles have zero spin, $C_S = 1$ and $C_A = 0$, while for particles with a spin of 1/2, $C_S = 1/4$, and $C_A = 3/4$.

No comparable formula is given for scattering by a shielded coulomb potential, but by analogy with the distinguishable particle case, we should presumably replace each $\sin^2(\theta/2)$ in equations (4.106) and (4.107) by $[\sin^2(\theta/2) + 1/4\ \theta_o^{\ 2}]$ and also replace each $\cos^2(\theta/2)$ by $[\cos^2(\theta/2) + 1/4\ \theta_o^{\ 2}]$. Thus, after setting

$$\frac{e_1^{\ 2}}{\hbar V_r} = \frac{1}{2}\sqrt{\alpha/x} \qquad (4.108)$$

we obtain a scattering cross section for identical particles of

$$\sigma_s(V_r,\theta) = \left(\frac{e_1^{\,2}}{mV_r^{\,2}}\right)^2 \left\{\left[\sin^2\left(\frac{\theta}{2}\right) + \frac{\theta_o^{\,2}}{4}\right]^{-2} + \left[\cos^2\left(\frac{\theta}{2}\right) + \frac{\theta_o^{\,2}}{4}\right]^{-2} + \frac{K\cos\left(\frac{1}{2}\sqrt{\frac{\alpha}{x}}\ln\left[\frac{\sin^2(\theta/2)+\theta_o^{\,2}/4}{\cos^2(\theta/2)+\theta_o^{\,2}/4}\right]\right)}{\left[\sin^2\left(\frac{\theta}{2}\right) + \frac{\theta_o^{\,2}}{4}\right]\left[\cos^2\left(\frac{\theta}{2}\right) + \frac{\theta_o^{\,2}}{4}\right]}\right\} \tag{4.109}$$

The constant K in the above equation takes on the value (+2) for identical, spinless particles, and the value (-1) for particles with spin of ½. When $x \gg \alpha$, the term

$$\cos\left(\frac{1}{2}\right)\sqrt{\frac{\alpha}{x}}\ \ln\left[\frac{\sin^2\ (\theta/2) + \theta_o^{\,2}/4}{\cos^2\ (\theta/2) + \theta_o^{\,2}/4}\right]$$

is essentially a constant, unity, independent of V_r (or x) and θ. On the other hand, for $x \ll \alpha$, this term will oscillate rapidly, as a function of V_r (or x) between ± 1, and the term which it multiplies will integrate over V_r or x to zero. Thus as a first approximation, we might let

$$\sigma_s(V_r,\theta) = \left(\frac{e_1^{\,2}}{mV_r^{\,2}}\right)^2 \left\{\left[\sin^2(\theta/2) + \frac{\theta_o^{\,2}}{4}\right]^{-2} + \left[\cos^2(\theta/2) + \frac{\theta_o^{\,2}}{4}\right]^{-2} + \frac{K}{\left[\sin^2(\theta/2) + \frac{\theta_o^{\,2}}{4}\right]\left[\cos^2(\theta/2) + \frac{\theta_o^{\,2}}{4}\right]}\right\} \qquad x \gg \alpha$$

$$= \left(\frac{e_1^{\,2}}{mV_r^{\,2}}\right)^2 \left\{\left[\sin^2(\theta/2) + \frac{\theta_o^{\,2}}{4}\right]^{-2} + \left[\cos^2(\theta/2) + \frac{\theta_o^{\,2}}{4}\right]^{-2}\right\} \qquad x \ll \alpha$$

$$\tag{4.110}$$

For energy loss calculations, we should limit the range of the center-of-mass scattering angle, θ, to $0 \leqslant \theta \leqslant \pi/2$. In any collision, there will always be one particle emerging with a value of θ in this interval, as well as one particle whose C.M. scattering angle is $(\pi-\theta)$. With identical particles, we cannot tell, of course, which is the original test particle; however, we may arbitrarily identify as the emergent test particle that particle whose orbit most nearly resembles that which the test particle would have followed had there not been any collision. This identification we accomplish by limiting θ to values of $\pi/2$ or less. In order to put this limit on θ in the energy loss equation, we need to switch from the polar angles χ and ν used previously to the new polar angles θ and ν', with equation (4.10) replaced by

$$d\Omega' = \sin\theta \, d\theta \, d\nu' \tag{4.111}$$

For identical particles, then, equation (4.38) becomes

$$L = \frac{2\pi}{v^2} \left(\frac{m}{2\pi kT}\right)^{3/2} \int_0^\infty V e^{-\frac{mV^2}{2kT}} \int_{|V-v|}^{V+v} V_r^2 \int_0^{\pi/2} \int_0^{2\pi} \frac{m}{2}\left[v^2 - v'^2\right] \sigma_s(V_r,\theta) \sin\theta \, d\theta \, d\nu' \, dV_r \, dV \tag{4.112}$$

with $\sigma_s(V_r,\theta)$ taken from equation (4.110). The energy loss was expressed, however, in equation (4.40) in terms of the angle χ or

$$v^2 - (v')^2 = -\frac{2M}{m+M} V_m V_r (\cos\varphi + \cos\chi)$$

So, if we substitute for V_m, and incorporate the condition that M=m, we get in place of equation (4.43)

$$v^2 - (v')^2 = \tfrac{1}{2}(v^2 - V^2)\left[1 + \frac{\cos\chi}{\cos\varphi}\right] \tag{4.113}$$

In order to express $\cos\chi$ in terms of θ and ν', we need to replace the $\hat{e}_1$, $\hat{e}_2$, $\hat{e}_3$ coordinate system of Figure 26 with the system $\hat{e}_1'$, $\hat{e}_2'$, $\hat{e}_3'$ defined so that

$$\underline{W}_1 = W_1 \, \hat{e}_3{}' \tag{4.114}$$

and

$$\hat{e}_2 = \hat{e}_2{}' \tag{4.115}$$

If we measure the angle ν' from $\hat{e}_1{}'$ axis, then

$$\underline{W}_1{}' = (W_1 \sin\theta \cos\nu')\,\hat{e}_1{}' + (W_1 \sin\theta \sin\nu')\,\hat{e}_2{}' + (W_1 \cos\theta)\,\hat{e}_3{}' \tag{4.116}$$

From Figure 26 we have

$$\cos\chi = \frac{\underline{W}_1{}' \cdot \hat{e}_3}{W_1} \tag{4.117}$$

and since

$$\hat{e}_3 = (\sin\varphi)\,\hat{e}_1{}' - (\cos\varphi)\,\hat{e}_3{}' \tag{4.118}$$

we also have

$$\cos\chi = \sin\theta \sin\varphi \cos\nu' - \cos\theta \cos\varphi \tag{4.119}$$

so that equation (4.113) becomes

$$v^2 - (v')^2 = \tfrac{1}{2}(v^2 - V^2)\,[(1 - \cos\theta) + \sin\theta \tan\varphi \cos\nu'] \tag{4.120}$$

The integration of ν' from 0 to 2π causes the second term inside the square brackets in equation (4.120) to drop out and we have, in place of equation (4.44),

$$L(v) = L_1(v) + L_2(v) + K\,L_3(v) \tag{4.121}$$

where

$$L_1(v) = \frac{4\pi^2}{v^2}\left(\frac{m}{2\pi kT}\right)^{3/2}\left(\frac{e_1{}^2}{m}\right)^2 m \int_0^\infty \frac{dV_r}{V_r{}^2} \int_{|V_r - v|}^{V_r + v} V[v^2 - V^2]\, e^{-\frac{mV^2}{2kT}} \int_0^1 \frac{(1-y)\,dy\, dV}{[1 + (\theta_0{}^2/2) - y]^2} \tag{4.122}$$

$$L_2(v) = \frac{4\pi^2}{v^2}\left(\frac{m}{2\pi kT}\right)^{3/2}\left(\frac{e_1^{\,2}}{m}\right) m \int_0^{\infty} \frac{dV_r}{V_r^2}$$

$$\int_{|V_r - v|}^{V_r + v} V[v^2 - V^2]\, e^{-\frac{mV^2}{2kT}} \int_0^1 \frac{(1-y)dy\ dV}{[1+(\theta_o^2/2)+y]^2} \qquad (4.123)$$

$$L_3(v) \doteq \frac{4\pi^2}{v^2}\left(\frac{m}{2\pi kT}\right)^{3/2}\left(\frac{e_1^{\,2}}{m}\right) m \int_{2e_1^2/\hbar}^{\infty} \frac{dV_r}{V_r^2} \qquad (4.124)$$

$$\int_{|V_r - v|}^{V_r + v} V[v^2 - V^2]\, e^{-\frac{mV^2}{2kT}} \int_0^1 \frac{(1-y)dy\ dV}{[1+(\theta_o^2/2)+y^2][1+(\theta_o^2/2)-y^2]}$$

The integrations over y give us

$$\int_0^1 \frac{(1-y)\ dy}{\left(1 + \frac{\theta_o^2}{2} - y\right)^2} = \frac{-1}{1 + \frac{\theta_o^2}{2}} + \ln\left(\frac{1 + \frac{\theta_o^2}{2}}{\frac{\theta_o^2}{2}}\right) \qquad (4.125)$$

$$\int_0^1 \frac{(1-y)\ dy}{\left(1 + \frac{\theta_o^2}{2} + y\right)^2} = \frac{1}{1 + \frac{\theta_o^2}{2}} - \ln\left(\frac{2 + \frac{\theta_o^2}{2}}{1 + \frac{1}{2}\theta_o^2}\right) \qquad (4.126)$$

and

$$\int_0^1 \frac{(1-y)\ dy}{\left(1 + \frac{\theta_o^2}{2} - y\right)\left(1 + \frac{\theta_o^2}{2} + y\right)} = \left(\frac{2 + \frac{1}{2}\theta_o^2}{2 + \theta_o^2}\right) \ln\left(\frac{2 + \frac{\theta_o^2}{2}}{1 + \frac{\theta_o^2}{2}}\right)$$

$$+ \left(\frac{\frac{\theta_o^2}{2}}{2 + \theta_o^2}\right) \ln\left(\frac{\frac{\theta_o^2}{2}}{1 + \frac{\theta_o^2}{2}}\right) \qquad (4.127)$$

Thus, the three analogs to equation (4.47) are

$$L_1(v) = \frac{2\sqrt{\pi}}{m}\left(\frac{e_1{}^2}{v}\right)^2 \left\{ \int_0^{\alpha} x^{-3/2} \left[\ln\left(\frac{\beta^2 + x^2/2}{\beta^2}\right) - \frac{x^2/2}{\beta^2 + x^2/2} \right] e^{-(R^2+x)} \right.$$

$$[2Rx^{1/2} \cosh(2Rx^{1/2}) - (1+x) \sinh(2Rx^{1/2})] \; dx$$

$$+ \int_{\alpha}^{\infty} x^{-3/2} \left[\ln \; \frac{\gamma + x/2}{\gamma} \;\; \frac{x/2}{\gamma + x/2} \right] e^{-(R^2+x)}$$

$$\left. [2Rx^{1/2} \cosh(2Rx^{1/2}) - (1+x) \sinh(2Rx^{1/2})] \; dx \right\} \qquad (4.128)$$

$$L_2(v) = \frac{2\sqrt{\pi}}{m}\left(\frac{e_1{}^2}{v}\right)^2 \left\{ \int_0^{\infty} x^{-3/2} \left[\frac{x^2/2}{\beta^2 + x^2/2} - \ln\left(\frac{\beta^2 + x^2}{\beta^2 + x^2/2}\right) \right] e^{-(R^2+x)} \right.$$

$$[2Rx^{1/2} \cosh(2Rx^{1/2}) - (1+x) \sinh(2Rx^{1/2})] \; dx$$

$$+ \int_{\alpha}^{\infty} x^{-3/2} \left[\frac{x/2}{\gamma + x/2} - \ln\left(\frac{\gamma + x}{\gamma + x/2}\right) \right] e^{-(R^2+x)}$$

$$\left. [2Rx^{1/2} \cosh(2Rx^{1/2}) - (1+x) \sinh(2Rx^{1/2})] \; dx \right\} \qquad (4.129)$$

$$L_3(v) = \frac{2\sqrt{\pi}}{m}\left(\frac{e_1{}^2}{v}\right)^2 \int_{\alpha}^{\infty} x^{-3/2} \left[\left(\frac{\gamma+x}{2\gamma+x}\right) \ln\left(\frac{\gamma+x}{\gamma + x/2}\right) \right.$$

$$\left. + \left(\frac{\gamma}{2\gamma + x}\right) \ln\left(\frac{\gamma}{\gamma + x/2}\right) \right] e^{-(R^2+x)}$$

$$[2Rx^{1/2} \cosh(2Rx^{1/2}) - (1+x) \sinh(2Rx^{1/2})] \; dx \qquad (4.130)$$

Now, if we make the same assumptions concerning the magnitudes of α, β, and γ that were used for the case of the distinguishable particles, we obtain

$$L_1(v) \simeq \frac{2\sqrt{\pi}}{m}\left(\frac{e_1{}^2}{v}\right)^2 \left\{ \ln\left(\frac{1}{\beta^2}\right) \int_0^{\alpha} x^{-3/2} e^{-(R^2+x)} \right.$$

$$[2Rx^{1/2} \cosh(2Rx^{1/2}) - (1+x) \sinh(2Rx^{1/2})]\, dx$$

$$+ \ln(1/\gamma) \int_{\alpha}^{\infty} x^{-3/2} e^{-(R^2+x)}$$

$$\left. [2Rx^{1/2} \cosh(2Rx^{1/2}) - (1+x) \sinh(2Rx^{1/2})]\, dx \right\} \tag{4.131}$$

and

$$L_2(v) \ll L_1(v) \tag{4.132}$$

$$L_3(v) \ll L_1(v) \tag{4.133}$$

Equation (4.131) is identical to equation (4.54), given $M = m$, so that the energy loss for identical particles is the same as that for distinguishable particles, within the limits of our approximations.

REFERENCES

1. Bohm, D. *Quantum Theory.* (Englewood Cliffs, N.J.: Prentice-Hall, 1951).
2. Butler, S. T. and M. J. Buckingham. *Physical Review, 126,* 1 (1962).
3. Forsen, H.K. Proceedings of the International Working Sessions on Fusion Reactor Technology Oak Ridge National Laboratory CONF - 710624 (1971).
4. Galbraith, D. L. and T.Kammash. "Alpha Particle Heating of Mirror Plasma," *Technology of Controlled Thermonuclear Fusion Experiments and the Engineering Aspects of Fusion Reactors* CONF - 721111, U.S. AEC (1974).
5. Husseiny, A. A. and H. K. Forsen. *Physical Review A, 2,* 2019 (1970).
6. Kammash, T. "Charged Particle Heating of Fusion Plasmas," *Trans. American Nuclear Society, 17,* 39 (1973).
7. Kammash, T. and D. L. Galbraith. *Nuclear Fusion, 13,* 133 (1973).

8. Mott, N. F. and H. S. W. Massey. *The Theory of Atomic Collisions,* 3rd ed. (Oxford: Oxford University Press, 1965).
9. Rose, D. J. *Nuclear Fusion, 9,* 183 (1969).
10. Sigmar, D. J., and G. Joyce. *Nuclear Fusion, 11,* 447 (1971).
11. Sivukhin, D. V. *Reviews of Plasma Physics*, Consultant Bureau, Vol. 4, 93 (1966).
12. Spitzer, L. *Physics of Fully Ionized Gases.* (New York : Interscience, 1950).
13. Trubnikov, B. *Reviews of Plasma Physics.* Consultant Bureau, Vol. 1, 105 (1965).

CHAPTER 5

PLASMA HEATING BY RELATIVISTIC ELECTRONS

Another method that has been proposed to heat the plasma to thermonuclear temperatures is one that involves the use of relativistic electron beams. Unlike the case of energetic heavy charged particles discussed in the previous chapter, the interaction of relativistic electrons with plasma is dominated by collective effects. These effects can be assessed only through the nonlinear evolution of the plasma instabilities triggered by the incident electron beam. Binary collision theory is expected to account for the smaller portion of the energy exchange and thus might be viewed as constituting a lower limit on heating by this method. Since collective phenomena are beyond the scope of this text we shall examine this method of heating purely on the basis of binary collision theory and keep in mind that the results based on it are quite conservative. In the previous section we showed that for large collision velocities, $V_r \gg e_1 e_2/\hbar$, the scattering cutoff angle θ_o takes on its quantum value, $\theta_o = \hbar/\mu \lambda_D V_r$, while for small collision velocities, $V_r \ll e_1 e_2/h$, θ_o assumes its classical value $\theta_o = (2e_1 e_2)/(\mu\lambda_D V_r{}^2)$. In the slowing down of an electron in plasma this consideration is not important, since for kinetic energies of 1 keV or more the scattering is exclusively in the "quantum" regime. However, for the electrons, the question of relativistic effects does become important, since $\beta = v/c$ equals 0.1 for kinetic energy of 2.5 keV, and is up to 0.5 for a kinetic energy of about 80 keV.

In this chapter we shall describe the slowing-down process using relativistic mechanics, and, for the interaction of primary concern—electrons colliding with electrons—we shall carry the derivation through to obtain an expansion in the parameter (kT/mc^2) valid for electron energies up to about 10 MeV.

The rate of energy loss suffered by a test particle whose energy is E, while traveling through a cloud of charged particles having the distribution $F[\epsilon, \Omega(\psi)]$ is

$$\frac{dE}{dt} = -n \int_{Mc^2}^{\infty} \int_{-1}^{1} \int_{\Omega'} \gamma_r V_r \sigma_s(\gamma_r, \theta) F(\epsilon, \Omega) [E-E'] \, d\Omega \, d(\cos\psi) \, d\epsilon \tag{5.1}$$

where E′ is the energy of the test particle after a given collision, V_r is the velocity of the test particle relative to the field particle with which it is interacting

$$\gamma_r \equiv \frac{1}{\sqrt{1-\beta_r^2}} = \frac{1}{\sqrt{1-V_r^2/c^2}} \tag{5.2}$$

M is the mass of the field particle, ϵ is the energy (rest plus kinetic) of the interacting field particle, θ is the center-of-mass scattering angle of the test particle in the collision, and ψ is the initial angle between test and field particle velocities (measured in the laboratory coordinate system).

The function $\sigma_s(\gamma_r, \theta)$ is the scattering cross section applicable to the interaction. In order to evaluate dE/dt, we must substitute the appropriate functional forms of σ_s $F[\epsilon, \Omega(\psi)]$ and [E-E′] into equation (5.1) and then perform the indicated integrations.

The energy loss, E-E′, depends upon the center-of-mass scattering angle. However, in order to relate the C.M. coordinate system to the laboratory system, we also need some measure of the angle between the incident velocities of the two interacting particles. It will prove convenient to use the two angles φ and χ, shown in Figure 25, which give the direction of the pre- and post-collision velocities in the center-of-mass coordinate system. If the momentum of the test particle in the C.M. coordinate system is $\vec{b}$, where the four-vector $\vec{b}$ is

$$\vec{b} = \begin{pmatrix} b_x \\ b_y \\ b_z \\ i\epsilon_1 k \end{pmatrix} \tag{5.3}$$

then, if we define the z coordinate axis to lie along V_m, the velocity vector (in three-space) of the center of mass, we have

$$b_z = -b \cos \varphi \tag{5.4}$$

and, for the post-collision vector b′,

$$b_z' = b \cos \chi \tag{5.5}$$

The transformation between the four-momentum $\vec{p}$ as seen in the laboratory system and the center-of-mass four-momentum, $\vec{b}$, is given by Goldstein (1959)

$$\vec{p} = \begin{pmatrix} p_x \\ p_y \\ p_z \\ i\,E/c \end{pmatrix} = \begin{pmatrix} 1 & 0 & 0 & 0 \\ 0 & 1 & 0 & 0 \\ 0 & 0 & \frac{1}{\sqrt{1-\beta_m^2}} & \frac{-i\,\beta_m}{\sqrt{1-\beta_m^2}} \\ 0 & 0 & \frac{i\beta_m}{\sqrt{1-\beta_m^2}} & \frac{1}{\sqrt{1-\beta_m^2}} \end{pmatrix} \begin{pmatrix} b_x \\ b_y \\ b_z \\ i\,\epsilon_1/c \end{pmatrix} \tag{5.6}$$

where

$$\beta_m = V_m/c \tag{5.7}$$

and ϵ_1 and E are the energy of the test particle as measured, respectively, in the center-of-mass and the laboratory coordinate system. From equation (5.6) we see that the pre-collision energy, E, is

$$E = \frac{c\,\beta_m\,b_z}{\sqrt{1-\beta_m^2}} + \frac{\epsilon_1}{\sqrt{1-\beta_m^2}} \tag{5.8}$$

Similarly, the post-collision energy, E′, is given by

$$E = \frac{c\,\beta_m\,b_z'}{\sqrt{1-\beta_m^2}} + \frac{\epsilon_1'}{\sqrt{1-\beta_m^2}} \tag{5.9}$$

Since $\epsilon_1{}' = \epsilon_1$ and $b' = b$, we combine equations (5.8) and (5.9) with equations (5.4) and (5.5) to obtain

$$E - E' = -\frac{c\beta_m b}{\sqrt{1-\beta_m{}^2}} (\cos \varphi + \cos \chi) \tag{5.10}$$

An equivalent form for equation (5.10) can be obtained, using equation (5.8), by noting that

$$E = \frac{mc^2}{\sqrt{1-\beta^2}} \tag{5.11}$$

where

$$\beta = v/c \tag{5.12}$$

and $\vec{v}$ is the velocity of the test particle in the laboratory system, and that

$$\epsilon_1 = [(mc^2)^2 + b^2c^2]^{1/2} \tag{5.13}$$

We thus have, in equation (5.8),

$$E = \frac{mc^2}{\sqrt{1-\beta^2}} = -\frac{c\beta_m b \cos \varphi}{\sqrt{1-\beta_m{}^2}} + \frac{\sqrt{(mc^2)^2 + b^2c^2}}{\sqrt{1-\beta_m{}^2}} \tag{5.14}$$

Using this result in equation (5.10), we have

$$E - E' = \left[\frac{mc^2}{\sqrt{1-\beta^2}} - \frac{\sqrt{(mc^2)^2 + b^2c^2}}{\sqrt{1-\beta_m{}^2}}\right] \left(1 + \frac{\cos \chi}{\cos \varphi}\right) \tag{5.15}$$

The quantities β_m and b in Equation (5.15) are not, however, primary variables in our problem, and thus we must still find substitute expressions for them. β_m can be eliminated through the momentum portions of equation (5.6), applied to both test and target particles. It is easiest to see this from the inverse transformations. Using the angles ω and ψ shown in Figure 26, we have

$$\begin{pmatrix} -\,b\sin\varphi \\ 0 \\ b\cos\varphi \\ i\sqrt{(mc)^2+b^2} \end{pmatrix} = \begin{pmatrix} 1 & 0 & 0 & 0 \\ 0 & 1 & 0 & 0 \\ 0 & 0 & \gamma_m & i\beta_m\gamma_m \\ 0 & 0 & -i\beta_m\gamma_m & \gamma_m \end{pmatrix} \begin{pmatrix} -mv\gamma\sin\omega \\ 0 \\ mv\gamma\cos\omega \\ imc\gamma \end{pmatrix} \tag{5.16}$$

$$\begin{pmatrix} b\sin\varphi \\ 0 \\ -b\cos\varphi \\ i\sqrt{(Mc)^2+b^2} \end{pmatrix} = \begin{pmatrix} & \downarrow & \end{pmatrix} \begin{pmatrix} MV\Gamma\sin(\psi-\omega) \\ 0 \\ MV\Gamma\cos(\psi-\omega) \\ iMc\Gamma \end{pmatrix} \tag{5.17}$$

where

$$\gamma = \frac{1}{\sqrt{1-\beta^2}} = \frac{1}{\sqrt{1-v^2/c^2}} \tag{5.18}$$

$$\Gamma = \frac{1}{\sqrt{1-B^2}} = \frac{1}{\sqrt{1-V^2/c^2}} \tag{5.19}$$

$$\gamma_m = \frac{1}{\sqrt{1-\beta_m{}^2}} = 1/\sqrt{1-V_m{}^2/c^2} \tag{5.20}$$

We thus have

$$-\,mv\gamma\,\sin\omega + MV\Gamma\,\sin(\psi-\omega) = 0 \tag{5.21}$$

and

$$mv\gamma\gamma_m\cos\omega - mc\gamma\beta_m\gamma_m + MV\Gamma\gamma_m\cos(\psi-\omega)$$
$$-MC\Gamma\beta_m\gamma_m = 0 \tag{5.22}$$

Combining equations (5.21) and (5.22) to eliminate the angle ω gives

$$\beta_m^2 = \frac{m^2\ \beta^2\ \gamma^2 + M^2\ B^2\ \Gamma^2 + 2\,m\,M\,\beta\,B\,\gamma\,\Gamma\ \cos\psi}{(m\ \gamma + M\Gamma)^2} \tag{5.23}$$

The center-of-mass momentum, b, can be related to the relative velocity V_r (which is in turn a function of the angle ψ) by means of a transformation from the coordinate system riding with the target particle to the C.M. coordinate system. Let the velocity of the target particle in the C.M. system be w_t. Then, if

$$\gamma_t = \frac{1}{\sqrt{1-\beta_t^2}} = \frac{1}{\sqrt{1-W_t^2/c^2}} \tag{5.24}$$

we have

$$\begin{pmatrix} 0 \\ 0 \\ b \\ i\sqrt{(mc)^2 + b^2} \end{pmatrix} = \begin{pmatrix} 1 & 0 & 0 & 0 \\ 0 & 1 & 0 & 0 \\ 0 & 0 & \gamma_t & i\beta_t\gamma_t \\ 0 & 0 & -\beta\gamma_t & \gamma_t \end{pmatrix} \begin{pmatrix} 0 \\ 0 \\ m\ V_r\ \gamma_r \\ i\ m\ c\ \gamma_r \end{pmatrix} \tag{5.25}$$

Thus,

$$b = mc\ \left(\sqrt{(\gamma_r^2-1)}\right)\ \gamma_t - mc\gamma_r\sqrt{(\gamma_t^2 - 1)} \tag{5.26}$$

But the target particle has a momentum in the C.M. system of

$$b = M\ W_t\,\gamma_t = Mc\beta_t\,\gamma_t = M\,c\,\sqrt{(\gamma_t^2-1)} \tag{5.27}$$

Combining equations (5.26) and (5.27), we get

$$b^2 = \frac{m^2\ M^2\ c^2\ (\gamma_r^2-1)}{m^2 + M^2 + 2mM\gamma_r} \tag{5.28}$$

The relationship between the angle ψ between test particle and target particle velocities (in the laboratory system) and the relative velocity V_r is obtained through a coordinate transformation from the laboratory system to a system riding on the target particle. This transformation gives, for the test particle,

$$\begin{pmatrix} p_{rx} \\ 0 \\ P_{rz} \\ imc\gamma_r \end{pmatrix} = \begin{pmatrix} 1 & 0 & 0 & 0 \\ 0 & 1 & 0 & 0 \\ 0 & 0 & \Gamma & iB\Gamma \\ 0 & 0 & -iB\Gamma & \Gamma \end{pmatrix} \begin{pmatrix} -mv\gamma \sin\psi \\ 0 \\ mv\gamma \cos\psi \\ imc\gamma \end{pmatrix} \tag{5.29}$$

so that

$$mc\gamma_r = -mv\gamma\Gamma\cos\psi + mc\gamma\Gamma \tag{5.30}$$

or

$$\beta B\gamma\Gamma \cos\psi = \gamma\Gamma - \gamma_r \tag{5.31}$$

If we now use γ_r rather than the angle ψ as our independent variable, then equation (5.23) for β_m can be rewritten as

$$1 - \beta_m{}^2 = \frac{m^2 + M^2 + 2mM\gamma_r}{(m\gamma + M\Gamma)^2} \tag{5.32}$$

Using the results of equations (5.29) and (5.32) in equation (5.15), we find, after a little algebra, that

$$E - E' = \left[\frac{(mc^2\gamma)(M^2 + mM\gamma_r) - (Mc^2\Gamma)(m^2 + mM\gamma_r)}{m^2 + M^2 + 2mM\gamma_r}\right]\left(1 + \frac{\cos\chi}{\cos\varphi}\right) \tag{5.33}$$

For the special case of M=m, which covers the electron-electron interaction, equation (5.33) simplifies considerably:

$$\left(E - E'\right)_{m=M} = \tfrac{1}{2}mc^2(\gamma - \Gamma)\left(1 + \frac{\cos\chi}{\cos\varphi}\right) \tag{5.34}$$

The scattering cross section $\sigma_s(\gamma_r,\theta)$ is given by Akheizer and Berestetskii (1965) for electron-electron scattering as

$$\sigma_s(\gamma_r,\theta) = \left(\frac{e^2}{mc^2}\right) \frac{[2(\epsilon_1/mc^2)^2 - 1]}{4\beta_1^{\,4}(\epsilon_1/mc^2)^6} \tag{5.35}$$

$$\left\{ \frac{4}{\sin^4\theta} - \frac{3}{\sin^2\theta} + \frac{[(\epsilon_1/mc^2)^2 - 1]^2}{[2(\epsilon_1/mc^2)^2 - 1]^2} \left(1 + \frac{4}{\sin^2\theta}\right) \right\}$$

where

$$\epsilon_1 = m\ c^2\ \gamma_1 \tag{5.36}$$

$$\gamma_1 = \frac{1}{\sqrt{1-\beta_1^{\,2}}} = \frac{1}{\sqrt{1-W_1^{\,2}/c^2}} \tag{5.37}$$

and W_1 is the velocity of the test particle measured in the C.M. coordinate system. Noting that

$$mc\gamma_1 = \sqrt{(mc)^2 + b^2} \tag{5.38}$$

and that, from equation (5.28) with M=m,

$$b^2 = ½(mc)^2\ (\gamma_r - 1) \tag{5.39}$$

we have

$$\gamma_1 = \sqrt{(\gamma_r + 1)/2} \tag{5.40}$$

After a little bit of algebra, we can write

$$\sigma_s(\gamma_r,\theta) = \left(\frac{e^2}{mc^2}\right)^2 \frac{\gamma_r^{\,2}(\gamma_r+1)}{2(\gamma_r^{\,2} - 1)^2} \left\{ \frac{1}{\sin^4(\theta/2)} + \frac{1}{\cos^4(\theta/2)} \right.$$

$$- \frac{1}{\sin^2(\theta/2)\cos^2(\theta/2)} + \frac{(\gamma_r - 1)^2}{\gamma_r^{\,2}}$$

$$\left. \left(1 + \frac{1}{\sin^2(\theta/2)\cos^2(\theta/2)}\right) \right\} \tag{5.41}$$

The form of $\sigma_s(\gamma_r,\theta)$ given in equation (5.41) still leaves us with the familiar coulomb scattering difficulty of blowup for small angles. In fact, this formulation of σ_s blows up for both $\theta=0$ and $\theta=\pi$. We therefore resort to one of the standard solutions to this problem, the replacing of each $\sin^2(\theta/2)$ by the quantity $[\sin^2(\theta/2) + \frac{1}{4}\theta_o^2]$. In order to keep σ_s finite for $\theta=\pi$, we must similarly replace each $\cos^2(\theta/2)$ by $[\cos^2(\theta/2) + \frac{1}{4}\theta_o^2]$. After some manipulation, we arrive at

$$\sigma_s(\gamma_r,\theta) = 2\left(\frac{e^2}{mc^2}\right)^2 \frac{\gamma_r^2(\gamma_r+1)}{(\gamma_r^2-1)^2}\left\{\frac{1}{[1+(\theta_o^2/2)+\cos\theta]^2}\right.$$

$$+\frac{1}{[1+(\theta_o^2/2)-\cos\theta]^2}+\frac{(1-2\gamma_r)/\gamma_r^2}{[1+(\theta_o^2/2)+\cos\theta][1+(\theta_o^2/2)-\cos\theta]}$$

$$\left.+\frac{(\gamma_r-1)^2}{\gamma_r^2}\right\} \qquad (5.42)$$

The Maxwellian distribution for a relativistic gas is given by Synge (1957) as

$$F(\epsilon,\Omega) = \frac{c^2/kT}{2M^2K_2\left(\frac{Mc^2}{kT}\right)}\, e^{-\epsilon/kT}\, \frac{(\epsilon^2-M^2c^4)^{\frac{1}{2}}}{c^6}\,\epsilon \qquad (5.43)$$

where

$$\epsilon = Mc^2\Gamma \qquad (5.44)$$

and $K_2(x)$ is the modified Bessel function of second order:

$$K_2(x) = \int_0^\infty e^{-x\cosh y}\cosh(2y)\,dy \qquad (5.45)$$

Making use of equation (5.44) and taking the limit as $Mc^2/kT \gg 1$:

$$\lim_{x \gg 1} K_2(x) = \sqrt{\frac{\pi}{x}}\, e^{-x}$$

we get (setting M=m)

$$F(\epsilon,\Omega)\ d\epsilon \simeq \frac{(mc^2/kT)^{1/2}}{\sqrt{2\pi}}\ \Gamma\sqrt{\Gamma^2-1}\ e^{-\frac{mc^2}{kT}(\Gamma-1)}\ d\Gamma \qquad (5.47)$$

This completes the derivation of the various functions which appear in equation (5.1). We have now replaced the variables of integration in that equation, ψ and ϵ, by the quantities γ_r and Γ. We note from equation (5.31) that

$$\gamma\Gamma - (\gamma^2-1)^{1/2}\,(\Gamma^2-1)^{1/2} \leqslant \gamma_r \leqslant \gamma\Gamma + (\gamma^2-1)^{1/2}\,(\Gamma^2-1)^{1/2} \qquad (5.48)$$

and that

$$d(\cos\psi) = -\frac{d\gamma_r}{\beta B\gamma\Gamma} = -\frac{d\gamma_r}{(\gamma^2-1)^{1/2}(\Gamma^2-1)} \qquad (5.49)$$

and note that, from equation (5.2), we also have

$$V_r = c\,\frac{(\gamma_r^{\,2}-1)^{1/2}}{\gamma_r} \qquad (5.50)$$

The solid angle, $d\Omega'$, into which the test particle is scattered, is given by

$$d\Omega' = \sin\theta\ d\theta\ d\nu \qquad (5.51)$$

where θ is the center-of-mass scattering angle defined previously, and the angle ν is shown in Figure 41. From that figure, we readily find that

$$\cos\chi = -\cos\theta\ \cos\varphi + \sin\theta\ \sin\varphi\ \cos\nu \qquad (5.52)$$

Since the angle ν appears in the integrand of equation (5.1) only in the factor $(1 + \cos\chi/\cos\varphi)$ of $(E - E')$, we can easily integrate over this variable, obtaining

$$\int_0^{2\pi} \left(1 + \frac{\cos \chi}{\cos \varphi}\right) d\nu = 2\pi(1 - \cos \theta) \tag{5.53}$$

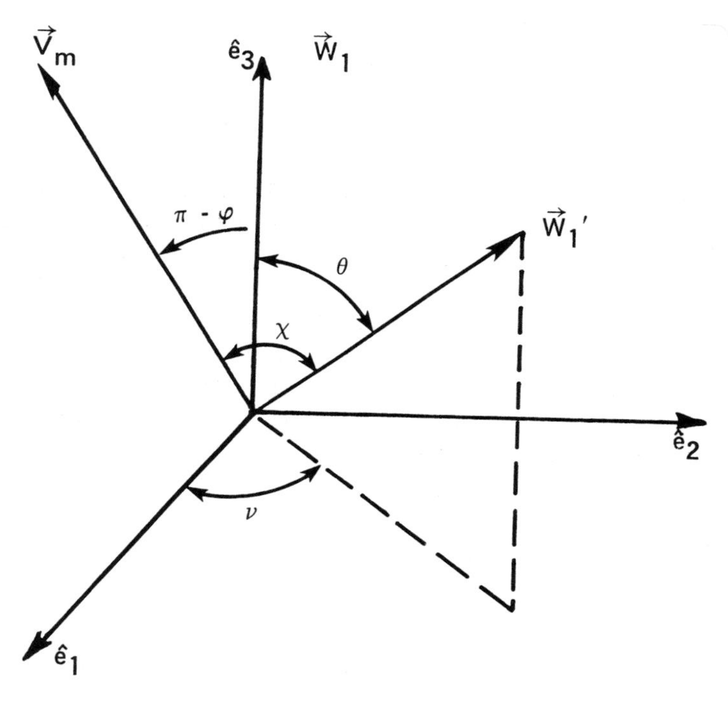

Figure 41. Angles in the C.M. system.

Thus, when all the substitutions are made into equation (5.1), we obtain

$$\frac{dE}{dt} = -\sqrt{2\pi}\,\frac{nce^4}{(\gamma^2-1)^{1/2}}\,\frac{(mc^2)^{1/2}}{(kT)^{3/2}} \int_1^{\infty} \Gamma(\gamma-\Gamma)\, e^{-\frac{mc^2}{kT}(\Gamma-1)}$$

$$\int_{\gamma\Gamma+(\gamma^2-1)^{1/2}(\Gamma^2-1)^{1/2}}^{\gamma\Gamma+(\gamma^2-1)^{1/2}(\Gamma^2-1)^{1/2}} \frac{\gamma_r^{\,2}(\gamma_r+1)}{(\gamma_r^{\,2}-1)^{3/2}} \int_0^{\pi/2} \left\{ \frac{1}{[1+\frac{1}{2}\theta_o^{\,2} - \cos\theta]^2}\right.$$

$$+ \frac{1}{[1 + \frac{1}{2}\theta_o^2 + \cos\theta]^2} + \frac{(1 - 2\gamma_r)/\gamma_r^2}{[1 + \frac{\theta_o^2}{2} - \cos\theta][1 + \frac{\theta_o^2}{2} + \cos\theta]}$$

$$+ \frac{(\gamma_r - 1)^2}{\gamma_r^2} \Bigg\} \quad (1 - \cos\theta) \sin\theta \, d\theta \, d\gamma_r \, d\Gamma \qquad (5.54)$$

The integration over θ in equation (5.54) has been restricted to the interval $0 \leqslant \theta \leqslant \pi/2$ since, for electron-electron scattering, the test and target particles are indistinguishable; if the test particle actually scatters by more than $\pi/2$, the target particle will appear at some C.M. angle less than $\pi/2$, and vice versa. Since we are interested in the degradation of the energy carried by an energetic particle, we therefore arbitrarily call whichever electron emerges from the collision with a C.M. angle θ less than $\pi/2$ the "test" particle. The θ integrations of equation (5.54) are readily performed to yield

$$\int_o^{\pi/2} (1 - \cos\theta) \sin\theta \, d\theta = \int_o^1 (1 - y) \, dy = 1/2 \qquad (5.55)$$

$$\int_o^{\pi/2} \frac{(1 - \cos\theta) \sin\theta \, d\theta}{[1 + \frac{1}{2}\theta_o^2 - \cos\theta]^2} = - \frac{1}{1 + \theta_o^2/2} - \ln\left(\frac{\frac{1}{2}\theta_o^2}{1 + \frac{1}{2}\theta_o^2}\right) \qquad (5.56)$$

$$\int_o^{\pi/2} \frac{(1 - \cos\theta) \sin\theta \, d\theta}{[1 + \frac{1}{2}\theta_o^2 + \cos\theta]^2} = \frac{1}{1 + \frac{1}{2}\theta_o^2} - \ln\left(\frac{2 + \frac{1}{2}\theta_o^2}{1 + \frac{1}{2}\theta_o^2}\right) \qquad (5.57)$$

$$\int_o^{\pi/2} \frac{(1 - \cos\theta) \sin\theta \, d\theta}{[1 + \frac{1}{2}\theta_o^2 - \cos\theta][1 + \frac{1}{2}\theta_o^2 + \cos\theta]} = \frac{\frac{1}{2}\theta_o^2}{2(1 + \frac{1}{2}\theta_o^2)}$$

$$\ln\left(\frac{\frac{1}{2}\theta_o^2}{1 + \frac{1}{2}\theta_o^2}\right) + \frac{2 + \frac{1}{2}\theta_o^2}{2(1 + \frac{1}{2}\theta_o^2)} \ln\left(\frac{2 + \frac{1}{2}\theta_o^2}{1 + \frac{1}{2}\theta_o^2}\right) \qquad (5.58)$$

The quantity θ_o, which is given by

$$\theta_o = \hbar/\lambda_D\, b \tag{5.59}$$

where λ_D is the Debye shielding length, and b is the C.M., momentum of equation (5.28), is very small, so that the dominant contribution to equation (5.54) from the four integrals (5.55) through (5.58) comes from the logarithmic part of (5.56). Thus, we have

$$\frac{dE}{dt} \simeq \sqrt{2\pi}\, \frac{nce^4}{(\gamma^2-1)^2}\, \frac{(mc^2)^{1/2}}{(kT)^{3/2}} \int_1^\infty \Gamma(\gamma-\Gamma)e^{-\frac{mc^2}{kT}(\Gamma-1)} \int_{(\gamma_r)_{min}}^{(\gamma_r)_{max}} \frac{\gamma_r^{\,2}(\gamma_r+1)}{(\gamma_r^{\,2}-1)^{3/2}} \ln\left[\frac{\hbar^2}{m^2c^2\lambda_D^{\,2}(\gamma_r-1)}\right] d\gamma_r\, d\Gamma \tag{5.60}$$

If we invert the order of integration in (5.60), the integration can be performed without difficulty. Therefore, we rewrite (5.60) as

$$\frac{dE}{dt} \simeq \sqrt{2\pi}\, \frac{nce^4}{(\gamma^2-1)^{1/2}}\, \frac{(mc^2)^{1/2}}{(kT)^{3/2}} \int_1^\infty \frac{\gamma_r^{\,2}(\gamma_r+1)}{(\gamma_r^{\,2}-1)^{3/2}} \ln\left[\frac{\hbar^2}{m^2c^2\lambda_D^{\,2}(\gamma_r-1)}\right] \int_{(\gamma_r)_{min}}^{(\gamma_r)_{max}} \Gamma(\gamma-\Gamma)e^{-\frac{mc^2}{kT}(\Gamma-1)}\, d\Gamma\, d\gamma_r \tag{5.61}$$

Now since

$$\int \Gamma(\gamma-\Gamma)e^{-\frac{mc^2}{kT}(\Gamma-1)}\, d\Gamma = \left(\frac{kT}{mc^2}\right)^3 \left[2 + \left(\frac{mc^2}{kT}\right)(2\Gamma-\gamma) + \left(\frac{mc^2}{kT}\right)^2 (\Gamma^2-\gamma\Gamma)\right] e^{-\frac{mc^2}{kT}(\Gamma-1)} \tag{5.62}$$

The energy loss equation becomes

$$\frac{dE}{dt} \simeq \frac{2\sqrt{2\pi}}{mc^2} \frac{nce^4}{(\gamma^2-1)^{1/2}} \left(\frac{kT}{mc^2}\right)^{1/2} \int_1^\infty \frac{\gamma_r^2(\gamma_r+1)}{(\gamma_r^2-1)^{3/2}} \ln\left[\frac{\hbar^2}{m^2c^2\lambda_D^2(\gamma_r-1)}\right]$$

$$e^{-\frac{mc^2}{kT}(\gamma\,\gamma_r-1)} \left\{ \left[\left(\frac{mc^2}{kT}\right) \gamma(2\gamma_r-1)(\gamma^2-1)^{1/2}(\gamma_r^2-1)^{1/2}\right.\right.$$

$$\left. + 2(\gamma^2-1)^{1/2}(\gamma_r^2-1)^{1/2}\right] \cosh\left[\frac{mc^2}{kT}(\gamma^2-1)^{1/2}(\gamma_r^2-1)^{1/2}\right]$$

$$- \left[\left(\frac{mc^2}{kT}\right)\left(\gamma^2\gamma_r(\gamma_r-1) + (\gamma^2-1)(\gamma_r^2-1)\right) + \gamma(2\gamma_r-1)\right.$$

$$\left.\left. + 2\left(\frac{kT}{mc^2}\right)\right] \sinh\left[\frac{mc^2}{kT}(\gamma^2-1)^{1/2}(\gamma_r^2-1)^{1/2}\right]\right\} d\gamma_r \qquad (5.63)$$

The logarithmic term in the above equation can be broken into two parts; *i.e.*,

$$\ln\left[\frac{\hbar^2}{m^2c^2\lambda_D^2(\gamma_r-1)}\right] = \ln\left(\frac{m}{kT}\left[\frac{\hbar}{m\lambda_D}\right]^2\right) + \ln\left[\frac{kT}{mc^2(\gamma_r-1)}\right] \qquad (5.64)$$

With

$$\lambda_D^2 = \frac{kT}{ne^2} \qquad (5.65)$$

the argument of the first of these two logarithms can be written as

$$\frac{m}{kT}\left(\frac{\hbar}{m\lambda_D}\right)^2 = \frac{n\hbar^2e^2}{m(kT)^2} = 1.058 \times 10^{-28} \frac{n}{(kT)^2} \qquad (5.66)$$

when n is given in electrons per cm^3, and kT is expressed in keV. The argument of the second logarithm in (5.64) is just the ratio of kT to the kinetic energy, in the C.M. coordinate system, of the test particle electron. Thus, while the quantity $m/kT(\hbar/m\lambda_D)^2$ will be on the order of 10^{-15}, the quantity $kT/mc^2(\gamma_r-1)$ will be on the order of 10^{-2} or 10^{-3}. This means that to a reasonable approximation, we may set

$$\ln\left[\frac{\hbar^2}{m^2c^2\lambda_D{}^2(\gamma_r-1)}\right] \simeq \ln\left[\frac{m}{kT}\left(\frac{\hbar}{m\lambda_D}\right)^2\right] \tag{5.67}$$

and, since the quantity on the right-hand side of equation (5.67) does not depend on γ_r, we may then remove it from the integral in (5.63).

With the logarithmic term out of the integrand, we can now perform an integration by parts on the cosh term in (5.63), converting this to a sinh term which can be combined with the sinh term already present. After this operation has been carried out we find that we have

$$\frac{dE}{dt} \simeq -\frac{2\sqrt{2\pi}}{mc^2}\, nce^4 \left(\frac{kT}{mc^2}\right)^{1/2} \ln\left[\frac{m}{kT}\left(\frac{\hbar}{m\lambda_D}\right)^2\right] \Big\{ 2\left(\frac{mc^2}{kT}\right)\gamma+2 \Big] \, e^{-\frac{mc^2}{kT}(\gamma-1)} - \frac{1}{(\gamma^2-1)^{1/2}} \int_1^\infty \frac{1}{(\gamma_r{}^2-1)^{1/2}} \Big\{ \left(\frac{mc^2}{kT}\right)\left[\gamma_r{}^2(\gamma_r+1)-\gamma^2\gamma_r\right]$$

$$-\gamma[4\gamma_r{}^2-1] - 2\left(\frac{kT}{mc^2}\right)[2\gamma_r+1] \Big\} \, e^{-\frac{mc^2}{kT}(\gamma\gamma_r-1)}$$

$$\sinh\left[\frac{mc^2}{kT}(\gamma^2-1)^{1/2}(\gamma_r{}^2-1)^{1/2}\right] d\gamma_r \Big\} \tag{5.68}$$

The integral in (5.68) can be carried forward most readily by expressing the sinh factor in terms of its component exponentials, and then transforming variables to obtain integrals of the form

$$\int f(x)\, e^{-ax}\, dx$$

The variable x thus must take the form of

$$x_{\pm} \equiv \gamma\, \gamma_r - 1 \pm \sqrt{\gamma^2 - 1}\, \sqrt{\gamma_r - 1} \tag{5.69}$$

Both x_+ and x_- are equal to $(\gamma - 1)$ for $\gamma_r = 1$. However, we note that while $\partial x_+/\partial \gamma_r$ is positive for all $\gamma_r \geqslant 1$, $\partial x_-/\partial \gamma_r$ starts out negative (equal to $-\infty$) at $\gamma_r = 1$, becomes zero for $\gamma_r = \gamma$, and is positive from there on out, becoming $(\gamma - \sqrt{\gamma^2 - 1})$ as $\gamma_r \to \infty$. Thus, while the γ_r to x_+ transformation is one-to-one on the interval of integration, we must divide the γ_r integral into two parts, one going from $\gamma_r = 1$ to $\gamma_r = \gamma$, and the other from $\gamma_r = \gamma$ to infinity, if we are to have a one-to-one transformation for each integral.

From equation (5.69), we find that

$$\gamma_r = \gamma(x_{\pm} + 1) \pm \sqrt{\gamma^2 - 1}\, \sqrt{(x_{\pm} + 1)^2 - 1} \tag{5.70}$$

where the $\pm$ ambiguity in front of the square roots is not immediately relatable to the $\pm$ subscript on x. However, since

$$\frac{\partial \gamma_r}{\partial x_{\pm}} = \gamma \pm \frac{(x_{\pm} + 1)\sqrt{\gamma^2 - 1}}{\sqrt{(x_{\pm} + 1)^2 - 1}} \tag{5.71}$$

and we have found previously that $\gamma_r = 1$ for $x_{\pm} = \gamma - 1$ and that at this point, $\partial x_{\pm}/\partial \gamma_r = \pm\infty$, so that

$$\left(\frac{\partial \gamma_r}{\partial x_{\pm}}\right)_{x_{\pm} = \gamma - 1} = 0$$

then the minus sign in equation (5.71), and thus in (5.70), must be used for the two integrals which include the point $\gamma_r = 1$. The third integral, which covers the interval $\gamma \leqslant \gamma_r \leqslant \infty$ through the γ_r to x-transformation, requires, as we saw above, that

$$\left(\frac{\partial \gamma_r}{\partial x_-}\right)_{x_- = \infty} = \frac{1}{(\partial x_-/\partial \gamma_r)_{\gamma_r = \infty}} = \frac{1}{\gamma - \sqrt{\gamma^2 - 1}}$$

which is possible only if we use the plus sign in (5.71) and (5.70). Therefore, we have

$$\int_1^\infty f(\gamma_r) e^{-\frac{mc^2}{kT}[\gamma\gamma_r - 1 + \sqrt{\gamma^2-1}\sqrt{\gamma_r^2-1}]} d\gamma_r =$$

$$\int_{\gamma-1}^\infty f(\gamma_r(x_+)) e^{-\frac{mc^2}{kT}x_+} \left[\gamma - \frac{(x_+ + 1)\sqrt{\gamma^2-1}}{\sqrt{(x_+ + 1)^2 - 1}}\right] dx_+ \qquad (5.72)$$

and

$$\int_1^\infty f(\gamma_r) e^{-\frac{mc^2}{kT}[\gamma\gamma_r - 1 - \sqrt{\gamma^2-1}\sqrt{\gamma_r^2-1}]} d\gamma_r =$$

$$\int_0^{\gamma-1} f(\gamma_r(x_-)) e^{-\frac{mc^2}{kT}x_-} \left[\frac{(x_- + 1)\sqrt{\gamma^2-1}}{\sqrt{(x_- + 1)^2 - 1}} - \gamma\right] dx_-$$

$$+ \int_0^\infty f(\gamma_r(x_-)) e^{-\frac{mc^2}{kT}x_-} \left[\gamma + \frac{(x_+ + 1)\sqrt{\gamma^2 - 1}}{\sqrt{(x_- + 1)^2 - 1}}\right] dx_- \qquad (5.73)$$

The function $f(\gamma_r)$ is, from equation (5.68),

$$f(\gamma_r) = \frac{1}{\sqrt{\gamma_r^2 - 1}} \left\{ \left(\frac{mc^2}{kT}\right) \gamma_r(\gamma_r^2 + \gamma_r - \gamma^2) - \gamma(4\gamma_r^2 - 1) - 2\left(\frac{kT}{mc^2}\right)(2\gamma_r + 1) \right\} \qquad (5.74)$$

A considerable simplification arises from the fact that

$$\sqrt{(\gamma_r^2 - 1)} = \sqrt{\left[\gamma(x+1) \pm \sqrt{\gamma^2-1}\sqrt{(x+1)^2 - 1}\right]^2 - 1} =$$

$$\left|\gamma\sqrt{(x+1)^2 - 1} \pm (x+1)\sqrt{\gamma^2 - 1}\right| \qquad (5.75)$$

Thus, the factor $1/\sqrt{\gamma_r^2 - 1}$ in $f(\gamma_r)$ is cancelled out by the numerator of the fraction which results from combining the two terms inside the square brackets in each of the integrals in equations (5.72) and (5.73).

If we define

$$I \equiv \tfrac{1}{2}\int_1^\infty f(\gamma_r) e^{-\frac{mc^2}{kT}[\gamma\gamma_r - 1 - \sqrt{\gamma^2-1}\sqrt{\gamma_r^2-1}]} d\gamma_r - \int_1^\infty \tfrac{1}{2} f(\gamma_r)\exp\left[-\frac{mc^2}{kT}(\gamma\gamma_r - 1 + \sqrt{\gamma^2-1}\sqrt{\gamma_r^2-1})\right] d\gamma_r \tag{5.76}$$

then we have

$$\frac{dE}{dt} \simeq -\frac{2\sqrt{2\pi}}{mc} ne^4 \left(\frac{kT}{mc^2}\right)^{1/2} \ln\left[\frac{m}{kT}\left(\frac{\hbar}{m\lambda_D}\right)^2\right] \left\{ 2\left[\left(\frac{mc^2}{kT}\right)\gamma + 2\right] e^{-\frac{mc^2}{kT}(\gamma-1)} - \frac{1}{\sqrt{\gamma^2-1}} I \right\} \tag{5.77}$$

when $f(\gamma_r)$ is given by equation (5.74). The function $I(\gamma)$ can be found, after a bit of cancellation, to be

$$\begin{aligned} I = \int_0^{\gamma-1} \frac{e^{-\frac{mc^2}{kT}x}}{\sqrt{(x+1)^2-1}} & \left\{ \left[\left(\frac{mc^2}{kT}\right)(4\gamma^3 - 3\gamma)\right] x^3 \right. \\ & + \left[\left(\frac{mc^2}{kT}\right)(12\gamma^3 + 2\gamma^2 - 9\gamma - 1) - (8\gamma^3 - \gamma)\right] x^2 \\ & + \left[\left(\frac{mc^2}{kT}\right)(8\gamma^3 + 4\gamma^2 - 6\gamma - 2) - (16\gamma^3 - 8\gamma) - \left(\frac{kT}{mc^2}\right)4\gamma\right] x \\ & \left. + \left[\left(\frac{mc^2}{kT}\right)\gamma^2 - (4\gamma^3 - \gamma) - \left(\frac{kT}{mc^2}\right)(4\gamma + 2)\right] \right\} dx \\ + \sqrt{\gamma^2-1} \int_{\gamma-1}^\infty e^{-\frac{mc^2}{kT}x} & \left\{ \left[\left(\frac{mc^2}{kT}\right)(4\gamma^2 - 1)\right] x^2 \right. \\ & + \left[\left(\frac{mc^2}{kT}\right)(8\gamma^2 + 2\gamma - 2) - 8\gamma^2\right] x + \left[\left(\frac{mc^2}{kT}\right)(2\gamma^2 + 2\gamma) \right. \\ & \left.\left. - (8\gamma^2) - \left(\frac{kT}{mc^2}\right)(4)\right] \right\} dx \end{aligned} \tag{5.78}$$

The second integral in (5.78) is readily evaluated, giving

$$\int_{\gamma-1}^{\infty} e^{-\frac{mc^2}{kT}x} \left\{ \left[\left(\frac{mc^2}{kT}\right)\left(4\gamma^2 - 1\right)\right] x^2 + \left[\left(\frac{mc^2}{kT}\right)\left(8\gamma^2 + 2\gamma - 2\right) - 8\gamma^2 \right] x + \left[\left(\frac{mc^2}{kT}\right)\left(2\gamma^2 + 2\gamma\right) - 8\gamma^2 - 4\left(\frac{kT}{mc^2}\right)\right] \right\} dx =$$

$$\left\{ \left[4\gamma^4 - \gamma^2 + 1 \right] - 6\left(\frac{kT}{mc^2}\right)^2 \right\} e^{-\frac{mc^2}{kT}(\gamma-1)} \qquad (5.79)$$

The first integral in (5.78), however, while perfectly well behaved, does not appear to integrate to any finite combination of tabulated functions. Therefore it seems that in order to proceed, we must either evaluate this integral numerically, or must resort to some approximation scheme.

The basic integral form we must consider is

$$I_o = \int_o^{\gamma-1} \frac{e^{-ax}}{\sqrt{x^2 + 2x}} \, dx \qquad (5.80)$$

since the first integral in (5.78) is composed of terms of this form and various derivatives of it with respect to the parameter a. This parameter, a, is given by

$$a \equiv mc^2/kT \qquad (5.81)$$

For electrons, $mc^2 \simeq 510$ keV, while the plasma electron temperature is likely to be considerably less than 100 keV. In fact, kT is probably on the order of 10 keV for most problems of interest. Thus, the quantity $1/a = (kT/mc^2)$ is a reasonable expansion parameter for our problem. In the integral of equation (5.80), if we replace the variable x by the new variable $y = ax$, we have

$$I_o = \int_o^{a(\gamma-1)} \frac{e^{-y}\, dy}{\sqrt{y(y+2a)}} = \frac{1}{\sqrt{2a}} \int_o^{a(\gamma-1)} \frac{e^{-y}\, dy}{\sqrt{y}\sqrt{1 + 1/2a\, y}} \qquad (5.82)$$

Since $a \gg 1$, we have $y/2a \ll 1$ wherever e^{-y} is greatly different from zero. We can therefore expand the term $(1 + 1/2a\ y)^{-1/2}$ in a power series, obtaining

$$I_o \simeq \frac{1}{\sqrt{2a}} \int_0^{a(\gamma-1)} \left[y^{-1/2} - \frac{1}{4a} y^{1/2} + \frac{3}{32a^2} y^{3/2} - \frac{5}{128a^3} y^{5/2} - \ldots \right] e^{-y} \, dy \tag{5.83}$$

Keeping only the lowest three orders in $1/a$, the first integral in equation (5.78) becomes

$$\int_0^{\gamma-1} \frac{e^{-\frac{mc^2}{kT}x}}{\sqrt{(x+1)^2-1}} \left\{ \left[\left(\frac{mc^2}{kT}\right)(4\gamma^3 - 3\gamma) \right] x^3 + \left[\left(\frac{mc^2}{kT}\right) (2\gamma^3 + 2\gamma^2 - 9\gamma - 1) \right. \right.$$

$$\left. - (8\gamma^3 - \gamma) \right] x^2 + \left[\left(\frac{mc^2}{kT}\right) (8\gamma^3 + 4\gamma^2 - 6\gamma - 2) - (16\gamma^3 - 8\gamma) \right.$$

$$\left. \left. - 4\gamma \left(\frac{kT}{mc^2}\right) \right] x + \left[\left(\frac{mc^2}{kT}\right) \gamma^2 - 4(\gamma^3 - \gamma) - \left(\frac{kT}{mc^2}\right) (4\gamma + 2) \right] \right\} dx$$

$$\cong \left(\frac{mc^2}{2kT}\right)^{1/2} \int_0^{\frac{mc^2}{kT}(\gamma-1)} e^{-y} \left\{ \left[\gamma^2 y^{-1/2} + \left(\frac{kT}{mc^2}\right) \left[- (4\gamma^3 - \gamma) y^{-1/2} \right. \right. \right.$$

$$\left. + (8\gamma^3 + \frac{15}{4}\gamma^2 - 6\gamma - 2) y^{1/2} \right] + \left(\frac{kT}{mc^2}\right)^2 \left[- (4\gamma + 2) y^{-1/2} \right.$$

$$\left. \left. - (15\gamma^3 - \frac{31}{4}\gamma) y^{1/2} + (10\gamma^3 + \frac{35}{32}\gamma^2 - \frac{15}{2}\gamma - \frac{1}{2}) y^{3/2} \right] \right\} dy \tag{5.84}$$

We now have three readily evaluable integrals; they are

$$\int_0^{a(\gamma-1)} y^{-1/2} e^{-y} \, dy = \sqrt{\pi} \; \Phi\left(\sqrt{a(\gamma-1)}\right) \tag{5.85}$$

$$\int_0^{a(\gamma-1)} y^{1/2} e^{-y} \, dy = \tfrac{1}{2}\sqrt{\pi} \; \Phi\left(\sqrt{a(\gamma-1)} - \sqrt{a(\gamma-1)}\right) e^{-a(\gamma-1)} \tag{5.86}$$

and

$$\int_0^{a(\gamma-1)} y^{3/2} e^{-y} \, dy = \frac{3}{4}\sqrt{\pi}\, \Phi\left(\sqrt{a(\gamma-1)}\right) - \left[\frac{3}{2} + a(\gamma-1)\right] \sqrt{a(\gamma-1)}\, e^{-a(\gamma-1)} \tag{5.87}$$

where $\Phi(z) = \frac{2}{\sqrt{\pi}} \int_0^z e^{-t^2} \, dt$ is the standard error function.

It may appear at first glance that our expansion in terms of 1/a is defeated, since each integral of the form

$$\int_{0}^{a(\gamma-1)} e^{\frac{2n+1}{2}} e^{-y} \, dy \tag{5.88}$$

contains a term proportional to $e^{(2n+1)/2}$ while the expansion process illustrated in equation (5.84) multiplies the integral by one factor of order $a^{-(2n+1/2)}$. However, we note that the integrals of the form of (5.88) are functions only of the combination $a(\gamma-1)$, so that the term $a^{2n+1/2}$ is multiplied by $(\gamma-1)^{2n+1/2}$; it is additionally multiplied by $e^{-a(\gamma-1)}$. Thus, if $a(\gamma-1) \gg 1$, the exponential causes these terms to become negligible, leaving only the error functions. The terms containing $e^{-a(\gamma-1)}$ will be non-negligible only when $a(\gamma-1) \ll a$.

From equation (5.78), therefore, if we utilize (5.79), (5.84), and (5.85) through (5.87), we obtain

$$\begin{aligned} I \simeq \sqrt{\frac{\pi}{2}} \left(\frac{mc^2}{kT}\right)^{1/2} \Bigg\{ & \gamma^2 + \left(\frac{kT}{mc^2}\right) \left(\frac{15}{8}\gamma^2 - 2\gamma - 1\right) + \left(\frac{kT}{mc^2}\right)^2 \\ & \left(\frac{105}{128}\gamma^2 - \frac{23}{4}\gamma - \frac{19}{8}\right) \Bigg\} \Phi\left(\sqrt{\frac{mc^2}{kT}(\gamma-1)}\right) - \frac{1}{\sqrt{2}} \left(\frac{kT}{mc^2}\right)^{1/2} \\ & \Bigg\{ \left(8\gamma^3 + \frac{15}{4}\gamma^2 - 6\gamma - 2\right) + \left(\frac{kT}{mc^2}\right) \Bigg[\left(\frac{105}{64}\gamma^2 - \frac{7}{2}\gamma - \frac{3}{4}\right) \\ & + \left(10\gamma^3 + \frac{15}{32}\gamma^2 - \frac{15}{2}\gamma - \frac{1}{2}\right) \left(\frac{mc^2}{kT}\right) (\gamma - 1) \Bigg] \Bigg\} e^{-\frac{mc^2}{kT}(\gamma-1)} \\ & + \sqrt{\gamma^2 - 1} \left\{ (4\gamma^4 - \gamma^2 + 1) - 6\left(\frac{kT}{mc^2}\right)^2 \right\} e^{-\frac{mc^2}{kT}(\gamma-1)} \end{aligned} \tag{5.89}$$

This result may now be incorporated into equation (5.77) to give us our final form for dE/dt. However, let us first simplify the notation somewhat by introducing the dimensionless parameter

$$R \equiv \sqrt{\frac{mc^2}{kt}(\gamma-1)} \tag{5.90}$$

We note that R^2 is just the ratio of the kinetic energy, $mc^2(\gamma-1)$ of the test particle electron to the temperature, kT, of the plasma electrons. In the non-relativistic limit,

$$R \rightarrow \sqrt{\frac{mv^2}{2kT}} \qquad \text{for } \beta = \frac{v}{c} \ll 1$$

In terms of R, then, we have

$$\frac{dE}{dt} \simeq \frac{2\pi ne^4}{m\sqrt{\gamma+1}} \left(\frac{m}{kT}\right)^{\frac{1}{2}} \ln\left[\frac{m}{kT}\left(\frac{\hbar}{m\lambda_D}\right)^2\right] \Bigg\{ \Bigg[\gamma^2 \frac{\Phi(R)}{R}$$

$$- \frac{2\sqrt{2(\gamma+1)}}{\sqrt{\pi}} e^{-R^2} \Bigg] + \left(\frac{kT}{mc^2}\right) \Bigg[\left(\frac{15}{8}\gamma^2 - 2\gamma - 1\right) \frac{\Phi(R)}{R}$$

$$+ \frac{\sqrt{2(\gamma+1)}}{\sqrt{\pi}} (4\gamma^4 - \gamma^2 - 3)e^{-R^2} - \frac{15}{4\sqrt{\pi}} \gamma^2 e^{-R^2} \Bigg] + \left(\frac{kT}{mc^2}\right)^2$$

$$\Bigg[\left(\frac{105}{128}\gamma^2 - \frac{23}{4}\gamma - \frac{19}{8}\right) \frac{\Phi(R)}{R} - \frac{1}{\sqrt{\pi}} \left(\frac{105}{64}\gamma^2 - \frac{7}{2}\gamma - \frac{3}{4}\right) e^{-R^2}$$

$$- \frac{1}{\sqrt{\pi}} \left(10\gamma^3 + \frac{291}{32}\gamma^2 + \frac{1}{2}\gamma + \frac{1}{2}\right) R^2 e^{-R^2} \Bigg] \Bigg\} \tag{5.91}$$

Some further simplification is possible by remembering that those terms multiplied by e^{-R^2} are negligible if $R \gg 1$. Thus any time we can factor out $(\gamma-1)$ from a coefficient, we can multiply by (mc^2/kT) to obtain R^2 and move the whole term down one order farther in magnitude. Thus, for example,

$$\begin{aligned}
(4\gamma^4 - \gamma^2 - 3) &= (4\gamma^3 + 4\gamma^2 + 3\gamma + 3)(\gamma - 1) \\
&= \left(\frac{kT}{mc^2}\right) (4\gamma^3 + 4\gamma^2 + 3\gamma + 3) \\
&= \left(\frac{kT}{mc^2}\right) [(4\gamma^2 + 8\gamma + 11)(\gamma - 1) + 14] R^2 \\
&= \left(\frac{kT}{mc^2}\right) (14R^2) + \left(\frac{kT}{mc^2}\right) (4\gamma^2 + 8\gamma + 11) R^4
\end{aligned} \tag{5.92}$$

We are therefore entitled to write dE/dt, within the limits of our approximation, as

$$\frac{dE}{dt} \simeq \frac{2\pi ne^4}{m\sqrt{\gamma+1}} \left(\frac{m}{kT}\right)^{1/2} \ln\left[\frac{m}{kT}\left(\frac{\hbar}{m\lambda_D}\right)^2\right] \Bigg\{ \left[\gamma^2 \frac{\Phi(R)}{R} - \frac{2\sqrt{2(\gamma+1)}}{\sqrt{\pi}} e^{-R^2}\right]$$

$$+ \left(\frac{kT}{mc^2}\right) \left[\left(\frac{15}{8}\gamma^2 - 2\gamma - 1\right) \frac{\Phi(R)}{R} - \frac{15}{4\sqrt{\pi}} e^{-R^2}\right] + \left(\frac{kT}{mc^2}\right)^2$$

$$\left[\left(\frac{105}{128}\gamma^2 - \frac{23}{4}\gamma - \frac{19}{8}\right) \frac{\Phi(R)}{R} + \frac{14\sqrt{2(\gamma+1)}}{\sqrt{\pi}} R^2 e^{-R^2}\right.$$

$$\left. + \frac{1}{\sqrt{\pi}} \left(\frac{167}{64} - \frac{915}{32} R^2\right) e^{-R^2}\right] \Bigg\} \tag{5.93}$$

While equation (5.93) carries dE/dt out to three orders in (kT/mc^2), we must remember that when we removed the logarithm from the integrand through the approximation of (5.67), we threw away a term which was down by a factor of 40 or less from the term retained. Therefore, even for (mc^2/kT) no larger than about 5, the terms in (5.93) which are multiplied by $(kT/mc^2)^2$ are probably not very meaningful.

The results of equation (5.93) are, however, valid for almost any test particle velocity, including those which are highly relativistic. The approximation procedure obviously breaks down if γ approaches the magnitude of (mc^2/kT), but, for $kT \simeq 10$ keV, this would require 10 MeV electrons. It may be of interest, however, to observe the form equation (5.93) takes for only slightly relativistic electrons, of perhaps 100 keV or less kinetic energy. In this case, with

$$\beta \equiv v/c$$

and

$$R_o \equiv \sqrt{\frac{mv^2}{2kT}}$$

we find, keeping the lowest two orders in

$$\frac{dE}{dt} \simeq \frac{2\pi ne^4}{m}\left(\frac{m}{2kT}\right)^{1/2} \ln\left[\frac{m}{kT}\left(\frac{\hbar}{m\lambda_D}\right)^2\right]\Bigg\{\Bigg[\frac{\Phi(R_o)}{R_o}$$

$$-\frac{4}{\sqrt{\pi}}e^{-R_o^2}\Bigg] + \frac{1}{2}\beta^2\Bigg[\left(1-\frac{9}{8R_o^2}\right)\frac{\Phi(R_o)}{R_o}$$

$$-\frac{1}{\sqrt{\pi}}\left(\frac{5}{2}+\frac{15}{4R_o}-6R_o^2\right)e^{-R_o^2}\Bigg]\Bigg\} \tag{5.94}$$

The zero order term in (5.94) is the standard non-relativistic result for our assumed (Born approximation) scattering cross section used in the previous chapter.

It is interesting and useful at this point to compare the above result with the slowing down of a relativistic electron in a neutral gas. The latter is given (classically) by the Bethe formula which gives the energy loss per unit path as result of ionizing collisions as

$$\left(\frac{dE}{ds}\right)_{ion} = \frac{2\pi e^4}{m_o v^2}\, n\left\{\ln\left[\frac{m_o v^2\, E}{I^2(1-\beta^2)}\right] - \beta^2\right\} \tag{5.95}$$

where m_o is the mass of the target electron, v is the velocity of the incident electron and I is the ionization energy. The above expression represents the loss due to "soft" collisions in which the maximum energy transfer is placed at half the incident energy. "Hard" collisions between electrons are ignored according to this theory which treats the interaction as occurring between an incident relativistic electron and an atomic electron which is initially free and at rest. For slightly relativistic electrons the logarithmic term in (5.95) can be expanded as follows

$$-\ln(1-\beta^2) = \beta^2 + \frac{1}{2}\beta^4 + \frac{1}{3}\beta^6 + \frac{1}{4}\beta^8 + \ldots \tag{5.96}$$

so that upon insertion in the equation in question it becomes

$$\frac{dE}{dt} = v\,\frac{dE}{ds} = \frac{4\pi ne^4}{m_o v}\left\{\ln\frac{2E}{I} + \frac{1}{2}\beta^4\right\} \tag{5.97}$$

having kept lowest order terms in beta only.

To effect the desired comparison we return to equation (5.94) and expand it for large values of

$$R_o = \sqrt{\frac{mv^2}{2kT}}$$

which represent the ratio of the kinetic energy of the incident electron to the temperature of the target electrons which we take to be fairly cold. The asymptotic series of the error function is given by

$$\Phi(R_o) \xrightarrow[R_o \gg 1]{} 1 - \frac{e^{-R_o^2}}{\sqrt{\pi}} \left\{ \frac{1}{R_o} - \frac{1}{2R_o^3} + \frac{3}{4R_o^5} - \cdots \right\} \tag{5.98}$$

which upon substituting in (5.94) and retaining the lowest order terms in beta we get

$$\frac{dE}{dt} = \frac{4\pi ne^4}{m_o v} \ln \left[\frac{\hbar\, \omega_p}{kT}\right] \left\{ 1 + \tfrac{1}{2}\beta^2 \right\} \tag{5.99}$$

In obtaining this result we have replaced

$$\left(\frac{4\pi n}{m_o}\right)^{1/2} e$$

by the plasma frequency, ω_p; and we observe that the characteristic energy ratio in this case is that of the "plasmon" energy, $\hbar\omega_p$, to the thermal energy kT. Moreover, we note that in the case of the fully ionized gas the power loss by the incident particle depends more strongly on beta ($\sim\beta^2$) than in the ionization case where the dependence is on the fourth power of beta. Since the target particle in both analyses is treated as free, the difference in the power loss is attributed to the fact that in the ionization collision the target is taken as stationary.

In order to further illustrate the remarks made in the opening paragraphs of this chapter concerning the validity of binary collision theory of plasma heating by relativistic electrons, we consider a simple example. Since the energy loss directly to the ions is so small we may ignore it in calculating the energy lost by an electron to the plasma via binary collisions. For a plasma of length "L" this energy loss is just

$$\Delta E = \int_0^L \left(\frac{dE}{ds}\right) ds \tag{5.100}$$

where we recall that $dE/ds = (1/v)(dE/dt)$. If binary collisions constituted the only loss mechanism then ΔE would be so small that dE/ds is essentially independent of s over any reasonable length "L." Therefore we can write

$$\Delta E = \frac{L}{v}\frac{dE}{dt} \tag{5.101}$$

For a plasma with an electron temperature of 6 keV, a density of 10^{14}, and length of one meter serving as the target of an incident electron of 100 keV energy we find that $\Delta E \simeq 2.7 \times 10^{-4}$ keV, and that the efficiency of this process as given by $\Delta E/E$ is indeed quite low. Ion heating, by this theory, takes place as result of collisions with the electrons of the plasma which are preferentially heated by the incident relativistic beam. Once again the relevant parameters are the thermalization time of the incident particles and the equilibration time between the electrons and the ions of the plasma.

REFERENCES

1. Akheizer, A. I. and V. B. Berestetski. *Quantum Electrodynamics.* (New York: Interscience, 1965).
2. Babykin, M. V. Zavoisky, A. A. Ivanov, and L. I. Rudakov. "Proc. IAEA Conference, Madison; Nuclear Fusion Supplement (1972), p. 75.
3. Goldstein, H. *Classical Mechanics.* (Addison-Wesley Publishing Co., Inc., 1959).
4. Kammash, T. and D. L. Galbraith. *Nuclear Fusion, 13,* 464 (1973).
5. Lovelace, R. V. and R. N. Sudan. *Phys. Rev. Letters, 27,* No. 19, 1256 (1971).
6. Synge, J. L. *The Relativistic Gas.* North-Holland Publishing Co., 1957.

CHAPTER 6

RADIOFREQUENCY (RF) HEATING OF FUSION PLASMAS

A number of methods which appear to have the potential of heating plasmas to thermonuclear temperatures have been examined in previous chapters where some have been singled out as especially suitable as supplementary heating methods. Another method which appears to be attractive and promising in this regard is radiofrequency heating and as we note from Figure 42 large amounts of power ($\geqslant$ 1 megawatt) are available at frequencies below a few gigahertz (GHz = 10^9 cycles per second).

The electron density of a typical laboratory plasma at the present is about 3×10^{13} cm^{-3}. The plasma frequency corresponding to this density is 50 GH. We note from Figure 42 that the available RF power at 50 GH is down by a factor of about 25 from the plateau. In a fusion reactor electron densities six times higher are expected and, to heat these plasmas, microwave power of the order of 100 megawatts will be needed for supplementary heating in bringing the plasma to ignition. This seems to imply that the highest frequencies for fusion applications presently are near or below the lower hybrid frequency [$\omega = (\omega_{pi}{}^2 + \Omega_i{}^2)^{1/2}$] thus excluding electron cyclotron heating as well as heating near the plasma frequency.

The lowest useful frequencies appear to be those associated with "Transit Time Magnetic Pumping" which are in the 100 KHz range. This is perhaps the best understood RF heating method and a rigorous treatment of the subject requires use of the kinetic theory of plasma and associated collective phenomena which are beyond the scope of this discussion. Nevertheless substantial insight, as well as order of magnitude estimates of

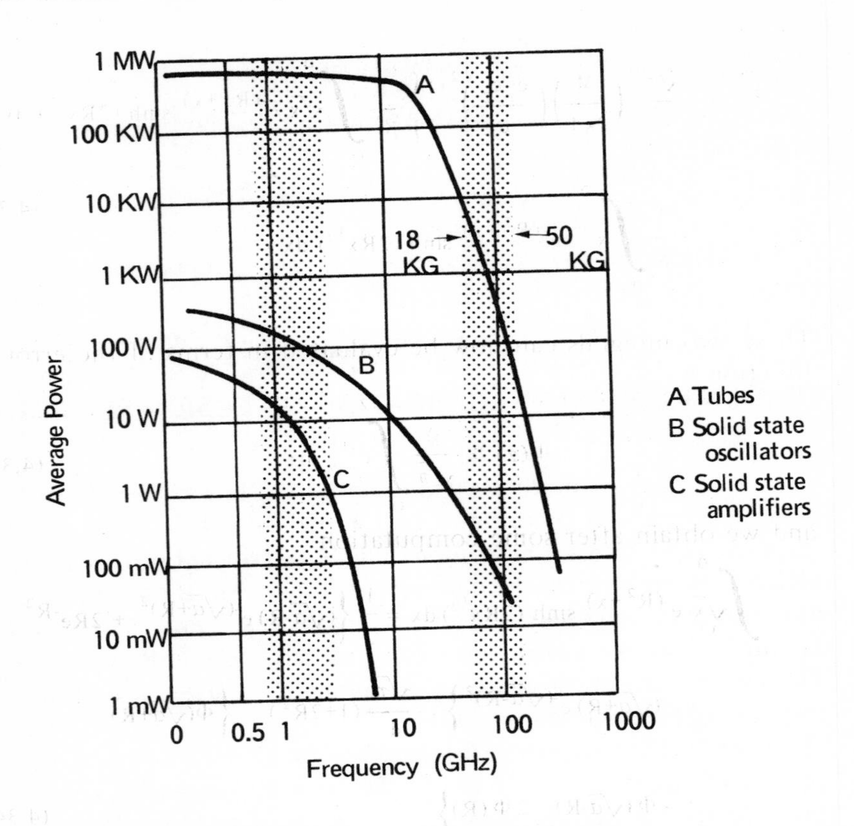

Figure 42. Microwave power from different sources.

heating can be gained using the single particle motion approach which we will adopt here.

Heating by magnetic pumping is based on the principle that if a plasma in a magnetic field is alternately compressed and expanded at a rate which is related to certain characteristic frequencies, the process will be thermodynamically irreversible and the plasma temperature will increase. The basic reason for this result is that because more work is done on the system during compression than is done by the system when it expands to its original volume without change in mass a net increase in the internal energy occurs. The alternate compression and expansion of the plasma is achieved by means of a rapidly oscillating magnetic field and thus the procedure is referred to as "magnetic pumping." We will see in equation (7.8) of Chapter 7 that constancy of the magnetic flux requires that

$$r^2 \; B = \text{constant}$$

which, when the particle gyrofrequency exceeds the collision frequency, leads to

$$B_p = B_{p_o} \, (r_o/r)^2$$

In this equation B_{p_o} represents the magnitude of the magnetic field trapped within the plasma of radius r_o before compression, and B_p and r are the corresponding values after the radial compression. If we assume that the length of the plasma remains constant during the radial compression then the volume V is proportional to r^2 and the above equation becomes

$$B_p = B_{p_o} \frac{V_o}{V} = \frac{\text{const}}{V} \tag{6.1}$$

Since the sum of the magnetic pressure inside the plasma and the plasma pressure must equal the pressure of the external magnetic field, equation (6.1) shows that an increase in the external magnetic field will be accompanied by a compression of the plasma; on the other hand as the field decreases the plasma expands. Therefore, an oscillating field will cause rapid alternate compression and expansion of the plasma.

Although both electrons and ions can be heated by magnetic pumping the interest is clearly in heating the ions. In the analysis of this heating method three characteristic times in addition to the ion gyromagnetic period are important. They are

1. the average collision time "t_c" which is the average time it takes an ion to undergo a large-angle collision as a result of encounters with both electrons and ions;
2. the oscillating time "t_o" of the radiofrequency field, *i.e.*, the reciprocal of the frequency;
3. the transit time "t_t" which is the time required for a typical ion to traverse the region over which the oscillating field is applied.

If all these times are of the same order of magnitude the analysis becomes quite complicated and difficult; however, the treatment is manageable in three limiting cases.

In the first case it is assumed that the oscillation time t_o is of the same order as the collision time t_c and both are much smaller than the transit time t_t; that is

$$t_o \simeq t_c \ll t_t$$

All these characteristic times are large compared to the ion's gyroperiod. The special case in which the gyroperiod is approximately equal to the oscillation time constitutes the basis for the "ion cyclotron resonance heating."

The second case corresponds to the situation in which the transit and oscillation times are approximately equal but they are much larger than the collision time, *i.e.*,

$$t_t \simeq t_o \gg t_c$$

Here the ion undergoes many collisions through its passage through the radiofrequency field and in the course of a single oscillation. The small effective mean free path of the ions corresponding to the short collision time may be due to high plasma densities, low temperature or cooperative plasma phenomena. In these circumstances the oscillating field produces perturbations in the plasma leading to waves which upon decay or saturation result in heating of the plasma particles. When the perturbation is associated with density variations, for example, low frequency (compared to the ion gyrofrequency) waves equivalent to sound waves can propagate in the plasma and the resulting heating is commonly referred to as "acoustic" heating. Such a wave is characterized by a frequency $\omega \simeq \sqrt{T_e/m_i}$ which is proportional to the electron temperature and inversely with the ion mass. It is also characterized as an "electrostatic" wave since the electric field of the perturbation E is parallel to its wave vector k. "Electromagnetic" waves where E is not parallel to k can also be induced in the plasma by the oscillating field and the slowest such wave is the so-called "Shear Alfven wave" which propagates along the externally applied magnetic field. These waves arise from the interaction of the magnetic field with the inertial properties of the plasma fluid and are not accompanied by pressure changes. They are characterized by a phase velocity $V_\varphi = \omega/k$ given by $V_\varphi = B/\sqrt{4\pi\rho}$ where B is the strength of the applied field and ρ is the mass density ($= n_e m_e + n_i m_i$) of the plasma.

6.1 TRANSIT TIME MAGNETIC PUMPING

The third case, which will be examined here in detail, corresponds to the situation where the plasma density is relatively low and the temperature is high. From equations (4.103) and (4.104) we recall that at low densities and high temperatures the ion collision time can be quite large, and if it is larger than both the transit and oscillation times which are of the same order then the condition

$$t_o \simeq t_t \ll t_c$$

constitutes the basis for "transit time heating." Under these circumstances the ions suffer essentially no collisions in traversing the magnetic pumping region, but they gain energy due to the radial compression induced by the oscillating field. We recall that for plasma densities and temperatures of thermonuclear interest the frequency of the oscillating field may be expected to be in the 50-100 KHz region with the oscillation time being somewhat shorter than the transit time in order for the ions to be heated rather than cooled.

Before proceeding to the detailed study of the interaction of charged particles with oscillating fields it would be desirable to obtain a rough relationship between the energy increase and the amplitude of the oscillating magnetic field from approximate considerations. We let B, the field strength in the magnetic pumping region at any time t, be given by

$$B = B_o + \Delta B \sin \omega t \tag{6.2}$$

where ΔB is the maximum amplitude of the oscillating field whose frequency is ω, and B_o is the strength of the field which confines the plasma. If we assume that B_o is spatially uniform then the field gradient at any instant t in, say, the axial direction z is roughly given by

$$\frac{dB}{dz} \simeq \frac{\Delta B}{\ell} \sin \omega t \tag{6.3}$$

where ℓ is the length of the pumping region. Because of the assertion that the transit time of an ion is approximately equal to the oscillation time, an ion moving in the axial direction will be subjected to a force associated with the field inhomogeneity given by

$$F = -\mu \frac{\partial B}{\partial z} \simeq -\mu \frac{\Delta B}{\ell} \sin \omega t \tag{6.4}$$

where μ is the magnetic moment defined by equation (1.8). Since the force is proportional to the time rate of change of the velocity of the ion in traversing the pumping region, then the transit time heating power, *i.e.*, the rate of energy increase, is proportional to F^2; thus

$$P_{tt} \simeq \text{const } (\mu \Delta B)^2 \tag{6.5}$$

However, the magnetic moment is equal to $W_\perp/B$ where $W_\perp$ is the particle energy perpendicular to the magnetic field. Since $W_\perp$ does not change appreciably equation (6.5) can be put in the form

$$P_{tt} \simeq \text{const } (\Delta B/B)^2 \tag{6.6}$$

The ratio $(\Delta B/B)$ is usually represented by M which is called the "modulation" of the field; it is more precisely defined by

$$M = B_{RF}/B_{DC} \tag{6.7}$$

where B_{RF} is the peak value of the radiofrequency field and B_{DC} is the strength of the direct current or confining magnetic field. We note from equation (6.6) and from subsequent calculations that the transit time heating rate is proportional to the square of the modulation.

6.2 GUIDING CENTER THEORY OF TRANSIT TIME HEATING

We consider in this section a transit time magnetic pumping which results from an electromagnetic wave ($\vec{E} \perp \vec{k}$) which is excited in the plasma by an array of external coils. We take the frequency of the oscillating field to be much smaller than the ion gyrofrequency, and the phase velocity ω/k to be comparable to the ion thermal velocity which is proportional to $\sqrt{T_i/m_i}$. For low-beta plasma, *i.e.*, a plasma whose pressure is much smaller than the pressure of the confining magnetic field, we would expect the phase velocity of the wave to be smaller than the Alfven velocity discussed earlier thus eliminating the possibility of an

Alfven wave propagation across the external field. Under these circumstances heating through collective processes is unlikely; rather collisionless heating of ions will take place and it can be evaluated using the concept of the motion of the guiding center. According to this approach, the motion of a charged particle in an electromagnetic field consists of gyrations around the magnetic field lines superimposed on drifts executed by its guiding center or its center of gyration. As pointed out in the previous section the ions (*i.e.*, their guiding centers) move under the force $F_z = -\mu\, \partial \widetilde{B}_z / \partial z$ where μ is the magnetic moment and $\widetilde{B}_z$ is the component of the oscillating magnetic field parallel to the static magnetic field B_o; the energy absorbed by the ion is then proportional to the square of this force.

In order to gain some insight into the mechanism of interaction between the particle and the field we use the simple physical model of a charged particle drifting through an electromagnetic field of the form

$$
\begin{aligned}
E_y &= \epsilon(x) \cos(kz - \omega t) \\
B_x &= -\frac{kc}{\omega}\, \epsilon(x) \cos(kz - \omega t) \\
B_z &= \frac{c}{\omega} \frac{\partial \epsilon}{\partial x} \sin(kz - \omega t)
\end{aligned}
\tag{6.8}
$$

The function $\epsilon(x)$ is arbitrary but assumed to change very slowly over a distance comparable to the ion Larmor radius ρ_i [see equation (1.11)], *i.e.*,

$$
\frac{1}{\rho_i} \left| \frac{1}{\epsilon} \frac{\partial \epsilon}{\partial x} \right| \ll 1 \tag{6.9}
$$

Although an electric field component parallel to the static magnetic field is needed to satisfy the condition of charge neutrality and the relative motion of electrons and ions along the field lines in the plasma, it has been omitted from (6.8) for simplicity. This is physically compatible with the self-consistent electromagnetic field in a low-beta plasma if it is assumed that the electron temperature T_e is much larger than the ion temperature T_i. From the point of view of general coupling of RF energy to the plasma the presence of an electric field $\vec{E} \parallel \vec{B}_o$ allows for a small "skin depth" or easier transmission which comes about as a result of the electrons moving along the magnetic lines of force and shorting out the $E_{\parallel}$ field. Of course if the plasma were a perfect conductor the skin depth will be zero. On the other hand if the polarization of E is approximately normal to B_o the electrons will move in the $\vec{E} \times \vec{B}$ direction and even if they do produce some current by this motion, this current is not able to short out the driving $E_{\perp}$ field. It is therefore these modes with approximately $\vec{E} \times \vec{B}$ which give rise to penetration at frequencies below the electron plasma frequency, $\omega_p = (4\pi ne^2/m_e)^{1/2}$, and these are the important modes for plasma heating.

The motion of a charged particle with charge e and mass m in an electric field $\vec{E}$ and a magnetic field $\vec{B}$ is given by the familiar Lorentz equation

$$m \frac{d\vec{v}}{dt} = e\left[\vec{E} + \frac{\vec{v} \times \vec{B}}{c}\right] \tag{6.10}$$

where c is the speed of light. The electric field felt by the particle in this case is simply E_y given by the first of equation (6.8) while the magnetic field in the z direction is the sum of the static field B_o and the component of the oscillating field B_z. The acceleration in the x direction is then given by

$$\dot{v}_x = \frac{e}{mc}\left[v_y B_o + v_y B_z\right]$$

which upon substitution from (6.8) becomes

$$\dot{v}_x = \Omega\, v_y + \frac{e}{m}\frac{v_y}{\omega} \sin(kz - \omega t) \tag{6.11}$$

having replaced eB_o/mc by the cyclotron frequency Ω. In a similar manner it is easy to show that

$$\begin{aligned} \dot{v}_y = &- \Omega\, v_x - \frac{e}{m}\frac{v_x}{\omega}\frac{\partial \epsilon}{\partial x} \sin(kz - \omega t) \\ &- \frac{e}{m}\left(\frac{k\, v_z}{\omega} - 1\right) \epsilon(x) \cos(kz - \omega t) \end{aligned} \tag{6.12}$$

and

$$\dot{v}_z = \frac{e}{m}\frac{k}{\omega} v_y\, \epsilon(x) \cos(kz - \omega t) \tag{6.13}$$

Since the electric field as expressed by $\epsilon(x)$ is arbitrary and restricted by (6.9), solutions of equations (6.11 - 6.13) can be obtained using iterative techniques. The zero order solutions are obtained from the above equations by setting the electric field equal to zero. These equations become

$$\dot{v}_{xo} = \Omega\, v_{yo}; \quad \dot{v}_{yo} = -\,\Omega\, v_{xo}; \quad \dot{v}_{zo} = 0 \qquad (6.14)$$

and if we let

$$\xi = v_{xo} + i\, v_{yo}$$

then the first two of equation (6.14) can be written as

$$\dot{\xi} + i\,\Omega\,\xi = \frac{d}{dt}(\xi\, e^{i\Omega t}) = 0 \qquad (6.15)$$

Integrating between the limits t = 0 and t = t we obtain

$$\xi(t)\, e^{i\Omega t} = \xi(0)$$

which upon separation of the real and imaginary parts yields

$$v_{xo}(t) = v_{xo}(0)\cos\Omega t + v_{yo}(0)\sin\Omega t \qquad (6.16)$$

$$v_{yo}(t) = v_{yo}(0)\cos\Omega t - v_{xo}(0)\sin\Omega t \qquad (6.17)$$

where $v_{xo}(0)$ and $v_{yo}(0)$ are the zero order velocities at t = 0. The third of equation (6.14) readily yields

$$v_{zo}(t) = v_{zo}(0) = v_{zo} \qquad (6.18)$$

which upon integration further gives

$$z(t) = z_o + v_{zo}\, t \qquad (6.19)$$

with z_o being the position at t = 0.

The first order velocity $\vec{v}_1$ (v_{1x}, v_{1y}, v_{1z}) is obtained by solving the first order acceleration equations which can be obtained from equations (6.11 - 6.13) by linearization. For example we rewrite equation (6.11) as

$$\dot{v}_{xo} + \dot{v}_{x1} = \Omega(v_{yo} + v_{y1}) + \frac{e}{m}\,\frac{v_{yo}}{\omega}\,\frac{\partial\epsilon}{\partial x}\sin(kz - \omega t)$$

having kept only v_{yo} in the last term to maintain the proper order (since $\partial\epsilon/\partial x$ is a first order quantity). In view of equation (6.14) the above equation reduces to

$$\dot{v}_{x1} = \Omega\, v_{y1} + \frac{e}{m}\,\frac{v_{yo}}{\omega}\,\frac{\partial\epsilon}{\partial x}\sin[k_{zo} + (kv_{so} - \omega)t]$$

having also utilized equation (6.19). If we now define

$$\alpha_o = k\, v_{zo} - \omega$$
$$\beta \;= k_{zo} + \alpha_o\, t \tag{6.20}$$

then the equation for $\dot{v}_{x1}$ can be expressed as

$$\dot{v}_{x1} = \Omega\, v_{y1} + \frac{e}{m}\frac{v_{yo}}{\omega}\frac{\partial\epsilon}{\partial x}\sin\beta \tag{6.21}$$

By the same token the other two components become

$$\dot{v}_{y1} = -\,\Omega\, v_{x1} - \frac{e}{m}\frac{v_{xo}}{\omega}\frac{\partial\epsilon}{\partial x}\sin\beta - \frac{e}{m}\frac{\alpha_o}{\omega}\,\epsilon(x)\cos\beta \tag{6.22}$$

$$\dot{v}_{z1} = \frac{e}{m}\frac{k}{\omega}\, v_{yo}\,\epsilon(x)\cos\beta \tag{6.23}$$

The procedure will be repeated in order to get the second order velocities; however that requires solving equations (6.21 - 6.23) for the first order velocities as a function of time. Once again we let

$$\xi_1 = v_{x1} + i\, v_{y1};\;\; \dot{\xi}_1 = \dot{v}_{x1} + i\, \dot{v}_{y1}$$

and upon substitution from equations (6.21) and (6.22) we find that

$$\dot{\xi}_1 + i\,\Omega\,\xi_1 = f(t) \tag{6.24}$$

where

$$f(t) = \frac{e}{m}\frac{\partial\epsilon}{\partial x}\frac{v_{yo}(t)}{\omega}\sin\beta - \frac{ie}{m}\frac{\partial\epsilon}{\partial x}\frac{v_{xo}(t)}{\omega}\sin\beta$$
$$-\,\frac{ie\alpha_o}{m\omega}\,\epsilon(x)\cos\beta \tag{6.25}$$

Multiplying equation (6.24) by $e^{i\Omega t}$ and integrating we obtain

$$\xi_1(t)\, e^{i\Omega t} - \xi_1(0) = \int_o^t f(u)\, e^{i\Omega u}\, du \tag{6.26}$$

and since the initial particle velocity is given by $v_{xo}(0)$ and $v_{yo}(0)$, we see that $\xi_1(0) = 0$. The integral in (6.26) cannot be performed without specifying $\epsilon(x)$ and its derivative; and since it is arbitrary and must be evaluated along the zero-order trajectory $\dot{x}(t) = v_{xo}(t)$

with $x = x_o$, $z = z_o$ for $t = 0$, we expand $\epsilon(x)$ and $\partial\epsilon/\partial x$ in a Taylor series about x_o. We therefore write

$$\epsilon(x) = \epsilon(x_o) + \partial\epsilon/\partial x \Big|_{x_o} (x - x_o) + \ldots$$

$$\partial\epsilon/\partial x = \frac{\partial\epsilon}{\partial x}\Big|_{x_o} + \frac{\partial^2\epsilon}{\partial x^2}\Big|_{x_o} (x - x_o) + \ldots \tag{6.27}$$

and will keep terms to first order in both expansions. Since we need $(x - x_o)$ we note that it is given by

$$x - x_o = \int_0^t v_{xo}(t')\, dt'$$

which upon substitution from (6.16) becomes

$$x - x_o = \frac{1}{\Omega} v_{xo}(o) \sin \Omega t + \frac{1}{\Omega} v_{yo}(o) \,[1 - \cos \Omega t] \tag{6.28}$$

We return now to equation (6.26) and substitute in it equations (6.25), (6.27) after utilizing equation (6.28). The manipulations are lengthy but straightforward and yield for $v_{x1} = R_e\, \xi_1$, and $v_{y1} = I_m\, \xi_1$, the following

$$v_{x1}(t) = \frac{e}{m\omega\alpha_o} \frac{\partial\epsilon}{\partial x} (\cos\beta - \cos kz_o)(v_{xoo} \sin \Omega t - v_{yoo} \cos \Omega t)$$

$$+ \frac{e}{2m\omega\Omega} \frac{\partial\epsilon}{\partial x} (\sin\beta - \sin kz_o)(v_{xoo} \cos \Omega t + v_{yoo} \sin \Omega t)$$

$$+ \frac{e\alpha_o}{m\omega} \frac{\epsilon(x_o)}{\alpha_o^2 - \Omega^2} [\alpha_o \sin kz_o \sin \Omega t - \Omega \cos kz_o \cos \Omega t$$

$$+ \Omega \cos\beta] + \frac{e\alpha_o}{m\omega\Omega} \frac{\partial\epsilon}{\partial x} \frac{v_{yoo}}{\alpha_o^2 - \Omega^2} [\alpha_o \sin kz_o \sin \Omega t - \Omega kz_o \cos \Omega t$$

$$+ \Omega \cos\beta] + \frac{e\alpha_o}{2m\omega\Omega} \frac{\partial\epsilon}{\partial x} \frac{v_{xoo}}{\alpha_o^2 - 4\Omega^2} [2\Omega \sin \Omega t (\cos\beta + \cos kz_o)$$

$$- \alpha_o \cos \Omega t (\sin\beta - \sin kz_o)] - \frac{e\alpha_o}{2m\omega\Omega} \frac{\partial\epsilon}{\partial x} \frac{v_{yoo}}{\alpha_o^2 - 4\Omega^2}$$

$$[\alpha_o \sin \Omega t (\sin\beta + \sin kz_o) + 2\,\Omega \cos \Omega t (\cos\beta - \cos kz_o)] \tag{6.29}$$

$$v_{y1}(t) = \frac{e}{m\omega\alpha_o}\frac{\partial\epsilon}{\partial x}(\cos\beta - \cos kz_o)(v_{xoo}\cos\Omega t + v_{yoo}\sin\Omega t)$$

$$+ \frac{e}{2m\omega\Omega}\frac{\partial\epsilon}{\partial x}(\sin\beta - \sin kz_o)(-v_{xoo}\sin\Omega t + v_{yoo}\cos\Omega t)$$

$$+ \frac{e\alpha_o}{m\omega}\frac{\epsilon(x_o)}{(\alpha_o^2 - \Omega^2)}[\alpha_o \sin kz_o \cos\Omega t + \Omega\cos kz_o \sin\Omega t$$

$$- \alpha_o \sin\beta] + \frac{e\alpha_o}{m\omega\Omega}\frac{\partial\epsilon}{\partial x}\frac{v_{yoo}}{(\alpha_o^2 - \Omega^2}[\alpha_o \sin kz_o \cos\Omega t$$

$$+ \Omega\cos kz_o \sin\Omega t - \alpha_o\sin\beta] - \frac{e\alpha_o}{2m\omega\Omega}\frac{\partial\epsilon}{\partial x}\frac{v_{xoo}}{(\alpha_o^2 - 4\Omega^2)}$$

$$[2\Omega\cos\Omega t(\cos\beta - \cos kz_o) + \alpha_o\sin\Omega t(\sin\beta + \sin kz_o)]$$

$$+ \frac{e\alpha_o}{2m\omega\Omega}\frac{\partial\epsilon}{\partial x}\frac{v_{yoo}}{(\alpha_o^2 - 4\Omega^2)}[\alpha_o\cos\Omega t(\sin\beta - \sin kz_o)$$

$$- 2\Omega\sin\Omega t(\cos\beta + \cos kz_o)] \tag{6.30}$$

where it is understood that $\partial\epsilon/\partial x$ is evaluated at x_o, and v_{xoo} and v_{yoo} are $v_{xo}(o)$ and $v_{yo}(o)$. To obtain v_{z1} we substitute for v_{yo} from (6.17), for $\epsilon(x)$ from (6.27) and (6.28), and integrate. The result is

$$v_{z1}(t) = -\frac{ek}{2m\omega\alpha_o\Omega}\frac{\partial\epsilon}{\partial x}v_{\perp oo}^2(\sin\beta - \sin kz_o) + \frac{ek\epsilon(x_o)}{m\omega(\alpha_o^2 - \Omega^2)}v_{xoo}$$

$$\left\{-\Omega(\cos\beta\cos\Omega t - \cos kz_o) - \alpha_o\sin\beta\sin\Omega t\right\}$$

$$+ \frac{ek\epsilon(x_o)}{m\omega(\alpha_o^2 - \Omega^2)}v_{yoo}\left\{\alpha_o(\sin\beta\cos\Omega t - \sin kz_o)\right.$$

$$\left.- \Omega\cos\beta\sin\Omega t\right\} + \frac{ek}{2m\omega\Omega}\frac{\partial\epsilon}{\partial x}\left(\frac{v_{xoo}^2 - v_{yoo}^2}{\alpha_o^2 - 4\Omega^2}\right)$$

$$\left\{ \alpha_o (\sin\beta \cos 2\Omega t - \sin kz_o) - 2\Omega \cos\beta \sin 2\Omega t \right\}$$

$$- \frac{ek}{m\omega\Omega} \frac{\partial\epsilon}{\partial x} \left(\frac{v_{xoo} - v_{yoo}}{\alpha_o^2 - 4\Omega^2} \right) \left\{ - \alpha_o \sin\beta \sin 2\Omega t - 2\Omega (\cos\beta \right.$$

$$\left. \cos 2\Omega t - \cos kz_o) \right\} + \frac{ek}{m\omega\Omega} \frac{\partial\epsilon}{\partial x} \frac{v_{xoo} - v_{yoo}}{\alpha_o^2 - \Omega^2} \left\{ - \alpha_o \sin\beta \sin \Omega t \right.$$

$$\left. - \Omega (\cos\beta \cos \Omega t - \cos kz_o) \right\} + \frac{ek}{m\omega\Omega} \frac{\partial\epsilon}{\partial x} \frac{v_{yoo}^2}{\alpha_o^2 - \Omega^2}$$

$$\left\{ \alpha_o (\sin\beta \sin \Omega t - \sin kz_o) - \Omega \cos\beta \sin \Omega t \right\} \tag{6.31}$$

where $v_{\perp oo}^2 = v_{xoo}^2 + v_{yoo}^2$.

Following the procedure used in obtaining the first order terms we proceed to calculate the second order terms. The equations to be solved subject to the conditions $\dot{v}_2 = 0$, at $t = 0$ are:

$$\dot{v}_{x2} = \Omega v_{y2} + \frac{e}{m} \frac{v_{y1}}{\omega} \frac{\partial\epsilon}{\partial x} \sin\beta + \frac{e}{m\omega} v_{y0} \left(\frac{\partial^2\epsilon}{\partial x^2} x_1 \sin\beta + \frac{\partial\epsilon}{\partial x} kz_1 \cos\beta \right) \tag{6.32}$$

$$\dot{v}_{y2} = - \Omega v_{x2} - \frac{e}{m} \frac{v_{x1}}{\omega} \frac{\partial\epsilon}{\partial x} \sin\beta - \frac{e}{m} \frac{v_{xo}}{\omega} \left(\frac{\partial^2\epsilon}{\partial x^2} x_1 \sin\beta + \frac{\partial\epsilon}{\partial x} kz_1 \cos\beta \right)$$

$$- \frac{e}{m} \frac{k}{\omega} v_{z1}\, \epsilon(x) \cos\beta - \frac{e}{m} \frac{\alpha_o}{\omega} \left(\frac{\partial\epsilon}{\partial x} x_1 \cos\beta - \epsilon(x) kz_1 \sin\beta \right) \tag{6.33}$$

$$\dot{v}_{z2} = \frac{e}{m} \frac{k}{\omega} v_{y1}\, \epsilon(x) \cos\beta + \frac{e}{m} \frac{kv_{yo}}{\omega} \left(\frac{\partial\epsilon}{\partial x} x_1 \cos\beta - \epsilon(x)\, kz_1 \sin\beta \right) \tag{6.34}$$

where

$$x_1 = \int_o^t v_{x1}(t')dt'; \quad z_1 = \int_o^t v_{z1}(t')dt' \qquad \text{or}$$

$$x_1(t) = \frac{e\, v_{yoo}}{m\omega\alpha_o\Omega} \frac{\partial\epsilon}{\partial x} \cos kz_o \sin \Omega t + \frac{e\, v_{xoo}}{m\omega\alpha_o\Omega} \frac{\partial\epsilon}{\partial x} \cos kz_o (\cos \Omega t\text{-}1)$$

$$+ \frac{e\, v_{yoo}}{m\omega\alpha_o(\alpha_o{}^2\text{-}\Omega^2)} \frac{\partial\epsilon}{\partial x} [\alpha_o \sin kz_o - \alpha_o \sin \beta \cos \Omega t + \Omega \cos \beta \sin \Omega t]$$

$$+ \frac{e\, v_{xoo}}{m\omega\alpha_o(\alpha_o{}^2\text{-}\Omega^2)} \frac{\partial\epsilon}{\partial x} [- \Omega \cos kz_o + \alpha_o \sin \beta \sin \Omega t + \Omega \cos \beta \cos \Omega t]$$

$$- \frac{e\, \alpha_o\, \epsilon(x_o)}{m\omega\Omega(\alpha_o{}^2\text{-}\Omega^2)} [\alpha_o \sin kz_o (\cos \Omega t\text{-}1) - \Omega \cos kz_o \sin \Omega t]$$

$$+ \frac{e\Omega\epsilon(x_o)}{m\omega(\alpha o^2\text{-}\Omega^2)} \sin \beta \cdot \qquad (6.35)$$

$$z_1(t) = \frac{ek}{2m\omega\alpha_o\Omega} \frac{\partial\epsilon}{\partial x} v_{\perp oo}^{\;2} \left[\frac{\cos \beta - \cos kz_o}{\alpha_o} + t \sin kz_o \right] - \frac{ek\epsilon(x_o)}{m\omega(\alpha_o{}^2\text{-}\Omega^2)} v_{xoo}$$

$$\Big\{ 2\alpha_o\Omega (\sin \beta \cos \Omega t - \sin kz_o) - (\alpha_o{}^2 + \Omega^2) \cos \beta \sin \Omega t$$

$$- \Omega t (\alpha_o{}^2 - \Omega^2) \cos kz_o \Big\} - \frac{ek\epsilon(x_o)}{m\omega(\alpha_o{}^2 - \Omega^2)} v_{yoo} \Big\{ (\alpha_o{}^2 + \Omega^2)$$

$$(\cos \beta \cos \Omega t - \cos kz_o) + 2\, \alpha_o\, \Omega \sin \beta \sin \Omega t + \alpha_o t (\alpha_o{}^2 - \Omega^2) \sin kz_o \Big\}$$

$$- \frac{ek}{2m\omega\Omega} \frac{\partial\epsilon}{\partial x} \frac{v_{xoo}{}^2 - v_{yoo}{}^2}{(\alpha_o{}^2 - 4\Omega^2)^2} \Big\{ (\alpha_o{}^2 + 4\Omega^2) (\cos \beta \cos 2\Omega t - \cos kz_o)$$

$$+ 4\alpha_o\, \Omega \sin \beta \sin 2\Omega t + \alpha_o t (\alpha_o{}^2 - 4\Omega^2) \sin kz_o \Big\} + \frac{ek}{m\omega\Omega} \frac{\partial\epsilon}{\partial x}$$

$$\frac{v_{xoo}\, v_{yoo}}{(\alpha_o{}^2 - 4\Omega^2)^2} \Big\{ 4\, \alpha_o\, \Omega (\sin \beta \cos 2\Omega t - \sin kz_o) - (\alpha_o{}^2 + 4\Omega^2)$$

$$\cos \beta \sin 2\Omega t - 2\Omega t (\alpha_o{}^2 - 4\Omega^2) \cos kz_o \Big\} - \frac{ek}{m\omega\Omega} \frac{\partial\epsilon}{\partial x} \frac{v_{xoo}\, v_{yoo}}{(\alpha_o{}^2 \text{-}\Omega^2)^2}$$

$$\Big\{ 2\alpha_o\, \Omega (\sin \beta \cos \Omega t - \sin kz_o) - (\alpha_o{}^2 + \Omega^2) \cos \beta \sin \Omega t$$

$$- \Omega t\,(\alpha_o^2 - \Omega^2)\cos kz_o \Bigg\} \quad - \frac{ek}{m\omega\Omega}\frac{\partial\epsilon}{\partial x}\frac{v_{yoo}^2}{(\alpha_o^2 - \Omega^2)^2} \Bigg\} (\alpha_o^2 + \Omega^2)$$

$$(\cos\beta\cos\Omega t - \cos kz_o) + 2\alpha_o \sin\beta \sin\Omega t + \alpha_o t(\alpha_o^2 - \Omega^2)\sin kz_o \Bigg\} \tag{6.36}$$

Since the main objective is to calculate the rate of change of the particle energy as a result of its interaction with the oscillating field, we need to solve equations (6.32 - 6.34) to obtain the second order velocities. That is also lengthy and tedious; instead we shall write the general equation for the rate of energy change and then examine some of the terms which contribute to it. The equation in question is given by

$$\frac{d}{dt}(mv^2/2) = \vec{v}_1 \cdot \frac{d}{dt} m\vec{v}_1 + \vec{v}_2 \cdot \frac{d}{dt} m\vec{v}_o + \vec{v}_o \cdot \frac{d}{dt} m\vec{v}_2 + \ldots \tag{6.37}$$

where terms up to second order will be retained. We focus our attention on the rate of change of the particle's energy in the z direction which we can readily write as

$$\frac{d}{dt}\left(\frac{mv_z^2}{2}\right) = mv_{z1}\dot{v}_{z1} + mv_{oz}\dot{v}_{z2}. \tag{6.38}$$

since $m\dot{v}_{oz} = 0$ by equation (6.14). A more meaningful result, however, is not the one given by the above equation; rather it is the rate of change of particle energy averaged over the initial position z_o and over a Larmor period. Before carrying out the averaging procedure we calculate the first term on the right-hand side of equation (6.38) by substituting directly from equations (6.17) and (6.23). The result is quite long and we will not write it down; rather we will take the first term of the product and demonstrate the process of averaging on it before putting down the final result. The term in question is

$$(m\, v_z \dot{v}_{z1})_1 = \frac{e^2 k^2 v_{\perp oo}^4}{4m\omega^2\Omega^2\alpha_o}\left(\frac{\partial\epsilon}{\partial x}\right)^2 (\sin\beta\cos\beta - \sin kz_o \cos\beta)$$

and its average over the initial position z_o leads to the form

$$\left\langle m\, v_{z1}\,\dot{v}_{z1}\right\rangle_{1z_o} = \frac{1}{\pi/k}\int_0^{\pi/k} dz_o\,(m\, v_{z1}\,\dot{v}_{z1})_1 \tag{6.39}$$

We recall from equation (6.20) that β is a function of z_o so that the trigonometric functions can be readily combined and integrated to give

$$\left\langle m\, v_{z1}\, \dot{v}_{z1} \right\rangle_{1z_o} = \frac{e^2 k^2}{8m\omega^2 \Omega^2} \left(\frac{\partial \epsilon}{\partial x}\right)^2 v_{\perp oo}^{\;4} \left(\frac{\sin \alpha_o t}{\alpha_o}\right) \tag{6.40}$$

Since none of the terms in this expression has Larmor periodicity, averaging over the Larmor period will not alter the result. A close observation of $m\, v_{z1}\, \dot{v}_{z1}$ in equation (6.38) will in fact reveal that the only surviving term is the first one, namely that given by (6.40). Similarly, the only term in the product $m\, v_{z0}\, \dot{v}_{z2}$ that remains after the appropriate averaging is

$$\frac{1}{\pi/k} \int_0^{\pi/k} dz_o \frac{e^2 k^3}{4m\omega^2 \Omega^2 \alpha_o} \left(\frac{\partial \epsilon}{\partial x}\right)^2 v_{\perp oo}^{\;4}\, v_{zo} \left[\frac{\cos \beta - \cos kz_o}{\alpha_o} + t \sin kz_o\right] \sin \beta = \frac{e^2 k^3\, v_{\perp oo}^{\;4}\, v_{zo}}{8m\omega^2 \Omega^2 \alpha_o} \left(\frac{\partial \epsilon}{\partial x}\right)^2 \left(- \frac{\sin \alpha_o t}{\alpha_o} + t \cos \alpha_o t\right) \tag{6.41}$$

Substitution of this result and (6.40) in (6.38) yields the average rate of change of the particle energy along the external field, *i.e.*,

$$\left\langle \frac{d}{dt} \left(\frac{m\, v_z^{\,2}}{2}\right)\right\rangle = - \frac{e^2 k^2\, v_{\perp oo}^{\;4}}{8m\omega^2 \Omega^2} \left[\omega \frac{\sin \alpha_o t}{\alpha_o^{\,2}} - k v_{zo} t \frac{\cos \alpha_o t}{\alpha_o}\right] \tag{6.42}$$

where we note that $v_{\perp oo}$ is the initial perpendicular velocity and that only terms to second order in the small quantities

$$\rho_i^{\,2} \left| \frac{1}{\epsilon} \frac{\partial^n \epsilon}{\partial x^n} \right|$$

or $(\omega/\Omega_i)^n$ have been retained. It can also be shown that the average rate of change in the perpendicular energy is zero or

$$\left\langle \frac{d}{dt} \left(\frac{m\, v_\perp^2}{2}\right)\right\rangle = 0 \tag{6.43}$$

Since the above results have been obtained for a single particle we need, in the presence of many particles, to average over the distribution of the initial velocities. We shall assume that the particles have a Maxwellian distribution in both $v_{\perp o}$ and v_{zo}, *i.e.*,

$$f_i(v_o) = \left(\frac{m}{2\pi KT_\perp}\right) \exp\left(-\frac{m\, v_{\perp o}^2}{2KT_\perp}\right) g(v_{zo})$$

$$g(v_{zo}) = \left(\frac{m}{2\pi KT_\parallel}\right)^{1/2} \exp\left(-\frac{m v_{zo}^2}{2KT_\parallel}\right) \tag{6.44}$$

where we have dropped the second index from the initial velocities since it is no longer needed. In the above expression K is the Boltzmann constant and $T_\perp$ and $T_\parallel$ are the particle temperature perpendicular and parallel to the external field, respectively. From (6.42) and (6.44) the quantity of interest becomes

$$\left\langle\!\left\langle \frac{d}{dt}\left(\frac{m\, v_z^2}{2}\right)\right\rangle\!\right\rangle = -\frac{m}{4B_o^2}\,\frac{k^2c^2}{\omega^2}\left(\frac{2KT_\perp}{m}\right)^2\left(\frac{\partial\epsilon}{\partial x}\right)^2$$

$$\int_{-\infty}^{+\infty} g(v_{zo})\left[\omega\,\frac{\sin\alpha_o t}{\alpha_o^2} - k\, v_{zo}\, t\,\frac{\cos\alpha_o t}{\alpha_o}\right] \tag{6.45}$$

where we have replaced Ω be its equivalent $e\,B_o/mc$, and have utilized the result

$$\int_0^\infty v_{\perp o}^5 \exp\left(-\frac{m\, v_{\perp o}^2}{2KT_\perp}\right) d\,v_{\perp o} = 8\left(\frac{KT_\perp}{m}\right)^3 \tag{6.46}$$

We focus our attention on the integral in (6.45) where we note that the integrand is well behaved at the singularity $\alpha_o = 0$ since by L'Hospital's rule

$$\lim_{\alpha_o\to 0}\ \frac{\cos\alpha_o t}{\alpha_o} \sim \frac{\sin\alpha_o t}{1} = 0$$

$$\lim_{\alpha_o\to 0}\ \frac{\sin\alpha_o t}{\alpha_o^2} \sim \frac{\cos\alpha_o t}{2\alpha_o} \sim \frac{\sin\alpha_o t}{2} = 0 \tag{6.47}$$

The integration, nevertheless, is not trivial due to the presence of this singularity, and contour integration will be required. We recall from (6.20) that α_o $(= k\, v_{zo} - \omega)$ depends on v_{zo}, and we further note that we can write

$$\int_{-\infty}^{+\infty} d\, v_{zo}\, g(v_{zo}) \left[\frac{\omega \sin \alpha_o t}{\alpha_o^2} - k\, v_{zo}\, t\, \frac{\cos \alpha_o t}{\alpha_o} \right]$$

$$= \frac{1}{k} \int_{-\infty}^{+\infty} d\, v_{zo} \frac{[(\partial/\partial v_{zo})\, g\,(v_{zo})]\ \omega \sin [(k\, v_{zo} - \omega)t]}{(k\, v_{zo} - \omega)}$$

$$- \int_{-\infty}^{+\infty} d\, v_{zo}\, t\, g\,(v_{zo}) \cos [(k\, v_{zo} - \omega)t] \tag{6.48}$$

having performed an integration by parts on the first term subject to the condition $g(\infty) = g(-\infty) = 0$. The second integral in the above expression can be readily carried out after expressing the cos in terms of exponential functions, completing the squares and utilizing the following familiar value of the error function:

$$\int_{-\infty}^{+\infty} e^{-\xi^2}\, d\xi = \sqrt{\pi} \tag{6.49}$$

The result is

$$\int_{-\infty}^{+\infty} d\, v_{zo}\, g(v_{zo}) \cos [(k\, v_{zo} - \omega)t] = e^{-\left(\frac{k^2 KT_{\parallel}}{2m}\right) t^2} \cos \omega t \tag{6.50}$$

We return to the first integral in (6.48) and utilizing (6.44) and the definition of sin θ we can write

$$\frac{1}{k} \int_{-\infty}^{+\infty} \frac{[(\partial/\partial v_{zo})\, g\,(v_{zo})]\ \omega \sin [(k\, v_{zo} - \omega)t]}{(k\, v_{zo} - \omega)}\, dv_{zo}$$

$$= I_1 - I_2 \tag{6.51}$$

where

$$I_1 = - \frac{\omega\, e^{-i\omega t}}{2i\sqrt{2\pi}\, k} \left(\frac{m}{KT_{\parallel}}\right)^{3/2} e^{-\frac{k^2 t^2 KT_{\parallel}}{2m}}$$

$$\int_{-\infty}^{+\infty} \frac{v_{zo}\, e^{-\left(\sqrt{\frac{m}{2kT_{\parallel}}}\, v_{zo} - ikt\sqrt{\frac{KT_{\parallel}}{2m}}\right)^2}}{(k\, v_{zo} - \omega)}$$

$$I_2 = -\frac{\omega\, e^{i\omega t}}{2i\sqrt{2\pi}\, k}\left(\frac{m}{KT_{\|}}\right)^{3/2} e^{-\frac{k^2t^2KT_{\|}}{2m}}$$

$$\int_{-\infty}^{+\infty} \frac{v_{zo}\, e^{-\left(\sqrt{\frac{m}{2KT_{\|}}}\, v_{zo} + i k t \sqrt{\frac{KT_{\|}}{2m}}\right)^2}}{(k\, v_{zo} - \omega)} \tag{6.52}$$

In evaluating the above integrals we utilize the well-known Plemelj formula which states

$$\lim_{\epsilon\to 0} \int_{-\infty}^{+\infty} \frac{h(u)\, du}{u - (x \mp i\,\epsilon)} = P \int_{-\infty}^{+\infty} \frac{h(u)\, du}{u - x} \mp i\pi$$

$$\int_{-\infty}^{+\infty} h(u)\delta\,(u - x)\, du \tag{6.53}$$

where P before the integral on the right-hand side of (6.53) denotes the principal value. By introducing the new independent variable $\xi = \sqrt{m/2KT_{\|}}\; v_{zo}$, I_1 takes on the form

$$I_1 = -\frac{\omega\, e^{-i\omega t}}{2i\sqrt{\pi}\, k^2}\left(\frac{m}{KT_{\|}}\right) e^{-\frac{k^2t^2KT_{\|}}{2m}}$$

$$\int_{-\infty}^{+\infty} d\xi\, \frac{(\xi - ikt\sqrt{KT_{\|}/2m})\, e^{-(\xi - ikt\sqrt{KT_{\|}/2m})^2}}{[(\xi - ikt\sqrt{KT_{\|}/2m}) - (\sqrt{m/2KT_{\|}}\,\frac{\omega}{k} - ikt\sqrt{KT_{\|}/2m})]}$$

$$-\frac{\omega e^{-i\omega t}}{2i\sqrt{\pi}\, k^2}\left(\frac{m}{KT_{\|}}\right) e^{-\frac{k^2KT_{\|}t^2}{2m}}$$

$$\int_{-\infty}^{+\infty} d\xi\, \frac{ikt\sqrt{KT_{\|}/2m}\;\; e^{-(\xi - ikt\sqrt{KT_{\|}/2m})^2}}{[(\xi - ikt\sqrt{KT_{\|}/2m}) - (\sqrt{m/2KT_{\|}}\,\frac{\omega}{k} - ikt\sqrt{KT_{\|}/2m})]} \tag{6.54}$$

with a similar form for I_2. We consider next the first integral in the above expression and we apply to it the Plemelj formula of (6.53) on the basis that $\omega = \omega + i\gamma$, $\gamma > 0$. Furthermore we choose the contour shown in Figure 43 to evaluate the principal

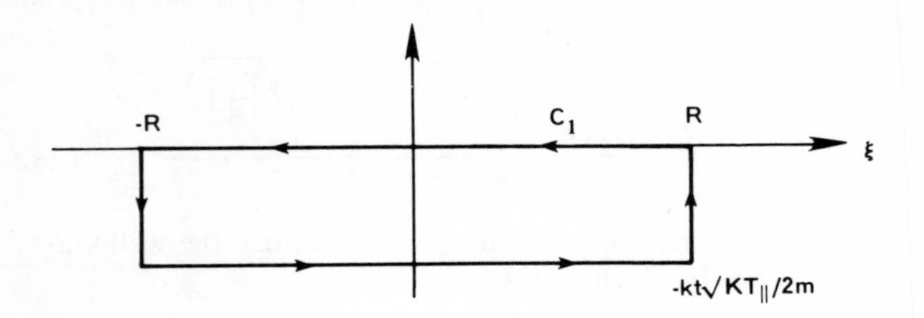

Figure 43. Contour for principal value integral.

value of the integral. The integral in question can be written as

$$\oint_{C_1} \frac{z\ e^{-z^2}}{z - a} = 2\pi i\ (a\ e^{-a^2}) \tag{6.55}$$

where $z = \xi + i\eta$, and

$$a = \sqrt{\frac{m}{2KT_{||}}}\ \frac{\omega}{k} - ikt\sqrt{\frac{KT_{||}}{2m}}$$

having invoked the residue theorem in connection with the contour which is taken to contain the singularity of interest. Carrying out the details and letting $R \to \infty$ we obtain

$$P\int_{-\infty}^{+\infty} \frac{(\xi - ikt\sqrt{KT_{||}/2m})\ e^{-(\xi - ikt\sqrt{KT_{||}/2m})^2}\ d\xi}{[(\xi - ikt\sqrt{KT_{||}/2m}) - (\sqrt{m/2KT_{||}}\ \frac{\omega}{k} - ikt\sqrt{KT_{||}/2m})]}$$

$$+ i\pi \int_{-\infty}^{+\infty} (\xi - ikt\sqrt{KT_{||}/2m})\ \exp\ (\xi - ikt\sqrt{KT_{||}/2m})$$

$$\delta\ (\xi - \sqrt{m/2KT_{||}}\ \frac{\omega}{k})\ d\xi = 2\pi i \left(\sqrt{\frac{m}{2KT_{||}}}\ \frac{\omega}{k} - ikt\sqrt{\frac{KT_{||}}{2m}} \right)$$

$$\exp \left\{ - \left(\sqrt{\frac{m}{2KT_{||}}}\ \frac{\omega}{k} - ikt\sqrt{\frac{KT_{||}}{2m}} \right)^2 \right\}$$

$$+ \int_{-\infty}^{+\infty} d\xi\ \frac{\xi\ e^{-\xi^2}}{\xi - (\sqrt{m/2KT_{||}}\ \frac{\omega}{k} - ikt\sqrt{KT_{||}/2m})} \tag{6.56}$$

A similar procedure can be followed to evaluate the second integral in (6.54) and upon combining the two results we find that

$$I_1 = \frac{\pi\omega}{2k^2}\left(\frac{m}{2\pi KT_{||}}\right)^{1/2}\left(-\frac{m}{KT_{||}}\frac{\omega}{k}\right)\exp\left[-\frac{m}{2KT_{||}}\left(\frac{\omega}{k}\right)^2\right]$$

$$-\frac{\omega\, e^{-i\omega t}}{2i\, k^2\sqrt{\pi}}\left(\frac{m}{KT_{||}}\right)e^{-\frac{k^2KT_{||}t^2}{2m}}$$

$$\int_{-\infty}^{+\infty} d\xi\, \frac{\xi\, e^{-\xi^2}}{\xi - [\sqrt{m/2KT_{||}}\,\frac{\omega}{k} - ikt\sqrt{KT_{||}/2m}]}$$

$$-\frac{\omega_t e^{-i\omega t}}{2k\sqrt{2\pi}}\left(\frac{m}{KT_{||}}\right)^{1/2}e^{-\frac{k^2KT_{||}t^2}{2m}}$$

$$\int_{-\infty}^{+\infty} d\xi\, \frac{e^{-\xi^2}}{\xi - [\sqrt{m/2KT_{||}}\,\frac{\omega}{k} - ikt\sqrt{KT_{||}/2m}]} \tag{6.57}$$

The procedure can be repeated in the evaluation of the integrals in I_2 which have the form

$$\oint_{C_2} \frac{z\, e^{-z^2}\, dz}{z - b} = 0; \qquad \oint_{C_2} \frac{e^{-z^2}\, dz}{z - b} = 0$$

with

$$b = \frac{m}{2KT_{||}}\,\frac{\omega}{k} + ikt\sqrt{KT_{||}/2m}$$

and where the contour c_2 is so chosen (as in Figure 44) as to exclude the singularity.

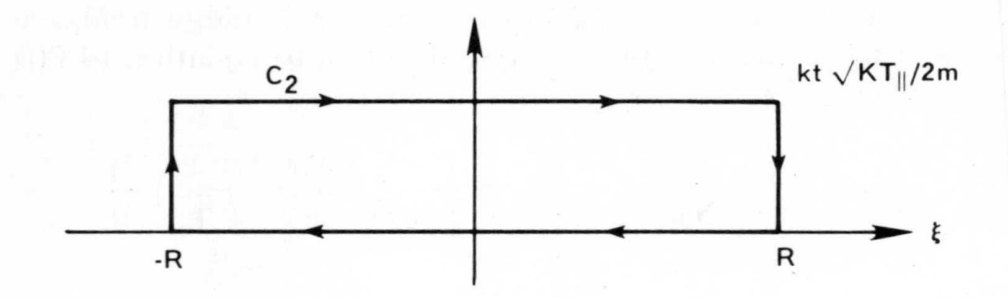

Figure 44. Contour for I_2 integrals.

The result is

$$I_2 = -\frac{\pi\omega}{2k^2}\left(\frac{m}{2\pi KT_{\|}}\right)^{\frac{1}{2}}\left(-\frac{m}{KT_{\|}}\frac{\omega}{k}\right)\exp\left\{-\frac{m}{2kT_{\|}}\left(\frac{\omega}{k}\right)^2\right\}$$

$$-\frac{\omega\, e^{i\omega t}\, e^{-\frac{k^2 KT_{\|} t^2}{2m}}}{2i\sqrt{\pi}\, k^2}\left(\frac{m}{KT_{\|}}\right)\int_{-\infty}^{+\infty}\frac{\xi\, e^{-\xi^2}\, d\xi}{[\xi-(\sqrt{m/2KT_{\|}}\;\frac{\omega}{k} + ikt\sqrt{KT_{\|}2m})]}$$

$$+\frac{\omega\, e^{i\omega t}\, e^{-\frac{k^2 KT_{\|} t^2}{2m}}}{2\sqrt{2\pi}\, k}\left(\frac{m}{KT_{\|}}\right)\int_{-\infty}^{+\infty}\frac{e^{-\xi^2}\, d\xi}{[\xi-(\sqrt{m/2KT_{\|}}\;\frac{\omega}{k} + ikt\sqrt{KT_{\|}/2m})]}$$

(6.58)

We now substitute (6.57) and (6.58) in (6.51); put the result in equation (6.45) and note that for large t we obtain

$$\left\langle\!\left\langle \frac{d}{dt}\left(\frac{m\, v_z^2}{2}\right)\right\rangle\!\right\rangle = -\frac{m}{4\, B_o^2}\;\frac{k^2 c^2}{\omega^2}\left(\frac{2KT_{\perp}}{m}\right)^2\left(\frac{\partial\epsilon}{\partial x}\right)^2\frac{\pi\omega}{kk}$$

$$\left[\frac{\partial}{\partial\, v_{zo}}\, g\,(v_{zo})\right]_{v_{zo}\,=\,\frac{\omega}{k}} \qquad (6.59)$$

having utilized the fact that

$$\left[\frac{\partial}{\partial v_{zo}} g(v_{zo})\right]_{v_{zo}=\frac{\omega}{k}} = \left(\frac{m}{2\pi KT_{\|}}\right)^{1/2}\left(-\frac{m}{KT_{\|}}\frac{\omega}{k}\right)e^{-\frac{m}{2KT_{\|}}\left(\frac{\omega}{k}\right)^2}$$

Moreover if we let for convenience $\zeta_o = \sqrt{m/2KT_{\|}}\ \omega/k$, then equation (6.59) assumes the form

$$\left\langle\!\left\langle \frac{d}{dt}\left(\frac{m\, v_z^2}{2}\right)\right\rangle\!\right\rangle = \frac{K}{\omega}\frac{T_{\perp}c^2}{B_o^2}\frac{T_{\perp}}{T_{\|}}\left(\frac{\partial\epsilon}{\partial x}\right)^2\sqrt{\pi}\,\frac{k}{|k|}\,\zeta_o\, e^{-\zeta_o^2} \tag{6.60}$$

Since at low beta, $\epsilon(x)$ in a first approximation is unaffected by the the presence of the plasma, the rate of increase of the ion kinetic energy is always positive and does not depend on the local value of the plasma density.

The RF power absorbed by the plasma per unit volume is given by

$$P = n(x)\frac{d}{dt}\left(\frac{mv^2}{2}\right) + \nabla\cdot\vec{\Phi}\left(\frac{mv^2}{2}\right) \tag{6.61}$$

where

$$\vec{\Phi}\left(\frac{mv^2}{2}\right) = n(x)\left(\frac{mv^2}{2}\right)\vec{v} \tag{6.62}$$

is the flux of ion kinetic energy, and n(x) is the local plasma density. Since no dependence on the y coordinate was assumed, and the dependence on the z coordinate vanishes after averaging over the initial position z_o, then only the x component of the energy flux need be considered, *i.e.*,

$$\Phi_x\left(\frac{mv_{\perp}^2}{2}\right) = n(x)\frac{m}{2}\Big\{(v_{x1}^2 + v_{y1}^2 + 2v_{xo}v_{x2} + 2v_{yo}v_{y2})\,v_{xo}$$
$$+ 2(v_{xo}v_{x1} + v_{yo}v_{y1})v_{x1} + (v_{xo}^2 + v_{yo}^2)v_{x2}\Big\} \tag{6.63}$$

$$\Phi_x\left(\frac{mv_z^2}{2}\right) = n(x)\frac{m}{2}\Big\{(v_{z1}^2 + 2v_{zo}v_{z2})\,v_{xo}$$
$$+ 2(v_{zo}v_{z1})\,v_{x1} + v_{zo}^2\,v_{x2}\Big\} \tag{6.64}$$

where again only terms not periodic in z_o are retained. After averaging over z_o and over a Larmor period, all the right-hand side of (6.63) and the right-hand side of (6.64) with the exception of the first term, namely $(v_{z1}^2 + 2v_{zo}v_{z2})\,v_{xo}(t)$, vanish. From (6.16) and (6.31) and after performing the above-mentioned averages we get

$$\left\langle v_{z1}^2\, v_{xo} \right\rangle = \frac{e^2 k^2 \epsilon(xo)\,(\partial\epsilon/\partial x)\, v_{\perp o}^4}{4m^2\omega^2\alpha_o\,(\alpha_o^2 - \Omega^2)} \sin \alpha_o t \tag{6.65}$$

By the same token one can also show that

$$\left\langle v_{zo}\, v_{z2}\, v_{xo} \right\rangle = -\frac{e^2 k^3\, \epsilon(xo)(\partial\epsilon/\partial x)\, v_{\perp o}^4\, v_{zo}}{8m^2\omega^2\alpha_o\,(\alpha_o^2 - \Omega^2)} \left\{ - t \cos \alpha_o t + \left(\frac{1}{\alpha_o} + \frac{2\alpha_o}{\alpha_o^2 - \Omega^2} \right) \sin \alpha_o t \right\} \tag{6.66}$$

and upon combining these two results the flux Φ_x becomes

$$\left\langle \Phi_x \left(\frac{m v_z^2}{2} \right) \right\rangle = n(x) \frac{e^2 k^2\, \epsilon(x)\,(\partial\epsilon/\partial x)\, v_{\perp o}^4}{8m\omega^2\alpha_o\,(\alpha_o^2 - \Omega^2)} \left\{ \left[1 - k v_{zo} \left(\frac{1}{\alpha_o} + \frac{2\alpha_o}{\alpha_o^2 - \Omega^2} \right) \right] \sin \alpha_o t + k v_{zo} t \cos \alpha_o t \right\} \tag{6.67}$$

If in addition we assume that the Doppler-shifted frequency $k v_{zo} - \omega = \alpha_o \ll \Omega_i$ is much smaller than the ion gyrofrequency then

$$\left\langle \Phi_x \left(\frac{m v_z^2}{2} \right) \right\rangle = n(x) \frac{e^2}{m} \frac{k^2 v_{\perp o}^4}{8\omega^2\Omega^2} \epsilon(x_o) \frac{\partial\epsilon}{\partial x} \left[\omega \frac{\sin \alpha_o t}{\alpha_o^2} - k\, v_{zo}\, t \frac{\cos \alpha_o t}{\alpha_o} \right] \tag{6.68}$$

$$\left\langle \Phi_x \left(\frac{m\, v_\perp^2}{2} \right) \right\rangle = 0 \tag{6.69}$$

We observe, therefore, that only parallel energy contributes to the energy flux and, as no contribution arises from terms proportional to v_{x2}, the flux of kinetic energy is not accompanied by any transport of mass. The origin of this energy flux can be made physically more intuitive as follows: as the ions move along the magnetic field lines under the slowly varying average force $F_z = -\mu \nabla B$, they are subject to a rapidly varying force $F_z = -v_y B_x$, and in turn acquire and lose energy according to the sign of v_y. An ion whose gyrocenter lies close enough to a plane $x =$ const will therefore cross this plane back and forth with different energies according to the sign of v_x.

Once again we average over the distribution function of the initial velocity (6.44) and note that the integrals involved are identical to the ones discussed earlier. The result is

$$\left\langle\left\langle \Phi_x \left(\frac{mv_z^2}{2}\right)\right\rangle\right\rangle = n(x) \frac{m}{4B_o^2} \frac{k^2 c^2}{\omega^2} \left(\frac{2KT_\perp}{m}\right)^2 \epsilon(x_o) \frac{\partial \epsilon}{\partial x} \frac{\pi\omega}{k|k|} \left[\frac{\partial g(v_{zo})}{\partial v_{zo}}\right]_{v_{zo} = \omega/k} \tag{6.70}$$

or

$$\left\langle\left\langle \Phi_x \left(\frac{mv_z^2}{2}\right)\right\rangle\right\rangle = - n(x) \frac{KT_\perp}{\omega B_o^2} \frac{T_\perp}{T_{||}} \epsilon(x_o) \frac{\partial \epsilon}{\partial x} \left[\sqrt{\pi} \frac{k}{|k|} \zeta_o e^{-\zeta_o^2}\right] \tag{6.71}$$

If we now substitute (6.60) and (6.71) into (6.61) we obtain the RF power absorbed by the plasma per unit volume

$$P = \frac{KT_\perp c^2}{\omega B_o^2} \frac{T_\perp}{T_{||}} \left\{ n(x) \left(\frac{\partial \epsilon}{\partial x}\right)^2 - \frac{\partial}{\partial x}\left[n(x) \epsilon(x_o) \frac{\partial \epsilon}{\partial x}\right]\right\} \left[\sqrt{\pi} \frac{k}{|k|} \zeta_o e^{-\zeta_o^2}\right] \tag{6.72}$$

This result shows that RF power is absorbed by the plasma only in a region where sufficiently high density gradient is present.

This does not mean however that only the few ions near the boundary or in the region where the external current flows can be heated at a very high rate. The rate of change of the ion kinetic energy is positive throughout the plasma as shown in (6.60), and heating occurs as a result of the existence of the energy flux (6.70) which is not accompanied by a mass transfer.

REFERENCES

1. Bekefi, G. *Radiation Processes in Plasmas.* (New York: John Wiley and Sons, 1966).
2. Cattanei, G. *Nuclear Fusion, 13,* 839 (1973).
3. Cattanei, G.. "A Physical Picture of Transit Time Magnetic Pumping," Culham Laboratory Report CLM-P326 (1972).
4. Dawson, J. M. and M. F. Uman. *Nuclear Fusion, 5,* 242 (1965).
5. Dologopolov, V. V. and K. N. Stepanov. *Nuclear Fusion, 3,* 205 (1963).
6. Piliya, A. D. and V. Y. Frenkel. *Soviet Phys. Tech. Phys., 9,* 1356 (1965).
7. *Proceedings of the First Topical Conference on R. F. Plasma Heating,* Texas Tech. University SR-2, Lubbock, Texas, July (1972).
8. Scott, A. *EDN* (October 1969).
9. Shohet, J. L. *Phys. Fluids, 16,* 1375 (1973).
10. Stepanov, K. N. *Soviet Physics, JETP, 18,* 826 (1964).
11. Stix, T. H. "Heating a Tokamak Plasma," Princeton University Report MATT-928 (1972).
12. Stix, T. H. "The Physics of R. F. Heating," Princeton University Report MATT-929 (1972).
13. Stix, T. H. *The Theory of Plasma Waves.* (New York: McGraw-Hill, 1962).
14. Wort, D. J. H. BNES, Nuclear Fusion Reactors (Proc. Int. Conf. Culham) (1969), p. 517.

CHAPTER 7

ADIABATIC COMPRESSION AND IGNITION OF FUSION REACTORS

We recall from Chapter 2 that the ignition temperature for a D-T fusion reactor is about 4.5 keV. None of the current heating methods has thus far resulted in ignition, and in the case of ohmic heating, as in Tokamaks, we recall that this method could lead to a saturation temperature which is short of ignition. This saturation for a constant-current heating is shown in Figure 45, and we observe that for a density of $n = 5 \times 10^{14}$ the temperature achieved is about 1 keV smaller than needed.

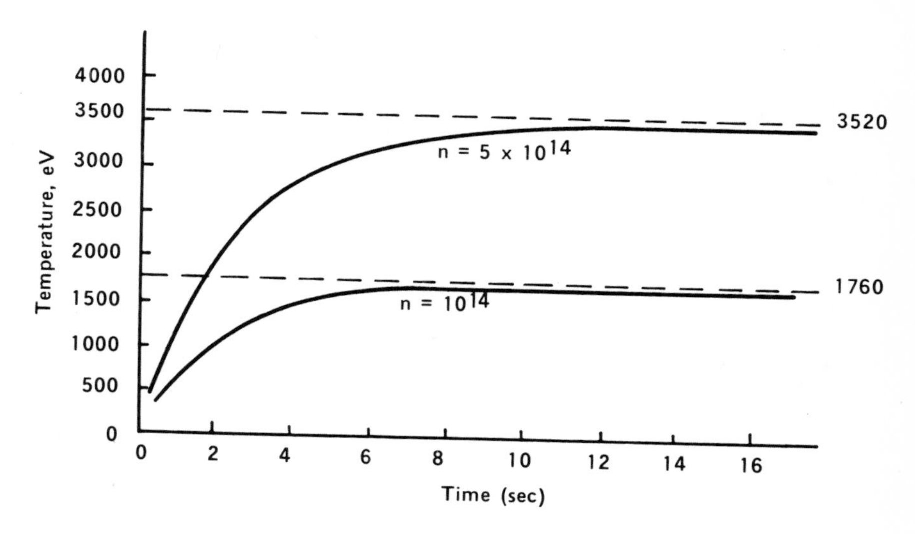

Figure 45. Ohmic heating at various densities.

We recall further that the limitations on ohmic heating are mainly due to two factors: the decrease of plasma resistivity with increasing temperature ($R \propto T_e^{-3/2}$); and the increase of bremsstrahlung losses with increasing temperature ($P_b \propto T_e^{1/2}$). Moreover, this method of heating heats the electrons preferentially, and the ions of the plasma are heated by the electrons through collisions. Therefore it is important that the equilibration time τ_{ei} (which is the time it takes to pass the heat from the electrons to ions) be shorter than the characteristic heating time shown in Figure 45. A plot of τ_{ei} vs electron temperature is shown in Figure 46, and we note that at low densities the equilibration time is quite long. This characteristic time can be computed directly from

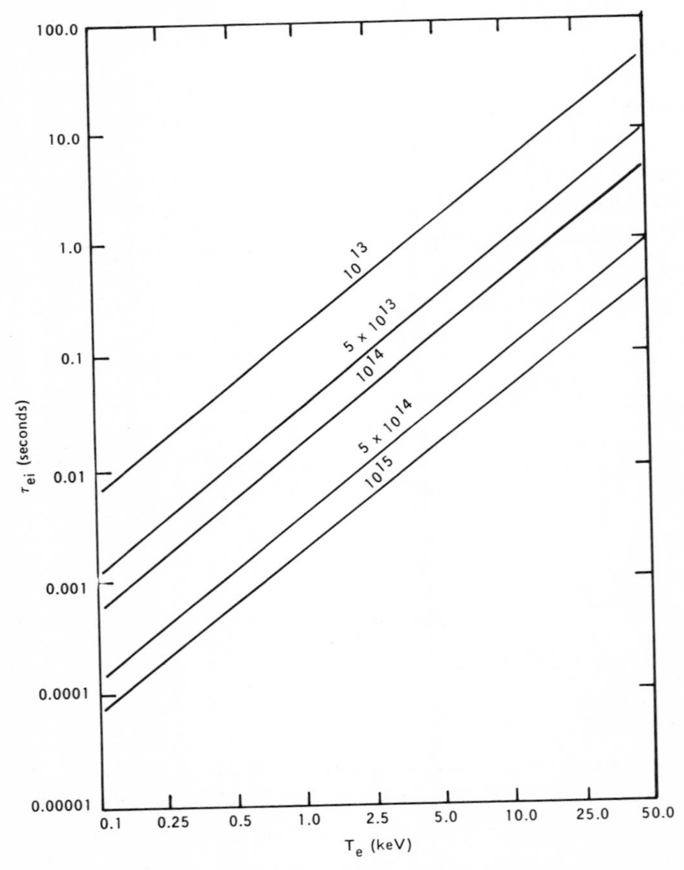

Figure 46. Electron-ion equilibration times.

Equation (4.104) as it is applied to two species: electrons with temperature T_e colliding with ions with temperature T_i. In this chapter we will examine the use of adiabatic compression as a second stage heating to bring a Tokamak reactor to ignition temperature. Since this method involves changing the magnetic field, knowledge of the Tokamak field geometry is necessary. In discussion of the Tokamak geometry we will have the opportunity of examining in simple terms the operating principle of such a reactor.

7.1. TOKAMAK GEOMETRY

The Tokamak device is a toroidal machine in which the toroidal magnetic field B_o is generated by the current I_o as shown in Figure 47. The field has a value B_o at the major radius

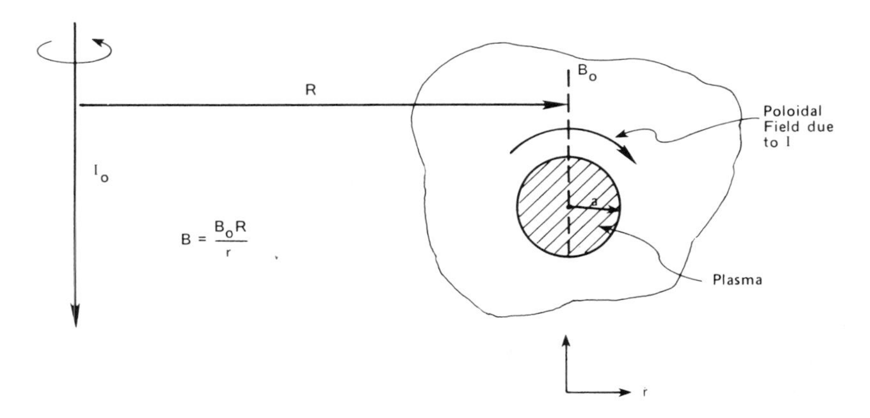

Figure 47. Tokamak geometry.

R which is the value of the radial distance r from the major axis of the plasma column. The plasma has a radius (the minor radius) "a" and in the plasma the field strength is smaller than B_o by an amount corresponding to the plasma pressure, NkT/V where N is the total number of particles confined, kT is the plasma temperature, and $V = 2\pi^2 a^2 R$ is the volume of the plasma ring. In addition to I_o, a toroidal current I is induced in the plasma to provide a "rotational transform" or, as we have seen in Chapter 1, a magnetic well with shear to provide stability. A small vertical magnetic field is also applied to provide equilibrium. Tokamaks have made use of heavy copper walls to provide eddy-current stabilization of the plasma column. If, as some recent experiments have indicated, it can be shown that a suitable vertical field can provide stable confinement without presence of copper walls, then very effective compressional heating should be possible. The simplest form of this heating is to increase the value of the vertical field in order to compress the major radius of the plasma. As this radius shrinks the plasma moves into a region of higher toroidal magnetic field causing the minor radius to shrink but not proportionally to the major radius. This type of adiabatic compression raises the temperature of the plasma and it will be called type B. Stronger heating by greater compression will be achieved if, simultaneously with the shrinking of the major radius, the strength of the toroidal field is increased. This is a more expensive method of heating since a large power source and power-handling capacity will be needed. Because of the greater expense one would want to minimize the amount of toroidal field increase, and for quantitative comparison we will choose an amount such that $B \sim R^{-2}$. We shall call this method type A.

7.2. DERIVATION OF THE VERTICAL FIELD

Following Mills (1970) we shall use the first law of thermodynamics to calculate the vertical magnetic field required for stable confinement. The first law states

$$dU = \delta Q - \delta W \tag{7.1}$$

where dU is the change in the (internal) energy of the system, δQ is the heat change and δW is the work done. Note that the minus sign before δW implies that when the system performs work its energy diminishes. For the case at hand the work is done (a) on the toroidal field current source, and (b) on the perpendicular field current source. The energy of the system is in (a_1) the toroidal field, (b_1) the poloidal field, and (c_1) in the plasma thermal energy 3/2 NkT. We have neglected the small energies in the vertical field and that corresponding to the cross product of the vertical field and the poloidal field.

Work on Toroidal Field Source

The work done on the toroidal field source can be written as

$$-\delta W = I_o \delta \Phi \tag{7.2}$$

where I_o is the current generating the toroidal field and $\delta\Phi$ is the change in the magnetic flux. The current I_o is related to the toroidal magnetic field through Ampere's law, *i.e.*,

$$\oint \vec{B} \cdot d\vec{\ell} = \mu_o I_o$$

or

$$B = \mu_o I_o / 2\pi r \tag{7.3}$$

But $\mu_o = 4\pi$, and $B = B_o$ at $r = R$; therefore

$$I_o = \frac{B_o R}{2}$$

and

$$B = \frac{B_o R}{r} \tag{7.4}$$

The magnetic flux can be expressed as the sum of two portions: that outside the plasma and the flux in the plasma, *i.e.*, we can write

$$\Phi = \int\int B drdz - B_o \pi a^2 + \Phi_{plasma} \tag{7.5}$$

where it has been assumed that B_o is constant throughout the plasma. This assumption is valid if $a \ll R$. Substituting for B from equation (7.4) the above equation becomes

$$\Phi = B_o R \int\int \frac{drdz}{r} - B_o \pi a^2 + \Phi_{plasma} \tag{7.6}$$

where the coordinates of Figure 47 have been used. The flux in the plasma is conserved and we can illustrate that by considering the following Maxwell equation

$$\vec{\nabla} \times \vec{E} = -\frac{\partial \vec{B}}{\partial t}$$

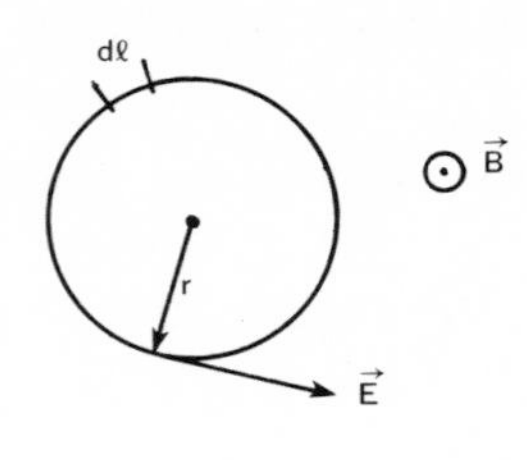

and assuming that B is uniform in space. Integrating the above equation we get

$$\int_s (\vec{\nabla} \times \vec{E}) \cdot \vec{d\ell} = -\int \frac{\partial \vec{B}}{\partial t} \cdot \vec{d\ell}$$

which by Stokes theorem becomes

$$\oint \vec{d\ell} \cdot \vec{E} = -\pi r^2 \frac{\partial \vec{B}}{\partial t} \tag{7.7}$$

or

$$\vec{E} = -\frac{r}{2} \frac{\partial \vec{B}}{\partial t}$$

We note therefore that time variation in the magnetic field gives rise to an electric field which in turn leads to plasma drift given by

$$v_D = \frac{dr}{dt} = \frac{|c\vec{E} x \vec{B}|}{B^2} = \frac{cE}{B} = -\frac{rc}{2B} \frac{\partial B}{\partial t}$$

This equation can be further written (using c = 1) as

$$2rB \frac{dr}{dt} + r^2 \frac{dB}{dt} = \frac{d}{dt} (r^2 B) = 0$$

Integration gives

$$r^2 B = \text{constant} = \pi r^2 B = \Phi_{plasma} \tag{7.8}$$

Returning to equation (7.6) we can now write

$$\delta\Phi = (R dB_o + B_o dR) \int\int \frac{drdz}{r} - \pi a^2 dB_o - 2B_o \pi a da$$

and upon substitution into equation (7.2) it becomes after utilizing (7.4)

$$-\delta W = (R^2 \frac{B_o dB_o}{2} + B_o{}^2 \frac{RdR}{2}) \int\int \frac{drdz}{r}$$

$$- \frac{2\pi^2 a^2 RB_o dB_o}{4\pi} - \frac{2B_o{}^2 \pi^2 a^2 Rda}{2\pi a} \qquad (7.9)$$

However, the plasma volume V is given by $2\pi^2 a^2 R$, and upon insertion of it in equation (7.9) and with slight rearrangement it can be put in the form

$$-\delta W = \tfrac{1}{2}(R^2 B_o dB_o + B_o{}^2 RdR) \int\int \frac{drdz}{r} - \frac{VB_o{}^2}{4\pi}\left(\frac{dB_o}{B_o} + \frac{2da}{a}\right) \qquad (7.10)$$

This result represents the work done on the toroidal field source, so we proceed now to the second component of the work, namely that done on the vertical field source.

Work on Vertical Field Source

The centering force per unit arc length required by the curvature is proportional to the product of the current and the magnetic field, *i.e.*, IB_X. This force balances three terms due to (1) gas pressure, (2) toroidal field, and (3) poloidal field. In order to calculate the first one we consider that segment of the plasma shown in Figure 48.

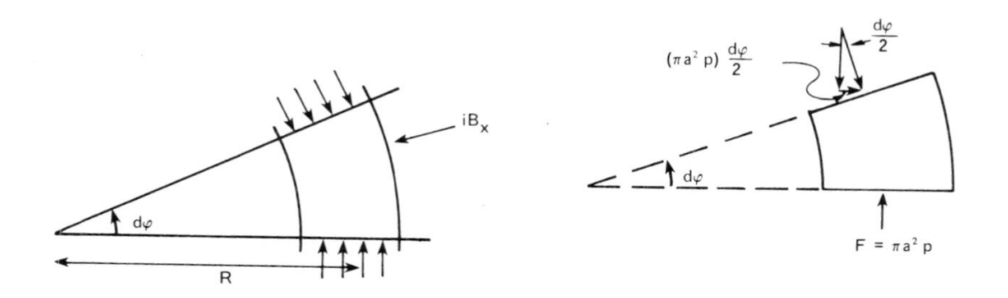

Figure 48. Segment of plasma.

Since the arc length ds = R$d\varphi$ then the force per unit arc length can be written as

$$\frac{dF_1}{ds} = \frac{\{\pi a^2 P \frac{d\varphi}{2}\}\,(2)}{R d\varphi} = \frac{\pi a^2 P}{R} \tag{7.11}$$

Substituting for the plasma pressure $P = NkT/V$ where $V = 2\pi^2 a^2 R$ is the plasma volume we obtain

$$\frac{dF_1}{ds} = \frac{NkT}{2\pi R^2} \tag{7.12}$$

where the subscript 1 denotes contribution from the first balance force.

In order to calculate the second balancing force we first recall the electromagnetic field tensor given by (see Longmire, 1963)

$$T_{ij} = \frac{1}{4\pi}\left\{\frac{E^2 + B^2}{2}\,\delta_{ij} - (E_iE_j + B_iB_j)\right\} \tag{7.13}$$

In the absence of an electric-field this tensor reduces to

$$T_{ij} = \frac{B^2}{8\pi}\,\delta_{ij} - \frac{B_iB_j}{4\pi} \tag{7.14}$$

where for a case in which the magnetic field is in the Z direction the components are

$$\begin{array}{cc} & \begin{array}{ccc} x & y & z \end{array} \\ T = \dfrac{1}{8\pi} & \begin{pmatrix} B^2 & 0 & 0 \\ 0 & B^2 & 0 \\ 0 & 0 & -B^2 \end{pmatrix} \begin{array}{c} x \\ y \\ z \end{array} \end{array} \tag{7.15}$$

We observe that there is compression (pressure) in the directions perpendicular to the magnetic field and tension ($-B^2/8\pi$) along the field itself. The magnetic field lines are, so to speak, like stretched rubber bands which repel one another. In a magneto static problem, the above tensor is usually balanced by the plasma pressure, and if the latter is a scaler then it can be represented by

$$P_{ij} = \frac{1}{V} \begin{pmatrix} NkT & 0 & 0 \\ 0 & NkT & 0 \\ 0 & 0 & NkT \end{pmatrix} \tag{7.16}$$

so that the equation of equilibrium can be written as

$$\frac{\partial}{\partial x_j}(P_{ij} + T_{ij}) = 0 \tag{7.17}$$

with repeated indices denoting summation. In cylindrical coordinates with the magnetic field in the azimuthal direction, the field tensor is given by

$$T = \frac{1}{8\pi} \begin{array}{c} \begin{array}{ccc} r & \varphi & z \end{array} \\ \begin{pmatrix} B^2 & 0 & 0 \\ 0 & -B^2 & 0 \\ 0 & 0 & B^2 \end{pmatrix} \end{array} \begin{array}{c} r \\ \varphi \\ z \end{array} \tag{7.18}$$

If we now assume that B is a function of r only (as it is with the toroidal field we are considering), and if we integrate equation (7.17) utilizing Gauss's theorem we get

$$\int n_j (P_{ij} + T_{ij})\, dA = 0 \tag{7.19}$$

where $\hat{n}$ is the outward normal to the surfaces on which the forces act. For the problem at hand we have already taken care of the contribution of the plasma pressure and we only need to consider

$$\int n_j T_{ij}\, dA = 0 \tag{7.20}$$

or the sum of electromagnetic forces.

If we once again consider a segment of the torus as shown in Figure 49 we note that only the azimuthal forces add to contribute to a radial force. This azimuthal force is the tensile force given in (7.18) and acts on the cross-sectional area of the plasma. We can therefore write

$$dF = \int \frac{B^2}{8\pi}\, dA = \frac{1}{8\pi} \int \frac{B_o{}^2 R^2}{r^2}\, dA = \frac{1}{8\pi} \int \frac{B_o{}^2 R^2}{(R + \ell \cos \theta)^2}\, dA$$

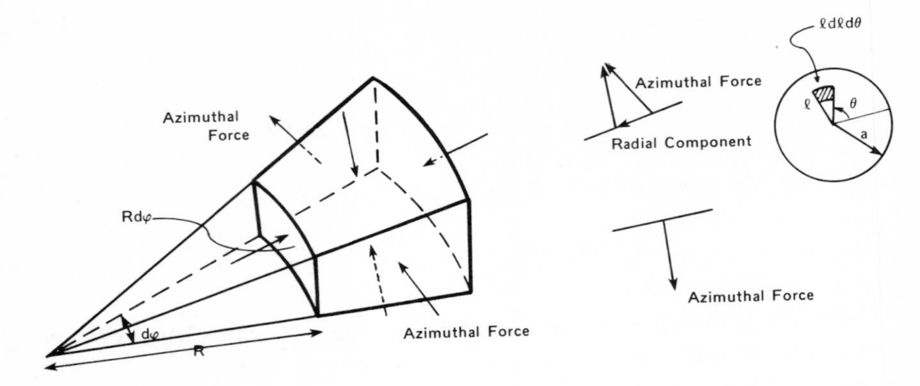

Figure 49. Segment of the torus.

and if we substitute dA = ℓdℓdθ,, then we get

$$dF = \frac{R^2 B_o^2}{8\pi} \int_o^{2\pi} \int_o^{a} \frac{\ell d\ell d\theta}{(R + \ell \cos\theta)^2} \tag{7.21}$$

Integration of the above equation yields

$$dF = \frac{R^2 B_o^2}{4} \left\{ \frac{1}{\sqrt{1-a^2/R^2}} - 1 \right\}$$

and if we now note that $(dF)_{rad} = 2(dF)\, d\varphi/2$ and divide the result by the arc length $ds = Rd\varphi$, we obtain

$$\frac{dF_2}{ds} = \frac{(dF)_{rad}}{Rd\varphi} = \frac{RB_o^2}{4} \left\{ \frac{1}{\sqrt{1-a^2/R^2}} - 1 \right\} \tag{7.22}$$

where the subscript 2 has been added to denote the second balancing force due to the toroidal field. For $a/R \ll 1$ the above result reduces to

$$\frac{dF_2}{ds} = \frac{B_o^2}{8} \frac{a^2}{R} \tag{7.23}$$

However the field in the plasma is not B_o^2 and the above equation should really be written as (see Figure 47)

$$\frac{dF_2}{ds} = \frac{B_o^2 - B_i^2}{8\pi} \frac{\pi a^2}{R} = \frac{NkT}{2\pi^2 a^2 R} \frac{\pi a^2}{R}$$

$$\frac{dF_2}{ds} = \frac{NkT}{2\pi R^2} \qquad (7.24)$$

where we have noted that the difference in the two fields is simply the plasma pressure.

We turn now to the third balancing force, namely that due to the poloidal field. This field is induced by the current I in the plasma, and if L is the inductance of the plasma ring then the energy $u = \frac{1}{2}LI^2$, and the force F_3 is given by

$$-F_3 = \frac{du}{dR} \qquad (7.25)$$

or

$$-F_3\,dR = du = \tfrac{1}{2}I^2\,dL + LI\,dI$$

but LI = potential difference = constant, thus

$$L\,dI + I\,dL = 0 \qquad \text{or} \qquad L\,dI = -I\,dL$$

The force F_3 cannot be calculated without the substitution in equation (7.25) of the proper expression for the inductance of a plasma loop. Before we calculate this quantity it would be desirable to focus our attention on the inductance of a coaxial system illustrated in Figure 50 in which the inner region carries the current I.

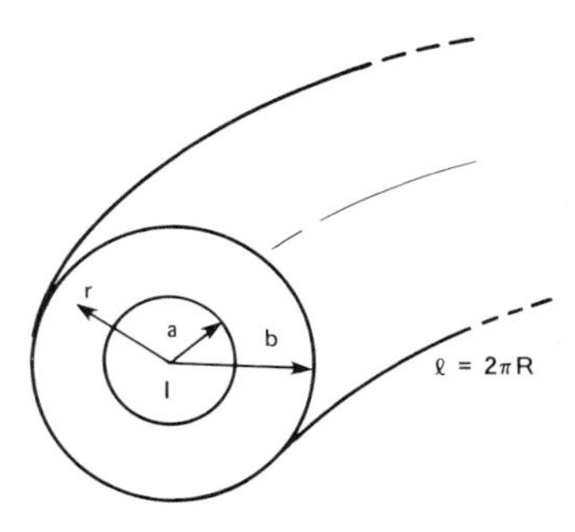

Figure 50. Coaxial system.

From Ampere's law we can once again write with the aid of Stokes' theorem

$$\int (\vec{\nabla} \times \vec{B}) \cdot d\vec{A} = 2\pi r B = 4\pi \int \vec{J} \cdot d\vec{A} = 4\pi \frac{I}{\pi a^2} \pi r^2 \qquad (7.26)$$

assuming that the current is uniformly distributed. The resulting magnetic field is given by

$$B = \frac{2I}{a^2} r$$

which, if we substitute in the expression for the flux, namely

$$\Phi = \int \vec{B} \cdot d\vec{A}' = \frac{2I}{a^2} \int_o^a \int_o^\ell r \, dr \, dz = I\ell$$

yields the internal inductance through the familiar relation $\Phi = IL$. The result is

$$L_{int} = \ell = 2\pi R \qquad (7.27)$$

For the calculation of the external inductance we first note from Equation (7.4) that

$$B = \frac{2I}{r}$$

so that the flux can be written as

$$\Phi = \int \vec{B} \cdot d\vec{A}' = 2\pi I \int_a^b \int_o^\ell \frac{drdz}{r} = 2 I \ln (b/a) \ell$$

which in turn yields

$$L_{ext} = 2\ell \ln (b/a) = 4\pi R \ln (b/a) \qquad (7.28)$$

The total inductance has the form

$$L_{tot} = 2\ell \left[\ln \frac{b}{a} + \frac{1}{2} \right] = 4\pi R \left[\ln \frac{b}{a} + \frac{1}{2} \right] \qquad (7.29)$$

or

$$L_{tot} = \mu_o R \left[\ln \frac{b}{a} + \frac{1}{2} \right] \qquad \text{Henry} \qquad (7.30)$$

For the plasma in a Tokamak device the symmetry invoked in the above derivation along with nonuniform distribution of the current density could lead to some corrections which Artsimovich (1972) represents by including the parameter i in the inductance, *i.e.*,

$$L_{tot} = 4\pi R \left[\ln \frac{b}{a} + \frac{i}{2}\right] \tag{7.31}$$

The value of i is typically 1.10.

If in place of the coaxial system we have a current-carrying loop whose radius is R and whose wire radius is a, then it can be shown that the inductance in Henrys is given by

$$L = \mu_o (2R - a) \left[(1 - \frac{k^2}{2})K(k) - E(k)\right] \tag{7.32}$$

where

$$k^2 = \frac{4R(R-a)}{(2R-a)^2} \tag{7.33}$$

and E(k) and K(k) are complete elliptic integrals of the first and second kind defined by

$$E(k) = \int_o^{\pi/2} \sqrt{1-k^2 \sin^2 \theta} \; d\theta \tag{7.34}$$

$$K(k) = \int_o^{\pi/2} \frac{d\theta}{\sqrt{1-k^2 \sin^2 \theta}} \tag{7.35}$$

If a/R is very small then k tends to unity and the elliptic integrals can be approximated by

$$K(k) \simeq \ln \left(\frac{4}{\sqrt{1-k^2}}\right)$$

$$E(k) = 1$$

Substitution of these results in equation (7.32) yields

$$L = \mu_o R \left[\ln\left(\frac{8R}{a}\right) - 2\right] \text{Henrys} = 4\pi R \left[\ln\left(\frac{8R}{a}\right) - 2\right] \tag{7.36}$$

Once again, for application to Tokamak systems the constant 2 is replaced by 1.75, and for the purposes of the present calculation we shall use as the inductance of the plasma ring:

$$L = 4\pi R \left\{ \ln\left(\frac{8R}{a}\right) - 1.75 \right\}$$

Equation (7.25) becomes with these substitutions

$$F_3 dR = \left\{ \frac{I^2 L}{2R} - 2\pi\, I^2\, R \left(\frac{1}{R} - \frac{1}{a} \frac{da}{dR} \right) \right\} dR$$

The force per unit arc length ds = $2\pi R dR$ then becomes

$$\frac{dF_3}{ds} = \frac{I^2 L}{4\pi R^2} + I^2 \left(\frac{1}{R} - \frac{1}{a} \frac{da}{dR} \right) \tag{7.37}$$

We now add the three balancing forces as given by equations (7.12), (7.24), and (7.37) and compute the vertical field by noting that

$$\delta W_{1,2,3} = 2\pi R I B_Z dR$$

where IB_Z is the force per unit length. The total balancing work is

$$\delta W_{1,2,3} = 2\pi R dR \left\{ \frac{NkT}{2\pi R^2} + \frac{NkT}{2\pi R^2} + \frac{I^2 L}{4\pi R^2} + I^2 \left(\frac{1}{R} - \frac{1}{a} \frac{da}{dR} \right) \right\} \tag{7.38}$$

from which the vertical field B_Z is found to be

$$B_Z = \frac{NkT}{\pi I R^2} + \frac{IL}{4\pi R^2} + I \left(\frac{1}{R} - \frac{1}{a} \frac{da}{dR} \right) \tag{7.39}$$

Thus far we have computed δW of equation (7.1), and in order to compute the heat change we need to obtain δU, the change in the internal energy. This consists of the energies in the toroidal field, the poloidal field, and in the plasma thermal energy.

Toroidal Field

In the toroidal field the energy can be written directly as

$$U_1 = \int \frac{B^2}{8\pi} \, dV$$

where the volume element dV = 2πrdrdz. Using equation (7.4) this energy integral assumes the form

$$U_1 = \iint \frac{B_o^2 R^2}{4r} \, drdz - \frac{B_o^2}{8\pi} V_{plasma} + \frac{B_i^2}{8\pi} V_{plasma}$$

or

$$U_1 = \frac{B_o^2 R^2}{4} \iint \frac{drdz}{r} - \frac{(B_o^2 - B_i^2)}{8\pi} V_{plasma}$$

The last term in the above equation is simply the total plasma pressure, hence

$$U_1 = \frac{B_o^2 R^2}{4} \iint \frac{drdz}{r} - NkT \qquad (7.40)$$

The remaining energy terms are $U_2 = ½LI^2$ (for poloidal field), and $U_3 = 3/2NkT$ (for plasma thermal energy), so that the total energy becomes

$$U_{.} = \frac{B_o^2 R^2}{4} \int \frac{drdz}{r} + ½LI^2 + ½NkT \qquad (7.41)$$

If we now take the variation on the above expression and recall the fact that LdI = -IdL we obtain

$$dU = ½(R^2 B_o dB_o + B_o^2 RdR) \int \frac{drdz}{r} - ½I^2 dL + ½NkdT$$

(7.42)

The total work is obtained by adding equations (7.38) and (7.9), and if the results along with (7.42) are substituted in equation (7.1) it finally becomes

$$\frac{\delta Q}{NkT} - \frac{2}{\beta}\left(\frac{dB_o}{B_o} + \frac{2da}{a}\right) - \frac{2dR}{R} = \frac{1}{2} \frac{dT}{T} \qquad (7.43)$$

where we have introduced for convenience the plasma beta as given by $\beta = 8\pi\ NkT/VB_o^2$. For "adiabatic" compression, *i.e.*, $\delta Q = 0$, the above equation yields

$$\frac{1}{2}\frac{dT}{T} = -\frac{2}{\beta}\left[\frac{dB_o}{B_o} + \frac{2da}{a}\right] - \frac{2dR}{R} \tag{7.44}$$

Before applying these results let us first examine the implication of the flux conservation in the plasma column. We recall from equation (7.8) that $\Phi_p = \pi a^2 B_i$ = constant; therefore we can write

$$\frac{dB_i}{B_i} = -\frac{2da}{a} \tag{7.45}$$

From the equilibrium condition

$$\frac{B_o^2 V}{8\pi} = \frac{B_i^2 V}{8\pi} + NkT$$

we can also write using (7.45)

$$8\pi NkT - 4\ B_i^2\ V\,(da/a) = (B_o^2 - B_i^2)dV + 2\ B_o dB_o V \tag{7.46}$$

If we now apply the variation on the volume $V = 2\pi^2 a^2 R$, *i.e.*,

$$\frac{dV}{V} = \frac{dR}{R} + \frac{2da}{a} \tag{7.47}$$

and substitute it into (7.46) utilizing the definition of β mentioned earlier we obtain after some rearrangement

$$\frac{dT}{T} = \frac{dR}{R} - \frac{2da}{a} + \frac{2}{\beta}\left(\frac{dB_o}{B_o} + \frac{2da}{a}\right) \tag{7.48}$$

This result can now be combined with equation (7.44) to yield

$$\frac{3}{2}\frac{dT}{T} + \frac{dR}{R} + \frac{2da}{a} = 0$$

which when combined with (7.47) further becomes

$$\frac{3}{2}\frac{dT}{T} + \frac{dV}{V} = 0$$

Integration of this equation gives

$$TV^{2/3} = \text{constant} \tag{7.49}$$

which we recognize as the adiabatic compression relation for a gas with three degrees of freedom. If we eliminate da/a, *i.e.*, the plasma dimensions, from equations (7.44) and (7.48) we get

$$\left(3 - \frac{\beta}{2}\right) \frac{dT}{T} = 2 \frac{dB_o}{B_o} - 2(1-\beta) \frac{dR}{R} \tag{7.50}$$

and if we eliminate T between these same two equations we get

$$2 \frac{dB_o}{B_o} + \frac{5}{3} \beta \frac{dR}{R} + \left(4 - \frac{2\beta}{3}\right) \frac{da}{a} = 0 \tag{7.51}$$

7.3 APPLICATION

The above relations will now be used to examine the heating of plasma via type A ($B \propto R^{-2}$) and type B ($B \propto R^{-1}$). For type A, $dB_o/B_o = -2\ dR/R$, and if we put this in equation (7.50) we get

$$\frac{dT}{T} = - \frac{(2 - 2\beta/3)}{(1 - \beta/6)} \frac{dR}{R} \tag{7.52}$$

while if we put the type A field variation in equation (7.51) we get

$$\left(1 - \frac{5}{12} \beta\right) \frac{dR}{R} = \left(1 - \frac{\beta}{6}\right) \frac{da}{a} \tag{7.53}$$

For type B variation we have $dB_o/B_o = -\ dR/R$ and if we put this in equations (7.50) and (7.51) respectively we obtain

$$\frac{dT}{T} = - \frac{(4 - 2\beta)}{(3 - \beta/2)} \frac{dR}{R} \tag{7.54}$$

$$\left(1 - \frac{5\beta}{6}\right) \frac{dR}{R} = \left(2 - \frac{\beta}{3}\right) \frac{da}{a} \tag{7.55}$$

The summary of these results is shown in Table 17.

Table 17

Effect of Adiabatic Compression

Type	*A*	*B*
Definition	$\frac{dB_o}{B_o} = -2\frac{dR}{R}$	$\frac{dB_o}{B_o} = -\frac{dR}{R}$
Effect on plasma geometry	$(1 - 5\beta/12)\frac{dR}{R} = (1 - \beta/6)\frac{da}{a}$	$(1 - 5\beta/6)\frac{dR}{R} = (2 - \beta/3)\frac{da}{a}$
Effect on plasma temperature	$\frac{dT}{T} = \frac{(2 - 2\beta/3)\frac{dR}{R}}{(1 - \beta/6)}$	$\frac{dT}{T} = -\frac{(4 - 2\beta)}{(3 - \beta/2)}\frac{dR}{R}$

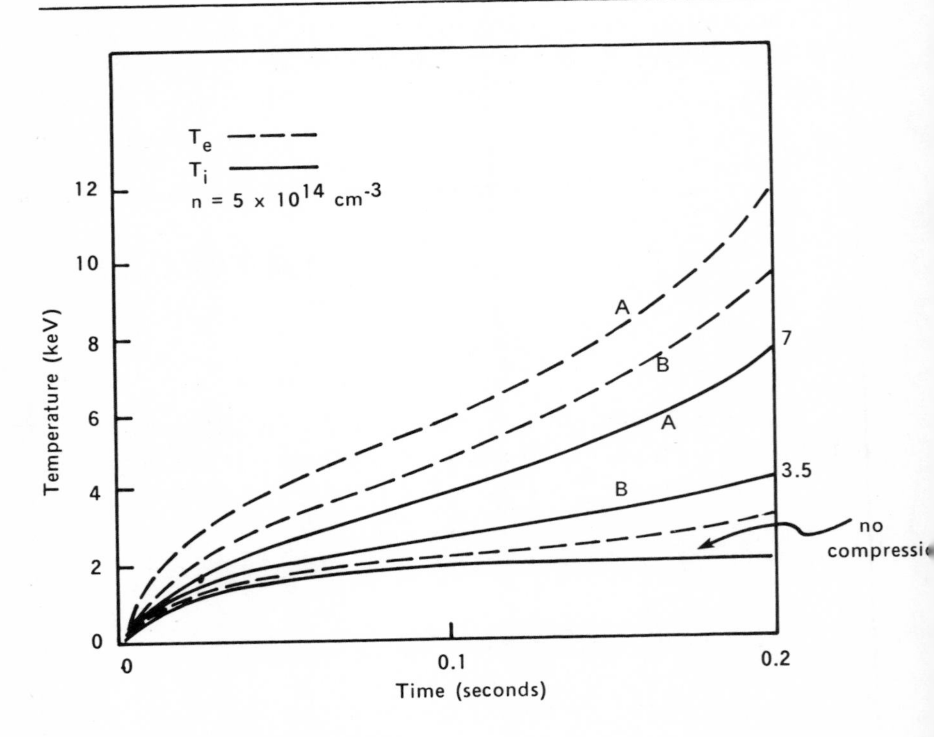

Figure 51. Tokamak compression.

Note that for type A compression as $\beta \to 0$, $dT/T = -2\ dR/R$ which leads to TR^2 = constant, while in the same limit for type B, the result is $TR^{4/3}$ = constant. These results have been applied to a Tokamak reactor containing a plasma of 90 cm radius and at a temperature of 20 eV. The particle confinement time is taken as 0.05 second and the compression takes place between 0.1 and 0.2 seconds. Three cases were followed: (1) with no compression at all, (2) type B compression, and (3) type A compression. The results are shown in Figure 51 for a D-T plasma for which the ignition temperature is about 4.2 keV. We observe that without compression the ion temperature is limited to a few hundred volts; with type B it reaches 3 keV and with type A, 7 keV. Clearly type B heating is inadequate, but type A is more than sufficient.

REFERENCES

1. Artsimovich, L. A. "Tokamak Devices," *Nuclear Fusion, 12,* 215 (1972).
2. Longmire, C. L. *Elementary Plasma Physics.* (New York: Interscience Publishers, 1963).
3. Mills, R. G. *Fusion Technology*. (Las Vegas, Nevada: Energy 70, 1970) p8.
4. Mills, R. G. "Time-Dependent Behavior of Fusion Reactors," Princeton Plasma Physics Lab. Matt - 728 (1970).
5. Ramo, S., J. R. Whinnery, and T. VanDuzen. *Fields and Waves in Communication Electronics.* (New York: Wiley and Sons, 1967).

CHAPTER 8

DYNAMICS AND CONTROL OF FUSION REACTORS

Although it is not clear at this time whether the first generation fusion reactors will be pulsed or steady state systems, it is expected that some of them at some point will operate in a steady state mode. For such systems (and even for quasi-steady state systems), the question of control is very important. To answer this question we must examine the dynamic equations which govern the temporal behavior of the system and assess its response to perturbations in such important plasma parameters as density and temperature. Even though the precise prediction of the plasma behavior is still beyond our reach, sufficient progress in the physics of confinement of high temperature plasma has been made to allow us to make sound predictions regarding stability and control of these systems, especially low beta toroidal devices such as Tokamak. In this chapter we will first study the thermal stability and control of an idealized system (as done by Ohta, *et al.*, 1971) and then re-examine the problem for a more realistic system in which all the important parameters are included.

8.1 THERMAL INSTABILITY AND FEEDBACK STABILIZATION

We consider a D-T toroidal reactor operating in steady state and containing a plasma where electron and ion temperatures are the same. The steady state parameters will first be determined and then we shall examine the stability of this system as a result of perturbations in particle density and temperature. We will also discuss feedback stabilization for such a reactor when it is operated in an unstable mode. The dynamic equations of interest are:

1) particle balance

$$\frac{dn}{dt} = -\frac{n}{\tau_n} + S \tag{8.1}$$

2) energy balance

$$\frac{d(nT)}{dt} = n^2 f(T) - \frac{nT}{\tau_E} + ST_s \tag{8.2}$$

where n is the particle density in particles/cm^3, T is the temperature in keV, and S is the rate of injected fuel (cm^{-3}sec^{-1})–injected at an energy of $3T_s$ (keV). The particle and energy confinement times are τ_n and τ_E, respectively, and f(T) is a temperature-dependent function consisting of two terms. The first comes from the power produced by the fusion reaction and deposited in the plasma, and the second represents power loss due to bremsstrahlung. Cyclotron radiation will be ignored in this analysis since it will be assumed that the system does not operate at very high temperatures; this effect will be included in the next section. In the above equations, we note the distinction between particle confinement time and energy confinement time since one represents particle diffusion from the system and the other represents energy (diffusion) conduction out of the system, and the two as we shall see later are not necessarily the same. If Q_c is the energy carried by the charged fusion product (3.5 MeV alpha in the case of a D-T reaction), then for equal amounts of D and T, the function f(T) can be written as:

$$f(T) = \frac{Q_c \langle\sigma v\rangle}{12} - 1.12 \times 10^{-15}\ T^{1/2} \tag{8.3}$$

where the last term represents the bremsstrahlung loss (see Chapter 1), and the numbers are obtained with the knowledge that 3/2 nT is the energy in the plasma as given in equation (8.2). By setting the left-hand sides of equations (8.1) and (8.2) equal to zero, we get the steady state conditions. The particle conservation yields:

$$\frac{n_0}{\tau_{n_0}} = S_0 \tag{8.4}$$

where the subscript zero denotes steady state conditions. From equation (8.2) with d(nT)/dt = 0, we also get

$$n_0 \tau_{E_0} f(T_0) = T_0 \left[1 - \frac{T_{S_0}}{T_0} \frac{\tau_{E_0}}{\tau_{n_0}} \right]$$

or

$$n_0 \tau_{E_0} = \frac{T_0}{f(T_0)} \left[1 - \frac{T_{S_0}}{T_0} \frac{\tau_{E_0}}{\tau_{n_0}} \right] = \xi_2 \frac{T_0}{f(T_0)} = F(T_0) \tag{8.5}$$

where

$$\xi_2 = \left[1 - \frac{T_{S_0}}{T_0} \frac{1}{R} \right], \quad R = \frac{\tau_{n_0}}{\tau_{E_0}} \tag{8.6}$$

When $T_S = 0$, a steady state is maintained by charged particle heating; in order to attain a steady state at lower τ_E, high energy injection of fuel ($T_S \neq 0$) is needed. We shall call the first method of heating "charged particle heating" and the second "injection heating." Equation (8.5) for both these methods is shown in Figure 52.

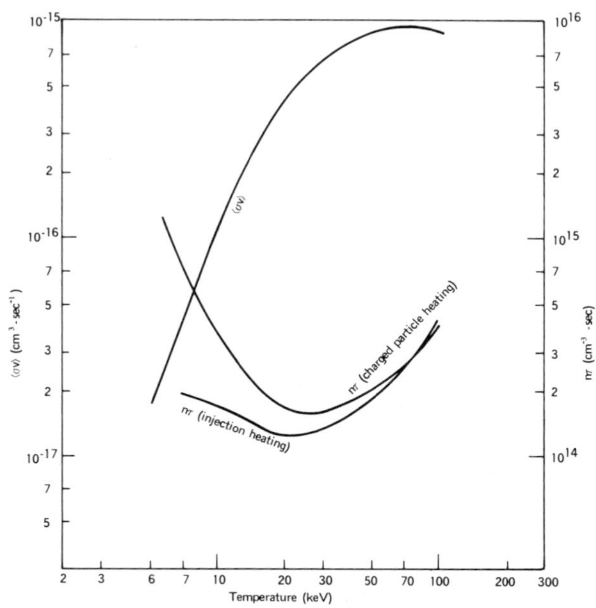

Figure 52. Reaction rate and $n\tau$ for injection and charged particle heated systems.

In order to examine the stability of the system, we linearize equations (8.1) and (8.2) by writing $n = n_0 + n'$, $T = T_0 + T'$, retaining only lowest order terms in perturbed quantities (*i.e.*, n' and T') since they are assumed to be much smaller than their equilibrium counterparts.

The result for equation (8.1) is:

$$\frac{dn'}{dt} = -\frac{n'}{\tau_{n_0}} - n_0 n' \frac{\partial}{\partial n} \tau_n^{-1} \Big|_0 - n_0 T' \frac{\partial}{\partial T} \tau_n^{-1} \Big|_0$$

and if we now define

$$n_0 \frac{\partial}{\partial n} \tau_n^{-1} \Big|_0 = \frac{1}{\tau_{n_1}}$$

$$T_0 \frac{\partial}{\partial T} \tau_n^{-1} \Big|_0 = \frac{1}{\tau_{T_1}} \tag{8.7}$$

then the linearized particle balance equation becomes

$$\frac{1}{n_0} \frac{dn'}{dt} = -\left(\frac{1}{\tau_{n_0}} + \frac{1}{\tau_{n_1}}\right) \frac{n'}{n_0} - \frac{1}{\tau_{T_1}} \frac{T'}{T_0} \tag{8.8}$$

In order to do the same with equation (8.2), we first note that

$$f(T) = f(T_0) + T' \frac{\partial}{\partial T} f \Big|_0 + \ldots$$

$$\frac{1}{\tau_E} = \frac{1}{\tau_{E_0}} + n' \frac{\partial}{\partial n} \frac{1}{\tau_E} \Big|_0 + T' \frac{\partial}{\partial T} \frac{1}{\tau_E} \Big|_0 + \ldots$$

and if we let the injection term be represented by its steady state value, *i.e.*, $ST_s = (n_0/\tau_{n_0})\, T_{s_0}$, then we obtain after some algebra

$$\frac{1}{n_0} \frac{dn'}{dt} + \frac{1}{T_0} \frac{dT'}{dt} = \left(\frac{1}{\tau_1} - \frac{1}{\tau_{E_0}}\right) \frac{n'}{n_0} + \left(\frac{1}{\tau_2} - \frac{1}{\tau_{E_0}}\right) \frac{T'}{T_0} \tag{8.9}$$

where we have let

$$\frac{1}{T_0} \left[2nf - nT \frac{\partial}{\partial n} \tau_E^{-1}\right]_0 = \frac{1}{\tau_1}$$

$$\frac{1}{n_0} \left[n^2 f - nT \frac{\partial}{\partial T} \tau_E^{-1}\right]_0 = \frac{1}{\tau_2} \tag{8.10}$$

If we now assume that both n′ and T′ vary with time as $e^{\gamma t}$, then by substituting in equations (8.8) and (8.9) we find that the condition for the existence of non-trivial solutions leads to

$$\gamma^2 + B\gamma + C = 0 \tag{8.11}$$

where the constants B and C are given by

$$B = \left(\frac{1}{\tau_{n_0}} + \frac{1}{\tau_{n_1}}\right) - \left(\frac{1}{\tau_2} - \frac{1}{\tau_{E_0}}\right) - \frac{1}{\tau_{T_1}} \tag{8.12}$$

$$C = -\left(\frac{1}{\tau_{n_0}} + \frac{1}{\tau_{n_1}}\right)\left(\frac{1}{\tau_2} - \frac{1}{\tau_{E_0}}\right) + \frac{1}{\tau_{T_1}}\left(\frac{1}{\tau_1} - \frac{1}{\tau_{E_0}}\right) \tag{8.13}$$

Since γ is in general complex, the plasma would be stable if $\mathrm{Re}\gamma < 0$, or if $B > 0$ and $C > 0$. Before we can investigate these stability conditions, we must first establish the dependence of both confinement times on density and temperature. We assume for the present that this dependence takes the form $\tau_E, \tau_n \sim n^{\ell}T^m$ and with it equations (8.7) and (8.10) become

$$\frac{1}{\tau_{n_1}} = -\frac{\ell}{\tau_{n_0}}$$

$$\frac{1}{\tau_{T_1}} = -\frac{m}{\tau_{n_0}}$$

$$\frac{1}{\tau_1} = \frac{1}{T_0}\left[2nf + \frac{T\ell}{\tau_E}\right]_0$$

$$\frac{1}{\tau_2} = \frac{1}{n_0}\left[n^2\frac{\partial f}{\partial T} + \frac{nm}{\tau_E}\right]_0$$

which, upon insertion in (8.12) and (8.13), lead to the following stability conditions:

$$B > 0 \Rightarrow \left(\frac{1}{\tau_{n_0}} - \frac{\ell}{\tau_{n_0}}\right) > -\frac{m}{\tau_{n_0}} + \left(n_0^2\frac{\partial f(T_0)}{\partial T_0} + \frac{n_0 m}{\tau_{E_0}} - \frac{n_0}{\tau_{E_0}}\right)\frac{1}{n_0} \tag{8.14}$$

$$C > 0 \Rightarrow -\frac{2mf}{T_0}\frac{n_0}{\tau_{n_0}} - \frac{m}{\tau_{n_0}}\frac{(\ell-1)}{\tau_{E_0}} > \left[\frac{1-\ell}{\tau_{n_0}}\right]\left\{n_0 \left.\frac{\partial f}{\partial T}\right|_0 + \frac{m}{\tau_{E_0}} - \frac{n_0}{F_0}\right\} \tag{8.15}$$

where from equation (8.5) we have used $n_0 \tau_{E_0} = F_0$. Equation (8.14) can be further simplified to read

$$\frac{1}{f}\left(1 - F_0 \left.\frac{\partial f}{\partial T}\right|_0\right) > \frac{F_0}{\xi_2 T_0}\left[m - \frac{1 - \ell + m}{R}\right] \tag{8.16}$$

and once again we recall from equation (8.5) that $fF_0 = \xi_2 T_0$, which upon differentiation can further be written as

$$f\frac{dF_0}{dT_0} + F_0\frac{df}{dT_0} = \xi_2 + T_0\frac{d\xi_2}{dT_0}$$

$$= \left(1 - \frac{\tau_{E_0}}{\tau_{n_0}}\frac{T_S}{T_0}\right) + T_0\left(\frac{\tau_{E_0}}{\tau_{n_0}}\frac{T_S}{T_0^2}\right)$$

$$= 1$$

Therefore, we note that

$$f\frac{dF_0}{dT_0} = 1 - F_0\frac{df}{dT_0}$$

or

$$\frac{dF_0}{dT_0} = \frac{1}{f}\left[1 - F_0\frac{df}{dT_0}\right] \tag{8.17}$$

which upon substitution into (8.16) leads to

$$\frac{dF_0}{dT_0} > \frac{F_0}{\xi_2 T_0}\left(m - \frac{1 - \ell + m}{R}\right) \tag{8.18}$$

Similar manipulations on equation (7.15) result in its final form, *i.e.*,

$$\frac{dF_0}{dT_0}(1 - \ell) > \frac{2mF_0}{T_0} \tag{8.19}$$

Equations (8.18) and (8.19) constitute the stability criteria for confinement times that vary with density and temperature as $n^\ell T^m$. We now apply these conditions to three examples:

1) $\ell = m = 0$ or τ_E = constant;

2) $\ell = 0$, $m = -1$, or τ_E and τ_n proportional to T^{-1}, which, we recall from Chapter 1, represents Bohm diffusion; and

3) $\ell = -1$, $m = \frac{1}{2}$, or $\tau_E, \tau_n \sim n^{-1} T^{\frac{1}{2}}$, which we also recall to represent classical diffusion.

For case 2) we get from equation (8.18),

$$\frac{T_0}{F_0} \frac{\partial F_0}{\partial T_0} > - \frac{1}{\xi_2 R} = \frac{T_0}{T_s - T_0}$$

and from (8.19)

$$\frac{\partial F_0}{\partial T_0} > 0 \tag{8.20}$$

and since the last condition automatically encompasses the first one, it can be regarded as the stability criterion for the case of constant confinement times. For Bohm confinement (2), the stability criteria are:

$$\frac{T_0}{F_0} \frac{\partial F_0}{\partial T_0} > - \frac{1}{\xi_2}$$

$$\frac{T_0}{F_0} \frac{\partial F_0}{\partial T_0} > - 2 \tag{8.21}$$

while for the classical diffusion case, they are:

$$\frac{T_0}{F_0} \frac{\partial F_0}{\partial T_0} > \frac{1}{2\xi_2} (1 - 5/R)$$

$$\frac{T_0}{F_0} \frac{\partial F_0}{\partial F_0} > \frac{1}{2} \tag{8.22}$$

These results are summarized in Table 18.

The above results indicate that reactor plasmas are stable when the derivative of the function $F_0 = F(T_0)(n_0 \tau_E$ at equilibrium) with respect to T_0 is greater than some constant, C, times $F(T_0)/T_0$ or simply

$$\frac{dF_0}{dT_0} > C \frac{F_0}{T_0}$$

Table 18

τ_E	Stability Criterion	T_c (keV) (D-T) CPH*	IH**
τ_E = constant	$\frac{T_0}{F_0}\frac{\partial F_0}{\partial T_0} > 0$	28	21
$\tau_E \sim T^{-1}$ $\ell = 0$ $m = -1$	$\frac{T_0}{F_0}\frac{\partial F_0}{\partial T_0} > -2, -\frac{1}{\xi_2}$	14	5
$\tau_E \sim T^{\frac{1}{2}}n^{-1}$ $\ell = -1$ $m = \frac{1}{2}$	$\frac{T_0}{F_0}\frac{\partial F_0}{\partial T_0} > \frac{1}{2},$ $\frac{1}{2\xi_2}(1 - 5/R)$	42	33

$^*R = 1, \xi_2 = 1, T_s = 0$

$^{**}T_s = 50, R = 10$

The function F(T) has one minimum at $T = T_m$ and increases with T for $T > T_m$ as shown in Figure 52. A critical temperature T_c is obtained when

$$\frac{dF_0}{dT_0} = C\frac{F_0}{K_0}$$

and in case (1) this critical temperature is equal to T_m since $C = 0$. When $C > 0$, $dF_0/dT_0 > 0$, and $T_c > T_m$ while the opposite is true when $C < 0$. The values of the critical temperature for charged particle heating and injection heating for a D-T reactor are also shown in Table 18.

In a stable operating mode, if the temperature increases due to some perturbation, and even if the reaction rate increases at the same time, the perturbation on the temperature decreases because the change in the energy loss due to the dependence of τ_E on T overcomes the heating; and if the temperature decreases by perturbation, the energy loss decreases more rapidly than the reaction rate and the temperature reduction can thus be recovered. This fact suggests that the temperature dependence of τ_E can be viewed as some kind of feedback, and hence it may be used to control the instabilities. In what follows, we will examine the use of feedback stabilization for controlling fusion reactors.

It is clear that when the reactor is operating at a temperature $T < T_c$ some sort of stabilization is needed. Feedback stabilization will be applied as follows: (a) to measure the deviation from operating values of density or temperature, and (b) to suppress these deviations (temperature, density, or both). A linear feedback system will be assumed and the analysis is carried out by adding terms proportional to n'/n_0 and T'/T to equations (8.8) and (8.9), depending on what kind of control is adopted. The mechanism of stabilization may be viewed as resulting from the modification of the energy gain or loss dependence on density and temperature due to the addition of n'/n and T'/T_0 terms. In this portion of the analysis we shall assume τ_E and τ_n to be constant and discuss the stabilization of a plasma which would be unstable unless controls are applied. We may note that inclusion of a term proportional to n'/n_0 only in equations (8.8) and (8.9) does not lend to stabilization because the plasma is stable to density perturbation only. One can readily see this by noting that an addition of a term $\alpha\ n'/n_0$ to equations (8.8) and (8.9) results in a stability condition which requires that $\alpha > 1/\tau_{no}$ and $\alpha < 1/\tau_{no}$ simultaneously, a condition that cannot be met. We will, therefore, examine the following three cases by monitoring temperature deviation.

Density Control

This simply means that we take equation (8.8) and add to it the term $\alpha\ T'/T_0$, with the result

$$\frac{1}{n_0}\frac{dn'}{dt} = -\left(\frac{1}{\tau_{no}}+\frac{1}{\tau_{n1}}\right)\frac{n'}{n_0} - \left(\frac{1}{\tau_{T_1}} - \alpha\right)\frac{T'}{T_0} \tag{8.23}$$

The above equation along with equation (8.9) are now analyzed for stability in exactly the same manner as before. We find that for stability the conditions are:

$$B > 0 \Rightarrow \alpha \quad \frac{1}{\tau_2} - \frac{1}{\tau_{no}} - \frac{1}{\tau_{E_0}} \tag{8.24}$$

$$C > 0 \Rightarrow \alpha \left(\frac{1}{\tau_1} - \frac{1}{\tau_{E_0}}\right) < \frac{1}{\tau_{no}}\left(\frac{1}{\tau_{E_0}} - \frac{1}{\tau_2}\right) \tag{8.25}$$

Temperature Control

In this case, the term $\alpha T'/T_0$ is added to the linearized energy equation (8.9), and the process is repeated. The results are:

$$B > 0 \Rightarrow \alpha < \frac{1}{\tau_{E0}} - \frac{1}{\tau_2} + \frac{1}{\tau_{n0}}$$

$$C > 0 \Rightarrow \alpha < \frac{1}{\tau_{E0}} - \frac{1}{\tau_2}$$

which can be combined to give:

$$\alpha < \frac{1}{\tau_{E0}} - \frac{1}{\tau_2} \qquad (8.26)$$

Temperature and Density Control

Both equations (8.8) and (8.9) with the term $\alpha T'/T_0$ included are examined in this case. Feedback stabilization can, however, be applied by means of a source term s', which is proportional to T'/T, *i.e.*, we let

$$s = s_0 + s'$$

$$s' = \alpha_s T'(t - \Delta t) = \alpha n_0 \frac{T'(t - \Delta t)}{T_0} \qquad (8.27)$$

in which a delay time Δt is taken into account. Assuming T_s = constant, and $\gamma \Delta t \ll 1$, then γ can be obtained from

$$A\gamma^2 + B\gamma + C = 0 \qquad (8.28)$$

with

$$A = 1 + \left(\frac{T_s}{T_0} - 1\right) \alpha \Delta T$$

$$B = \frac{1}{\tau_{n0}} - \frac{1}{\tau_2} + \frac{1}{\tau_{E0}} - \left(\frac{T_s}{T_0} - 1\right) \alpha + \frac{\alpha \Delta t \xi_2}{\tau_{E0}}$$

$$C = \frac{1}{\tau_{n0}} \left(\frac{1}{\tau_{E0}} - \frac{1}{\tau_2}\right) - \frac{\alpha \xi_2}{\tau_{E0}} \qquad (8.29)$$

The solution of (8.28) is

$$\gamma_{1,2} = \frac{-B \pm \sqrt{B^2 - 4AC}}{2A}$$

which leads to the following stability conditions

$$A > 0, \quad B > 0, \quad C > 0$$

or

$$A < 0, \quad B < 0, \quad C < 0 \tag{8.30}$$

The stabilized region is the dashed region shown in Figure 53.

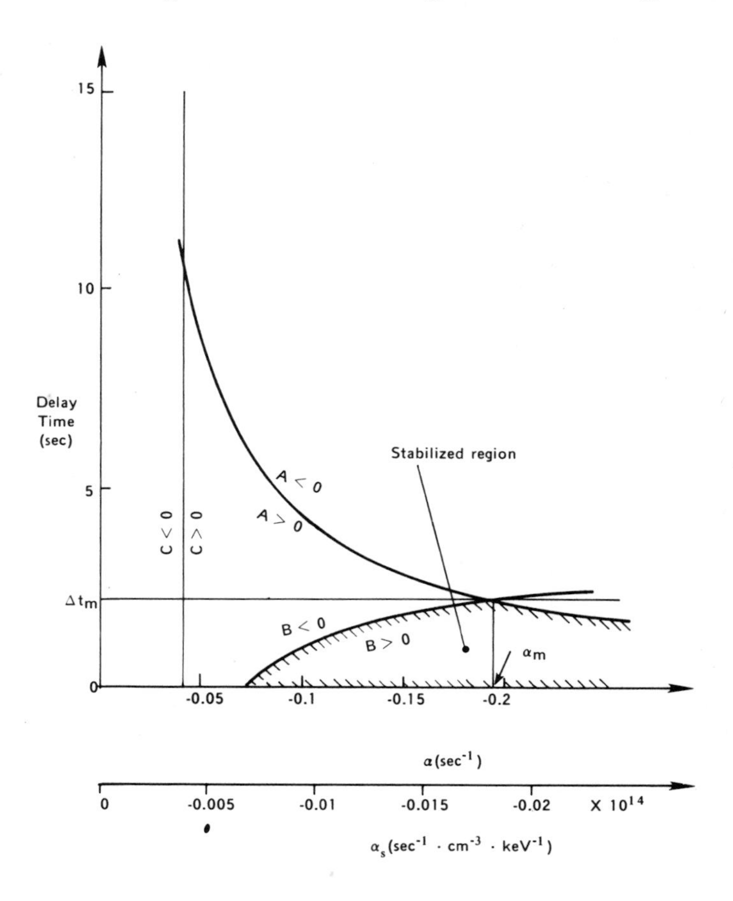

Figure 53. Regions of feedback stabilization.

8.2 THE DYNAMIC BEHAVIOR OF A LOW BETA TOKAMAK REACTOR

In this section we will follow the work of Horton and Kammash (1973) and examine the dynamics of a more realistic system in which the electrons and the ions of the plasma as well as the fusion reaction products are treated individually. The plasma will be assumed to be heated by alpha particles, and both bremsstrahlung and synchrotron radiation losses will be taken into account. The question of thermal stability of the equilibrium states will also be investigated through the time-dependent solutions of the dynamic equations. The analysis in the previous section was restricted to a very simple plasma model which does not distinguish between the fuel and the exhaust components of the plasma. In this analysis, we present results which follow all of the particle and energy components of the fusion plasma. Separating the plasma into its fuel and exhaust components and following the transfer of energy between the two components is important even for a qualitatively correct description.

The basic laws required to predict the fusion plasma behavior are the following:

1) nuclear reaction rates
2) confinement laws
3) charged particle heating rates
4) radiation loss rates.

Before turning to the dynamic equations and their solutions, we will present and discuss the formulas used to describe these four processes. The model we employ will be concerned mainly with variations in the confinement laws where the largest unknown remains. We will shortly note that fusion reactors can be designed over a wide range of "anomalous" confinement laws.

Once again, we restrict our consideration to a deuterium-tritium fueled reactor of the Tokamak type operated in a steady state. As a reference reactor, we take the device parameters given in Table 19. The quantities defined in the table are: 1) a—the minor radius in meters; 2) A = R/a—the aspect ratio with R being the major radius (see Figure 47); 3) B—the confining magnetic field strength in Webers/m^2 (one Weber/m^2 = 10 k.Gauss); and 4) q—the Tokamak safety factor [when its value exceeds unity it implies stability of plasma against MHD modes (see Chapter 1)]. This safety factor also implies a total plasma current given by $I = 5 \times 10^3\, aB/qA$ in kiloamperes.

Table 19

Reference Reactor

B = 60 kilogauss	a = 2 meters
A = R/a = 5	q = 2
R_e = 90%	D-T fuel

Nuclear Reaction Rate

Since we have chosen the D-T reaction as the fusion reaction of interest, we can write for the reaction rate:

$$\langle\sigma v\rangle = 3.7 \times 10^{-12}\, h(T_i)T_i^{-2/3} \exp(-20/T_i^{1/3})\ \mathrm{cm^3/sec} \quad (8.31)$$

where the deuterium and tritium are in Maxwell-Boltzmann distributions at the common temperature T_i expressed in keV. The above expression is the same as equation (2.32) except for the factor $h(T_i)$ which is unity for low temperatures and corrects the simple formula (2.32) in the T_i = 50-500 keV region to represent the $\langle\sigma v\rangle$ function with its maximum at $T_i \simeq 70$ keV. It has been found that this function has the form $h(x) = [1 + (x/70)^{1.3}]^{-1}$. Released in each D-T reaction are a 3.5 MeV alpha particle, which is confined by the magnetic field, and a 14.1 MeV neutron, which escapes from the plasma and is absorbed in the neutron blanket. For the reactor power output we take the energy from the 14.1 MeV neutron and do not include the energy multiplication obtained from breeding tritium in the neutron blanket. The reactor power is then given by

$$P_n = \tfrac{1}{4}\, n_i^2 \langle\sigma v\rangle Q_n V \quad (8.32)$$

where $V = 2\pi^2 a^3 A$ is the plasma volume and Q_n = 14.1 MeV. In the reference reactor, P_n varies from 0.1 to 100 gigawatts depending on the fueling rate, the confinement law, and other parameters.

Confinement Laws

A. Neoclassical Confinement

In a quiescent plasma (*i.e.*, with a thermal level of fluctuations), the confinement in the long-mean-free path regime (low collision plasma) of a low beta fusion reactor is determined by the loss rate of trapped particles as reviewed by Kadomtsev and Pogutse (1971). The particle confinement time, τ_p, is given by

$$\frac{1}{\tau_p} = \frac{\nu_e \rho_e^2}{a^2} \quad q^2 \ A^{3/2} \tag{8.33}$$

where

$$\nu_e = \sqrt{2\pi}\ C^4 \ \overline{Z} \ \ln \Lambda \ / \ 3 \ m_e^{1/2} \ T_e^{3/2} \tag{8.34}$$

is the electron-ion collision frequency and

$$\overline{Z} = \sum_{ions} n_i Z_i^2 \ / \ \sum_i n_i Z_i$$

is an average charge number which varies from 1 to 2 for D-T fusion plasma. The quantity ρ_e is the mean electron gyroradius whose values is given by $\rho_e = c\,(m_e T_e)^{1/2}/eB$.

With classical processes, that is, on the basis of two-body coulomb collisions, one expects that it will be the ions that transport most of the heat across the magnetic field. This is simply because the step length, which is the Larmor radius, is much larger for ions than for electrons. It is for a different reason that the ion heat transport across the magnetic field goes at a faster rate than the ion particle transport. When two particles of like charge and mass collide, then individual energies may be redistributed, but their center of gravity does not move. Moreover, transport of ions and electrons across the magnetic field must proceed at the same rate and, as we shall see shortly, this rate is down by the square root of the mass ratio from the rate of heat transport. If we, therefore, direct our attention to the classical heat transport processes, we can neglect the heat which is carried out bodily by each particle. We can make a qualitative estimate of the magnitude of the coefficient of thermal conductivity. The heat flow equation of interest is of the type

$$\vec{Q}_\perp = - K_\perp \nabla \ kT \tag{8.35}$$

where

$$K_\perp = n_i D_\perp$$

$$D_\perp \sim \frac{\lambda^2}{\tau} = \frac{\rho_{Li}^2}{\tau_{90}} \tag{8.36}$$

Here $D_\perp$ represents the particle diffusion coefficient for transport across the magnetic field [see equation (1.14)], ρ_{Li} is the ion Larmor radius in the toroidal magnetic field, and τ_{90} is the 90° deflection time for ion-ion collisions. In the long-mean-free path regime, there is a step length longer than ρ_{Li} which must be used. That is the thickness of the ion banana orbits as illustrated in Figure 54.

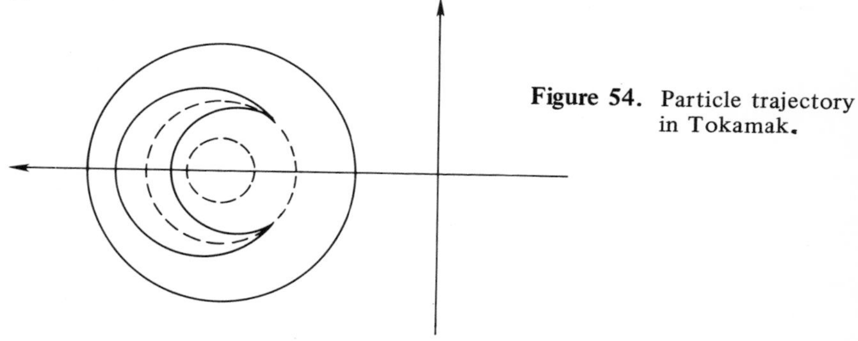

Figure 54. Particle trajectory in Tokamak.

Particle orbits in Tokamak geometry may be divided into two categories–passing particles and trapped particles. The trapped particles have velocity vectors close to 90° from the magnetic field direction and are reflected by the somewhat stronger magnetic field on the inside of the torus. The total thickness of the banana orbit at its fattest point is just $2^{3/2}$ times the Larmor radius in the local poloidal field $B_{pol}(r)$ divided by the square root of the local aspect ratio (R/r).

We can now make a new estimate of the heat conductivity in what is called the "neoclassical" or banana regime. To do this, we need to look briefly at the physics of magnetic mirror reflection (see Chapter 1). Consider a particle with magnetic moment $\mu = mv_\perp^2/2B$ situated between two magnetic mirrors. At the point of reflection, which we will call B_{ref}, the parallel velocity will go to zero, and $v_\perp^2$ at this point will be equal to v^2, where

$\frac{1}{2}mv^2$ is the total particle kinetic energy. The condition for reflection, or for trapping, is

$$\frac{mv_\perp^2}{2B} = \frac{mv^2}{2B_{ref}} \tag{8.37}$$

where B is the field strength midway between the mirrors. From the above equation we can write

$$\sin^2\theta = \frac{v_\perp^2}{v^2} = \frac{B}{B_{ref}} \sim \frac{R-r}{R} = 1 - r/R \tag{8.38}$$

$$\cos\theta \simeq \sqrt{r/R}$$

where R is the major radius of the torus and r is the minor radius coordinate of the particle. With this information, we can now estimate the neoclassical heat conductivity. The heat is carried mainly by the trapped ions. Therefore, we reduce n_i by the ratio of the trapped particle solid angle to 4π, *i.e.*, by

$$\int_0^{\pi/2} d(\cos\theta) = \cos\theta \simeq \sqrt{r/R}.$$

Also, we reduce the angle of deflection from 90° to $(\pi/2 - \theta)$, which is the critical angle for scattering into the loss cone. This effect reduces the collision time for 90° deflection, τ_{90}, to r/R times its earlier value. Putting it together, we have

$$K_\perp = \frac{n_i(r/R)^{1/2}\,[\rho_{Li}\,(B/B_p)\,(r/R)^{1/2}]^2}{\tau_{90}\,(r/R)} \tag{8.39}$$

Since $(r/R) = 1/A$, and the safety factor, $q = 1/A\ B/B_{pol}$ the thermal conductivity, $K_\perp$, becomes

$$K_\perp = \frac{n_i\,q^2\,\rho_i^2\,A^{3/2}}{\tau_i} = \nu_{ii}\,A^{3/2}\,q^2\,\rho_i^2 \quad (\rho_i \equiv \rho_{Li})$$

where ν_{ii} is the ion-ion collision frequency. The energy confinement time in the neoclassical regime can, therefore, be written as

$$\frac{1}{\tau_E} = \frac{K_\perp}{a^2} = \frac{\nu_{ii}\,A^{3/2}\,q^2\,\rho_i}{a^2} \tag{8.40}$$

If we now compare this with the neoclassical particle confinement time as given by equation (8.33), we find

$$\frac{\tau_p}{\tau_E} = \frac{\nu_i \rho_i^2}{\nu_e \rho_e^2} = \frac{\nu_i}{\nu_e} \frac{2T_i}{m_i \Omega_i^2} \frac{m_e \Omega_e^2}{2 T_e} = \frac{\nu_i}{\nu_e} \frac{T_i}{T_e} \frac{m_i}{m_e}$$

But from equation (8.34) we note that $\nu \sim m^{-1/2} T^{-3/2}$; thus the above ratio becomes

$$\frac{\tau_p}{\tau_E} = \sqrt{\frac{m_i T_e}{m_e T_i}}$$

and if the electron and ion temperatures are comparable this ratio becomes

$$\frac{\tau_p}{\tau_E} = \sqrt{\frac{m_i}{m_E}} \qquad (8.41)$$

Expressing the temperature in keV, the magnetic field in Webers/m^2, the density in particles per cm^3, and the minor radius in meters we can write

$$\frac{1}{\tau_p} = 9 \times 10^{-18}\, c_1 \frac{n_e}{T_e^{1/2}} \frac{q^2 A^{3/2}}{B^2 a^2} \quad (\mathrm{sec}^{-1})$$

$$\frac{1}{\tau_E} = 540 \times 10^{-18}\, c_1 \frac{n_i}{T_i^{3/2}} \frac{q^2 A^{3/2}}{B^2 a^2} \quad (\mathrm{sec}^{-1}) \qquad (8.42)$$

where c_1 is a constant employed to represent various effects, *e.g.*, the characteristic length for the density gradient $(a/n\ dn/dr)^2$ from the radial profiles.

B. Anomalous Confinement (Drift Wave Turbulence)

In the presence of density fluctuations of the type due to drift wave instabilities, studies (see Liu, Rosenbluth, and Horton, 1972) have given the particle confinement time as

$$\frac{1}{\tau_p} = c_2 \frac{c\, T_e}{e\, B} \frac{\rho_e}{a^3} (q\, A)^{1/2} = \frac{c_2}{10} \frac{T_e^{3/2}}{B^2\, a^3} (qA)^{1/2} \tag{8.43}$$

A low level of these fluctuations is predicted to occur in Tokamaks due to the relatively weak shear inherent in such devices. In this case the energy confinement time is approximately equal to the particle confinement time.

C. *Anomalous Confinement (Trapped Particle Turbulence)*

Because of the existence of localized magnetic mirrors in toroidal devices such as Tokamaks, the presence of trapped particles in these regions can lead to mirror-like instabilities referred to as trapped particle instabilities. In contrast with mechanisms A and B, where there is considerable experimental evidence for the basic mechanisms, the presence of trapped particle turbulence is basically only expected to occur in fusion reactor-size devices. For this mechanism of plasma loss, we know much less about the confinement time. We simply use the Bohm formula given in Chapter 1 and write

$$\frac{1}{\tau_p} = c_3 \frac{c\, T_e}{e\, B} \frac{1}{a^2} = 62.5\, c_3 \frac{T_e}{Ba^2} \tag{8.44}$$

where one theoretical estimate of c_3 is $c_3 \sim 1/A^2$.

Charged Particle Heating Rates

In the self-contained reactor, the charged particles produced from the nuclear reactions provide the heating for the incoming cold fuel. Hence, the rate of charged particle heating compared with the particle loss rate has an important influence over the reactor design as exemplified in the comparison of a mirror and toroidal reactor.

The alpha particles are born with a nearly monoenergetic distribution at 3.5 MeV. The alpha particle confinement in the torus for most of the confinement models A through C is sufficiently long to allow considerable broadening and thermalization of the alpha particle distribution function. In calculating the energy transfer rate from the alphas to the plasma, we will assume that the alphas have a Maxwellian distribution. In this case, we can simply use equation (4.103) in the "classical" limit and write

$$\frac{3}{2}\frac{d}{dt}(nT) = -8\sqrt{\pi}\, nn'(ee')^2 \ln\Lambda \left[\frac{mm'}{2(mT' + m'T)}\right]^{3/2} \times (T - T') \tag{8.45}$$

where primed quantities denote the target particles. With the appropriate substitutions in the above equation, we find that the rates of energy transfer from the alphas to the electrons and ions of the plasma are:

$$W_{\alpha e} = 2.7 \times 10^{-12} \frac{n_e n_\alpha (T_\alpha - T_e)}{T_e^{3/2}} \tag{8.46}$$

and

$$W_{\alpha i} = 1.3 \times 10^{-10} \frac{n_i n_\alpha (T_\alpha - T_i)}{(T_\alpha + 1.6\, T_i)^{3/2}} \tag{8.47}$$

Since the alpha particle heating rates differ only slightly for deuterium and tritium, we have represented the D-T ions by a single mean mass hydrogen ion, $m_i = (m_D + m_T)/2$, and have taken $T_i = T_D = T_T$. From equation (8.45) we can also write the rate of energy transfer from ions to electrons as

$$-W_{ei} = +W_{ie} = -5.1 \times 10^{-13} \frac{n_e n_i (T_e - T_i)}{T_e^{3/2}} \tag{8.48}$$

Radiation Loss Rates

The bremsstrahlung radiation emitted from the plasma electrons is given by (see Chapter 1)

$$P_B = 3.3 \times 10^{-15}\, n_e^2\, \bar{Z}\, T_e^{1/2} \text{ (keV/cm}^3\text{-sec)} \tag{8.49}$$

This component of radiant energy is in the ultra-violet to X-ray range and is usually absorbed by the liner walls of the device.

The calculation of the net synchrotron radiation loss rate is (as we saw in Chapter 2) a complicated problem and requires further research for a quantitative result valid over the full range of temperatures. At low temperatures ($T_e < 50$ keV) the dominant line-broadening mechanism responsible for self-absorption is the magnetic field inhomogeneity, and we use the result given by Rosenbluth (1970) in this range, *i.e.*,

$$P_S = 4.1 \times 10^3 \, n_e^{1/2} \, T_e^2 \, B^{5/2} \left(\frac{1-R_e}{aA}\right)^{1/2} \quad \text{keV/cm}^3\text{-sec} \tag{8.50}$$

where R_e is the average reflectivity to microwaves of the surfaces facing the plasma. At higher temperatures we neglect the effect of field inhomogeneities and use the result from relativistic line broadening given by (see Chapter 2)

$$P_S = 70 \, n_e^{1/2} \, T_e^{2.8} \, B^{5/2} \left(\frac{1-R_e}{aA}\right)^{1\,2} \quad \text{keV/cm}^3\text{-sec} \tag{8.51}$$

In the solutions discussed below we make a simple transition from the low temperature law to the high temperature one. The synchrotron radiation is nearly perfectly reflected from the metal liner; however, the vacuum and fueling ports and the "divertor" (through which the unspent fuel is diverted) are areas of low reflectivity. For the reference reactor, we take the mean reflectivity to be 90% and then consider the variation of the equilibrium temperature with R_e as shown for a typical case in Figure 55.

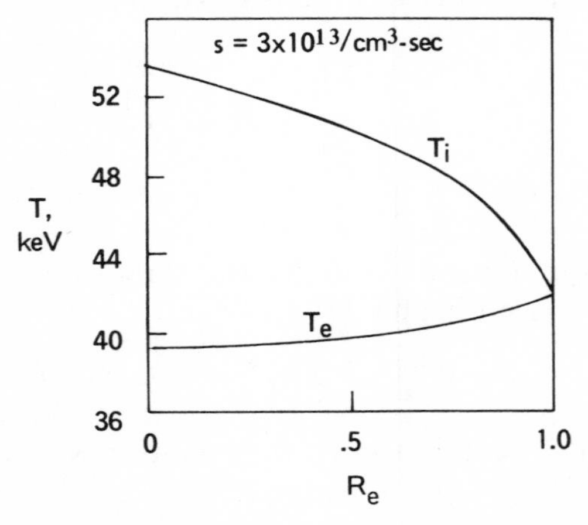

Figure 55. Variation of electron and ion temperature with reflection coefficient in Model B confinement.

Dynamic Analysis

Having obtained the information necessary, we proceed now to investigate the dynamics of a low beta Tokamak reactor by first writing down the appropriate dynamic equations. Unlike those in the previous section, where a highly simplified model was used, the equations we use here will be for every species involved where distinction is made between fuel and exhaust. We describe the fusion plasma by a mean concentration n_j and a mean temperature T_j for each species followed in the reactor. Combining the four mechanisms discussed above, we obtain equations giving the time rate of change of these mean plasma variables and solve the equations for our reactor model.

The confinement laws A-C are "ambipolar" and preserve charge neutrality, $n_e(t) = n_i(t) + 2n_\alpha(t)$, by having the same particle confinement time τ_p for each species. As a consequence of charge neutrality, the mean hydrogen mass approximation, and the assumption of a 50-50 deuterium-tritium fuel injection mixture, we can reduce the eight dynamic equations for the D,T, α, e system to the following five rate equations:

$$\frac{dn_i}{dt} = -\frac{n_i}{\tau_p} - \frac{1}{2} n_i^2 \langle\sigma v\rangle + s \tag{8.52}$$

$$\frac{dn_\alpha}{dt} = -\frac{n_\alpha}{\tau_p} + \frac{1}{4} n_i^2 \langle\sigma v\rangle \tag{8.53}$$

$$\frac{3}{2}\frac{d}{dt}(n_\alpha T_\alpha) = -\frac{3}{2}\frac{n_\alpha T_\alpha}{\tau_{E\alpha}} + \left(Q_\alpha + \frac{3}{5}T_i\right)\frac{n_i^2 \langle\sigma v\rangle}{4} - W_{\alpha e} - W_{\alpha i} \tag{8.54}$$

$$\frac{3}{2}\frac{d}{dt}(n_i T_i) = -\frac{3}{2}\frac{n_i T_i}{\tau_{Ei}} + W_{\alpha i} + W_{ei} - \frac{3}{4}T_i n_i^2 \langle\sigma v\rangle + sV_i \tag{8.55}$$

$$\frac{3}{2}\frac{d}{dt}(n_e T_e) = -\frac{3}{2}\frac{n_e T_e}{\tau_{Ee}} + W_{\alpha e} - W_{ei} - P_B - P_s + sV_e \tag{8.56}$$

where s is the fueling rate in ions/cm^3 sec, $V_{i,e}$ is the energy of the injected ions (electrons) which is taken as zero in this analysis, and Q_α is the alpha particle energy, 3.5 MeV. The transfer rates W_{ij} have already been given in equations (8.45-8.47).

Solutions to the above equations have been obtained numerically, and it has been found that these (time-dependent) solutions either evolve to a stable stationary reactor state or cool to the ignition point and then rapidly collapse to a negligible temperature. The distinction between the two cases is not always obvious in that in some cases there is a slow decay from a quasi-stationary state which can take 30 seconds or longer to lower the temperature to the ignition point. These results are discussed below for the three confinement laws, A-C, presented earlier.

Anomalous Reactor Model C (Bohm)

Since the Bohm confinement law, equation (8.44) with $c_3 = 1/16$, has become a widely used reference point in confinement studies, the results obtained were for this value of c_3. It has been noted, however, that the theoretical estimates that give a Bohm-like scaling predict smaller values of c_3. Steady state fusion plasmas are obtained with $c_3 = 1/16$ for magnetic fields in the range B = 80 to 150 kG, but at B = 60 kG for the reference reactor the confinement is inadequate. The time-dependent solutions approach the steady states on a second time scale with negligible overshoot. Steady state solutions are obtained for reactor parameters in the range

$$\begin{aligned} B &= 80 - 150 \text{ kG} \\ a &= 1.8 - 8 \text{ m} \\ s &= (6\text{-}30) \times 10^{14}/\text{cm}^3 \text{ sec} \end{aligned} \qquad (8.57)$$

for A = 5, q = 2, and R = 0.9 with not all the parameters in equation (8.57) taking their minimum value simultaneously. The solutions are characterized by burn-up fractions from 3-15%, ion and electron temperatures from 200-100 keV, plasma pressures with $\beta \approx 0.05$ to 0.5 and relatively large power outputs ($P_n >$ 10 gigawatts), which increase with the fifth power of the minor radius. For a given reactor power, the fuel injection rates are approximately a factor of ten greater than for confinement law B. The results from fifteen steady state solutions obtained in the parameter range given in (8.57) are summarized by

$$P_n = 0.069\ a^5\ B^{1.2} (s/10^{15})^{1.5}$$

$$\beta = 0.73\ P_n^{0.7}/a^{1.1}\ B^{1.6} \qquad (8.58)$$

where B, a, s are restricted to the range given in (8.57) and P_n is in gigawatts. The requirement of a maximum tolerable neutron power per unit wall area together with the rapid fifth power increase of P_n with minor radius imply an upper limit to the plasma radius which is relatively insensitive to the precise value of the wall load limit in neutron energy flux.

Anomalous Reactor Model B

For this reactor model, the anomalous confinement law given in equation (8.43), with $c_2 = 1.0$ was used. The time-dependent solutions approach the steady states through heavily damped oscillations with periods of the order of ten seconds. The steady states are stable to large perturbations in density or temperature up to the order of several hundred per cent. Steady state solutions are obtained for reactor parameters in the range

$$\begin{aligned} B &= 40 - 150 \text{ kG} \\ a &= 1\text{-}6 \text{ m} \\ s &= (1\text{-}30) \times 10^{13}/\text{cm}^3 \text{ sec} \end{aligned} \qquad (8.59)$$

for $A = 5$, $q = 2$, and $R_e = 0.9$ with not all the parameters above at their minimum value simultaneously. The steady state fusion plasmas obtained are characterized by burn-up fractions from 7-60%, ion and electron temperatures from 30-150 keV, plasma pressures with $\beta = 0.04\text{-}1.0$ and power outputs with $P_n = 1.0\text{-}100$ gigawatts. The reference reactor has $P_n = 3$ gigawatts for $s = 3 \times 10^{13}/\text{cm}^3$ sec. In this case the burn-up fraction is 10%, $T_i = 46$ keV, $T_e = 40$ keV, $\beta = 0.12$, the alpha particle concentration is 6% of the ion density, and the alpha temperature is 400 keV.

The results for the steady state power and beta obtained in the parameter range given in equation (8.59) are summarized by

$$P_n = 0.001\ a^{4.5}\ B^{1.8}\ (s/10^{13})^{1.3}$$

$$\beta = 1.6\ \frac{P_n^{0.7}}{a^{0.8}\ B^{1.6}} \qquad (8.60)$$

Figure 55 shows the effect of varying the reflection coefficient R_e in the reference reactor with $s = 3 \times 10^{13}/cm^3$ sec. The decrease in the ion temperature with increasing electron temperature is due to the shorter confinement time at high electron temperatures. The reactor power is relatively insensitive to the change in R_e, varying from 3.2 gigawatts at $R_e = 0$ to 2.4 gigawatts at $R_e = 1.0$.

Neoclassical Reactor Model A

For the neoclassical reactor, equations (8.52-8.56) are solved for the confinement law given in equation (8.42) with $c_1 = 1.0$. For this model the particle confinement time is on the order of an hour, and the energy confinement time is on the order of a minute. Thus, the burn-up fraction is essentially 100%, and the plasma has a high alpha particle concentration ($n_\alpha \simeq n_i$) in the time of a reaction period $\tau_R \simeq 30$ seconds. As a consequence, in the neoclassical reactor we note that the solutions do not reach a true steady state but rather evolve rapidly on a second time scale to a quasi-steady state which slowly decays on a 30-second time scale to the ignition temperature whereafter the fusion plasma is quickly lost. For the neoclassical plasma we must keep in mind that we are discussing the properties of the quasi-steady state as a function of the reactor parameters rather than a true steady state as in models B and C discussed above.

The neoclassical reactor states are characterized by relatively high plasma temperatures (greater than 100 keV) and lower fuel injection rates for the same reactor power as compared to the anomalously confined fusion plasma.

The reference reactor produces $P_n \simeq 3$ gigawatts for a source $s \simeq 3 \times 10^{12}/cm^3$ sec, and has $\beta = 0.5$ with the ion temperature $T_i \simeq 200$ keV and $T_e \simeq 100$ keV. Quasi-steady state solutions were obtained for reactor parameters in the range

$$B = 40\text{-}150 \text{ kG}$$
$$a = 1.5 - 8 \text{ m}$$
$$s = (1\text{-}30) \times 10^{12}/cm^3\text{-sec} \qquad (8.61)$$

for $A = 5$, $q = 2$, $R_e = 0.9$. For these parameters, the reactor power varies from 1-200 gigawatts, and $\beta \simeq 0.2\text{-}1.0$. The results obtained for the reactor power and beta are summarized by

$$P_n = 0.14\ a^3\ (s/10^{12})^{0.85}$$

$$\beta = 83\ P_n^{0.6} / a^{1.6}\ B^{2.5} \qquad (8.62)$$

for B, a, s in the range given in (8.61). In Figure 56 we see the effect on electron and ion temperature of varying the reflection coefficient R_e in the neoclassically confined plasma in the reference reactor with $s = 3 \times 10^{12}/cm^3$ sec.

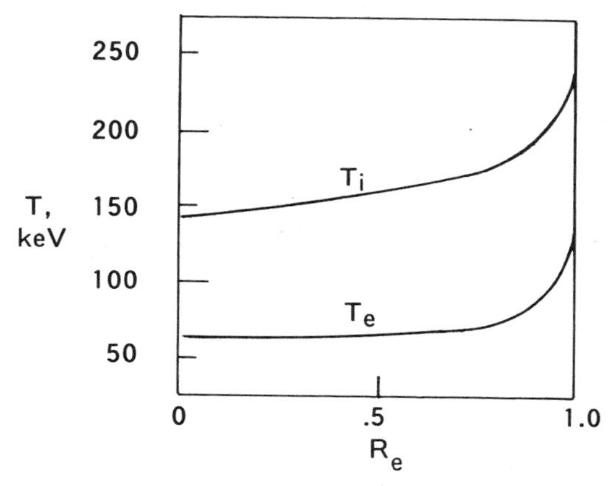

Figure 56. Electron and ion temperature as a function of R_e for model A confinement.

In this case the ion temperature follows the electron temperature and both make a significant increase for $R_e > 90\%$ due to the reflection of synchrotron radiation. In the neoclassical model, synchrotron radiation is the dominant energy loss mechanism until the reflection coefficient is essentially unity.

In Figures 57 and 58, we note the contrast in the fueling requirements and the plasma beta as a function of reactor power in the reference reactor for the neoclassical model (A) and the anomalous plasma (model B). For the reactor in this sample, the ion source at $s = 10^{13}/cm^3$ sec corresponds to an injection current of 1260 amperes.

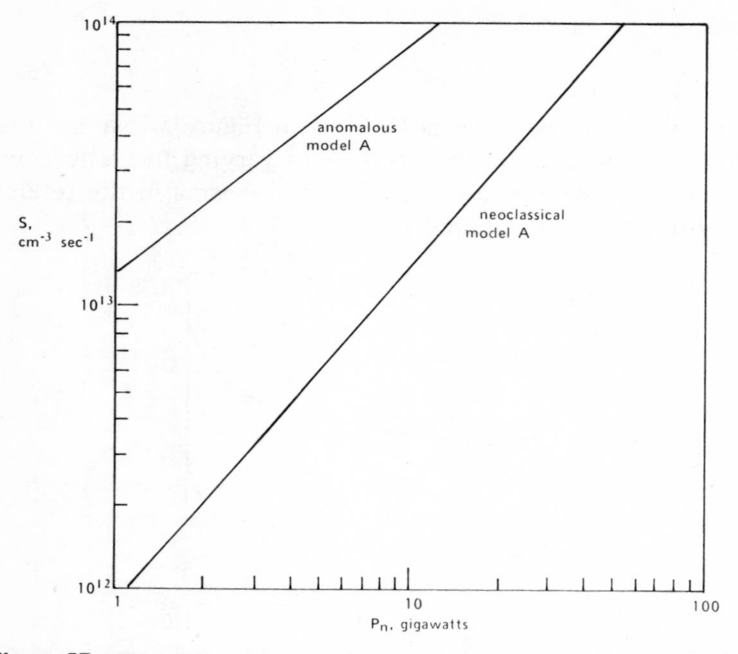

Figure 57. Injection rate versus neutron power for confinement model A and B.

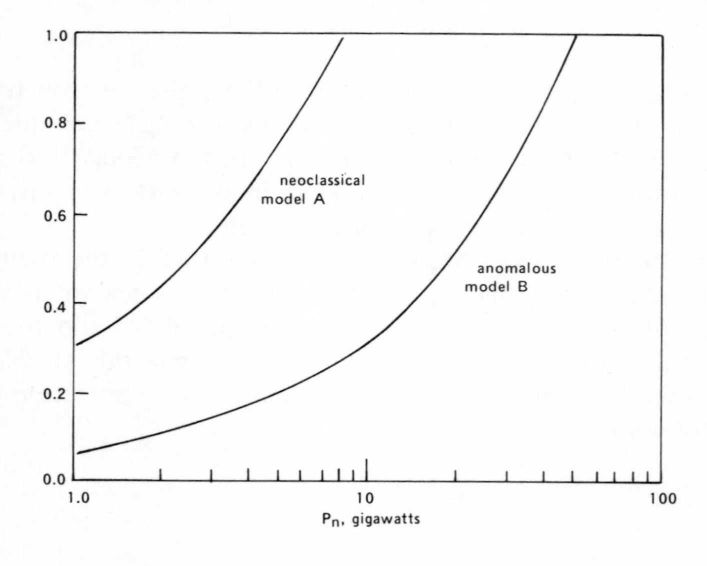

Figure 58. The plasma beta as a function of neutron power.

Summary and Conclusion

We have seen in the above analysis that plausible steady state fusion plasmas can be obtained over a wide range of anomalous confinement laws. Due to the increase of plasma loss with increasing temperature, the anomalous confinement laws provide a limit on the plasma beta which is not present with classical confinement. In the presence of a critical plasma beta such as from MHD stability requirements, the anomalous plasma confinement provides a self-regulating mechanism.

The neoclassically confined plasmas are characterized by confinement times that are greater than the nuclear reaction period, with the result of a complete burn-up of the fuel and a high alpha particle concentration ($n_\alpha \simeq n_i$). As a consequence, the plasma pressure is high, and relatively large magnetic fields are required to maintain a given plasma beta. Lacking an exhaust removal mechanism, the neoclassical solutions form a quasi-steady state which burns out over several reaction periods. From another point of view, the anomalous confinement rates allow a reactor to operate at higher thermonuclear reaction rates while maintaining the plasma pressure at less than a given critical plasma pressure by providing for fast removal of spent fuel.

Synchrotron radiation is the dominant energy loss mechanism in the neoclassical reactor, and the radiation law provides a stable upper temperature limit for the neoclassical reactor. Both anomalous models B and C possess a high degree of thermal stability, while the neoclassical model A appears stable within its quasi-steady state.

The lack of existence of thermal equilibrium states for the neoclassical reactor in this analysis is due mainly to one assumption, namely the alpha particle confinement time. We recall that no distinction was made between the alpha particle and ion confinement times although the energy confinement times were appropriately adjusted by incorporating the corresponding mass ratios. The question of alpha particle confinement in toroidal devices has not yet been theoretically resolved but some arguments based on simple classical concepts can be made to distinguish between ion and alpha particle diffusion rates. From equations (1.11) and (1.14) one can readily see that the classical diffusion coefficient varies directly with the square of the average Larmor radius and the collision frequency, a product which depends on the masses and charges of the particles involved. This fact in itself is sufficient to indicate that because of different mass and charge an alpha particle is expected to diffuse across magnetic fields at a different rate than that of a deuteron or a triton. In fact it has been shown (by Kammash and Galbraith, 1974) that thermal equilibrium states can exist in a neoclassically confined Tokamak reactor if $\tau_\alpha \leqslant \frac{1}{2} \tau_i$, and that the existence or nonexistence of an equilibrium depends upon the relationships between the various time constants and not upon their overall magnitude.

REFERENCES

1. Conn, R. W., D. G. McAlees, and G. A. Emmert. *Technology of Controlled Thermonuclear Fusion Experiments and the Engineering Aspects of Fusion Reactors* CONF - 721111 U.S. AEC (1974).
2. Horton, W. Jr., and T. Kammash. *Nuclear Fusion, 13,* 753 (1973).
3. Horton, W. Jr.,and T. Kammash. *Technology of Controlled Thermonuclear Fusion Experiments and the Engineering Aspects of Fusion Reactors,* CONF - 721111 U.S. AEC (1974).
4. Kadomtsev, B. B. and O. P. Pogutse. *Nuclear Fusion, 11,* 67 (1971).
5. Kammash, T. and D. L. Galbraith. *Nuclear Fusion, 13,* 133 (1973).
6. Kammash, T. and D. L. Galbraith. Unpublished results (1974).
7. Liu, C. S., M. N. Rosenbluth, and W. Horton. *Phys. Rev. Letters, 29,* 1489 (1972).
8. Mills, R. G. "Proceedings of the Symposium on Engineering Problems of Fusion Research," Los Alamos Scientific Laboratory, LA-4250 (1969), p. B1.
9. Ohta, M., H. Yamato, and S. Mori. *Plasma Physics and Controlled Nuclear Fusion Research (*Proc. 4th Int. Conf. Madison, 1971), *3,* IAEA, Vienna, 423 (1971).
10. Powell, C. and O. J. Hahn. *Nuclear Fusion, 12,* 667 (1972).
11. Rosenbluth, M. N. *Nuclear Fusion, 10,* 340 (1970).
12. Stacey, W. M. Jr. *Nuclear Fusion, 13,* 843 (1973).
13. Yamato, H., M. Ohta, and S. Mori. *Nuclear Fusion, 12,* 604 (1972).

CHAPTER 9

AN ENVIRONMENTAL ASPECT OF A FUSION POWER PLANT–THERMAL EFFICIENCY AND WASTE HEAT

In addition to being an important measure of how viable a power producing system is, the thermal efficiency of a power plant is also a measure of the waste heat rejected by the plant into the surroundings. In this chapter we shall calculate the thermal efficiency of mirror fusion reactors with direct conversion (a process to be discussed later) and compare it with conventional power plants. We shall see that unless certain component efficiencies are kept high this type of fusion reactor will barely be competitive with conventional (*e.g.*, coal-fired) plants which implies that it may be considered a "thermal pollutant." This is not an entirely true statement since, as we shall note in subsequent chapters, the rejected heat is high-grade heat and can be efficiently utilized. A thorough examination of the various component efficiencies as well as the overall reactor efficiency of the different approaches to fusion will be given in Chapter 16.

We recall from Chapter 1 that mirror machines are inherently lossy, *i.e.*, particles (ions, electrons, alphas, etc.) can readily escape the best practical mirror. In escaping, these particles carry with them an appreciable fraction of the plasma kinetic energy; to prevent a rapid cooling of the plasma, it is necessary to continually supply energy to it. Much of the energy supplied to the plasma is recoverable since it appears as kinetic energy of the escaping particles or possibly as radiation from the plasma; however, since no energy recovery system is 100 per cent efficient some of the fusion power must be siphoned off to replace the lost energy. Calculations generally show that this heating power must be of the same order of magnitude as the fusion power and in some cases may even be considerably greater. Thus, unless one can develop some highly efficient means of supplying and recovering the plasma heating power, the mirror machine concept may not prove suitable for a fusion power plant.

The following analysis is aimed at showing with minimal assumptions the efficiencies which are required of these processes if the mirror machine is to be competitive with other forms of power. The basis on which such a judgment is made is the thermal efficiency of the power plant. This can, with a few simplifying assumptions, be readily calculated and does not require any estimates of capital or fuel costs twenty or thirty years hence. As both demand for power and concern for the environment increase throughout the world it seems probable that no power system will be allowed to be built which rejects to the environment appreciably more heat per kilowatt of electrical power than competing systems so long as fuel remains available for the competitors.

9.1 ANALYSIS AND RESULTS

We consider the simple block diagram of the power flow in a mirror fusion power plant shown in Figure 59. It should apply to any type of fusion system with the qualifications, for example, that the "ion accelerator" box might require re-labeling to accommodate another type of plasma heating (*e.g.*, laser heating or relativistic electrons) and the direct converter might be missing in some systems, with P_ℓ going instead into the thermal converter. In Figure 59, the input power P_{in} required to maintain the system is divided into two parts. The first, P_a, is the auxiliary power needed to make up magnet losses, run pumps, supply cryogenic cooling, etc. This power is not recoverable since it appears as low grade heat. The second, P_i, is the power which must be supplied to the plasma heating device which in the mirror case may be an ion accelerator. If this device operates with efficiency ξ_i, then a power $\xi_i P_i$ is supplied to the plasma and is potentially recoverable.

The fusion power generated in the plasma is $P_n + P_c$, where P_n is the kinetic energy of the reaction product neutrons plus any additional power generated by the neutron-induced reactions in the blanket, and P_c is the kinetic energy of the charged particle products

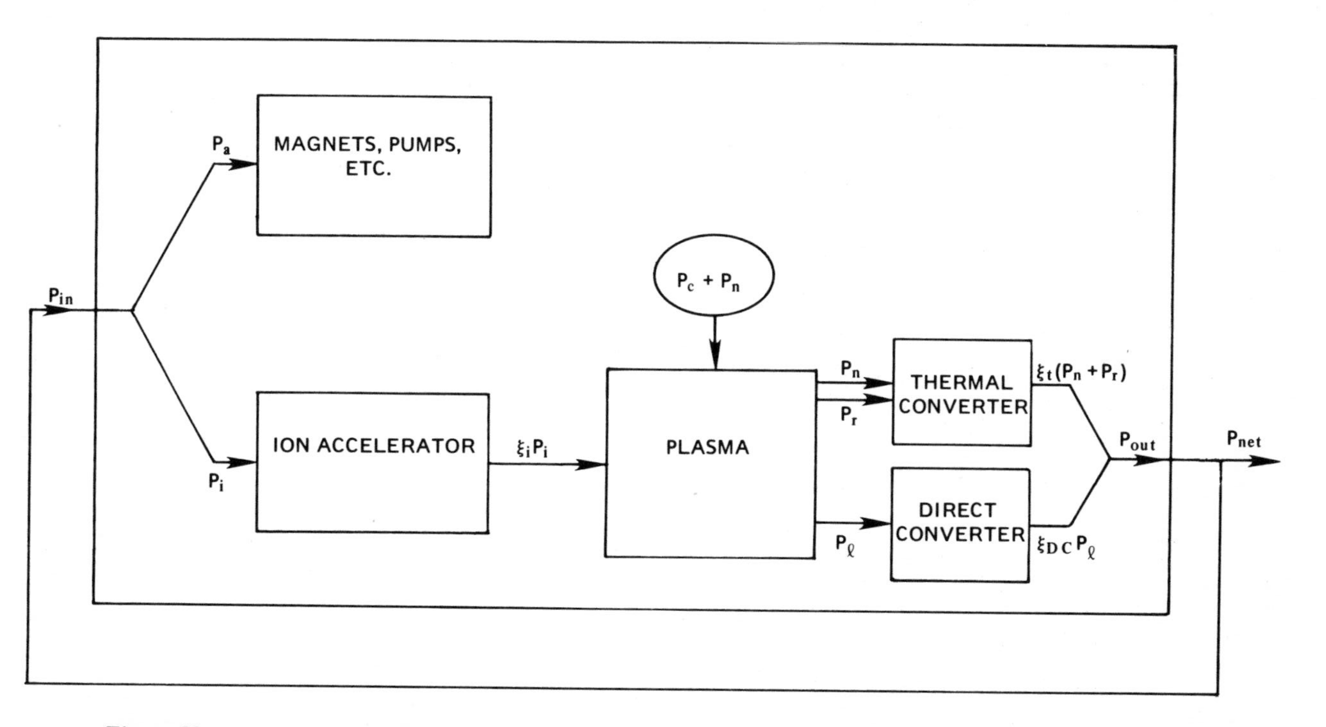

Figure 59. Power flow diagram.

of the fusion reaction. Since the neutrons do not react to any appreciable degree with the plasma, the energy available to heat the plasma is $\xi_i P_i + P_c$. Of this energy part will be converted into electromagnetic radiation, P_r, and the remainder is carried out as the kinetic energy, P_ℓ, of the escaping charged particles, primarily through the mirrors in a mirror machine. Hence we can write

$$P_\ell = \xi_i P_i + P_c - P_r \tag{9.1}$$

The neutron plus radiation power, $P_n + P_r$, appears as heat in the blanket and must be converted to electrical power by a thermal converter which operates with efficiency ξ_t. The "loss" power P_ℓ, however, is in the form of moving charged particles, and thus is already a form of electrical power. This can be converted to a usable form by a direct converter which operates with efficiency ξ_{DC}. The power which is generated by the plant is, therefore,

$$\begin{aligned} P_{out} &= \xi_t (P_n + P_r) + \xi_{DC} P_\ell \\ &= \xi_t (P_n + P_r) + \xi_{DC} (\xi_i P_i + P_c - P_r) \end{aligned} \tag{9.2}$$

and when the input power $P_{in} = P_i + P_a$ is subtracted from this, the net output power becomes

$$\begin{aligned} P_{net} &= P_{out} - P_{in} \\ &= \xi_t P_n + (\xi_t - \xi_{DC}) P_r + \xi_{DC} P_c - (1 - \xi_i \xi_{DC}) P_i - P_a \end{aligned} \tag{9.3}$$

Clearly, whenever P_{out} much exceeds P_{in}, the fusion reactor becomes a potential source of electrical power; by simply enlarging the system or paralleling a number of small systems, P_{out} can be made to reach any desired level. However, if P_{in} is only slightly smaller than P_{out}, the inevitable system losses are likely to result in a quantity of waste heat much larger than P_{net}. In an ecologically conscious age there is certain to be strong resistance to the construction of any power plant which rejects to the environment an excessive amount of waste heat.

Many people have considered that an appropriate figure of merit for a fusion reactor would be

$$Q_s \equiv P_{out}/P_{in} \tag{9.4}$$

This factor displays very well the break-even condition ($Q_s = 1$), but there is considerable debate as to how much larger than unity Q_s must be before the fusion reactor becomes feasible for an operating power plant. It appears that no such difficulties of interpretation would occur if, instead of Q_s, we looked at the overall thermal efficiency of the plant defined as

$$\epsilon \equiv P_{net}/(P_c + P_n)$$

The power plant in Figure 59 will, in its operating condition, receive no power from the outside. Therefore, the total energy released to the environment must be the fusion power, $P_c + P_n$. Of this, a power P_{net} appears as usable electric power, and the remainder (if not utilized) as waste heat. From equation (9.3) the plant thermal efficiency becomes

$$\epsilon = \frac{\xi_t P_n + (\xi_t - \xi_{DC}) P_r + \xi_{DC} P_c - (1 - \xi_i \xi_{DC}) P_i - P_a}{P_c + P_n} \qquad (9.5)$$

Since competing power plant systems will probably have thermal efficiencies quite close to ξ_t, the fusion plant will be at a disadvantage if ϵ falls much below this value, and may be quite attractive if ϵ can be made larger than ξ_t. To simplify the present calculations we shall assume that both P_r and P_a, the radiation power and auxiliary power, are negligible compared to other terms. We note that since presumably ξ_{DC} is greater than ξ_t, the effect of each of these powers is to reduce the plant thermal efficiency, ϵ. Nevertheless, for a sufficiently large system we may hope to make these quantities reasonably negligible, so that our error is perhaps only a few per cent. In this case the plasma (plus blanket) is represented by two parameters, Q and P_c/P_n. Q is the widely used "plasma" figure of merit, defined as the ratio of the fusion power produced in the plasma plus blanket to the power which must be added to the plasma externally in order to maintain its temperature, *i.e.*,

$$Q = \frac{P_c + P_n}{\xi_i P_i} \qquad (9.6)$$

The ratio P_c/P_n, of course, indicates how much of the fusion power must be processed by the thermal converter, and how much may be processed by the (hopefully) more efficient direct converter. When P_r and P_a are ignored equation (9.5) becomes

$$\epsilon = \frac{\xi_t}{1 + P_c/P_n} + \frac{(P_c/P_n)\xi_{DC}}{1 + P_c/P_n} - \frac{1 - \xi_i\xi_{DC}}{\xi_i Q} \tag{9.7}$$

Recent (Fokker-Planck) calculations have made it possible to compute the value of Q for many systems of interest to within a factor of two or better, and possibly to as close as 25%. The calculated values of the ratio P_c/P_n should, for a given blanket, be quite accurate. The efficiency of the thermal converter, ξ_t, is probably quite predictable and should fall within the range $0.45 \leqslant \xi_t \leqslant 0.55$.in the time period of interest for early fusion reactors. The efficiencies of the direct converter, ξ_{DC}, and of the injection system, ξ_i, are less accurately predictable since neither of these important devices yet exists in anything close to the form required for the operating power plant. Therefore, the best we can do at present is to determine through a study of ϵ the values which ξ_{DC} and ξ_i must have if the mirror fusion reactor is to constitute a feasible power plant.

We proceed now to calculate the plasma Q and the ratio P_c/P_n confining our attention to the following reactions involving deuterium:

$$\begin{aligned} D + D &\longrightarrow T(1.0) + p(3.02) \\ &\longrightarrow H_e(0.82) + n(2.45) \\ D + T &\longrightarrow H_e(3.52) + n(14.1) \\ D + H_e^3 &\longrightarrow H_e^4(3.6) + p(14.7) \end{aligned} \tag{9.8}$$

where the numbers in the brackets denote the energy of the particle. Following Post (1969) we denote by 1,2,3 and α the particles D, T, H_e^3 and H_e^4 and define:

λ_{11} = power multiplication factor for 2.45 MeV neutrons in the blanket

λ_{12} = power multiplication factor for 14.1 MeV neutrons in the blanket

π_{21} = probability that a 1.0 MeV triton will undergo D-T fusion before escaping

π_{31} = probability that a 0.82 MeV H_e^3 ion will undergo D-H_e^3 fusion before escaping.

Moreover, we will employ the following notation for convenience:

$$
\begin{aligned}
W_{12\alpha} &= 3.52 & W_{11p} &= 3.02 \\
W_{12n} &= 14.1 & W_{112} &= 1.0 \\
W_{11n} &= 2.45 & W_{13} &= W_{13p} + W_{13n} \\
W_{113} &= 0.82 & &= 14.7 + 3.6 = 18.3 \text{ MeV}
\end{aligned}
\tag{9.9}
$$

so that we can write

$$
\begin{aligned}
P_c = {} & n_1 n_{2i} \langle\sigma v\rangle_{12i} W_{12\alpha} + n_1 n_{3i} \langle\sigma v\rangle_{13i} W_{13} + n_1 n_{2r} \langle\sigma v\rangle_{12r} W_{12\alpha} \\
& + n_1 n_{3r} \langle\sigma v\rangle_{13r} W_{13} + \tfrac{1}{2} n_1^2 \langle\sigma v\rangle_{11n} W_{113} + \tfrac{1}{2} n_1^2 \langle\sigma v\rangle_{11p} \\
& (W_{11p} + W_{112})
\end{aligned}
\tag{9.10}
$$

Similarly, we can write for P_n,

$$
\begin{aligned}
P_n = {} & n_1 n_{2i} \langle\sigma v\rangle_{12i} W_{12n} \lambda_{12} + n_1 n_{2r} \langle\sigma v\rangle_{12r} W_{12n} \lambda_{12} \\
& + \tfrac{1}{2} n_1^2 \langle\sigma v\rangle_{11n} W_{11n} \lambda_{11}
\end{aligned}
\tag{9.11}
$$

We note, however, that

$$
\begin{aligned}
\tfrac{1}{2} n_1^2 \langle\sigma v\rangle_{11p} \pi_{21} &= n_1 n_{2r} \langle\sigma v\rangle_{12r} \\
\tfrac{1}{2} n_1^2 \langle\sigma v\rangle_{11n} \pi_{31} &= n_1 n_{3r} \langle\sigma v\rangle_{13r}
\end{aligned}
\tag{9.12}
$$

and with these relations equations (9.10) and (9.11) can be put in the form

$$
\begin{aligned}
P_c &= \tfrac{1}{4} n^2 \langle\sigma v\rangle_c W_c A_c \\
P_n &= \tfrac{1}{4} n^2 \langle\sigma v\rangle_n W_n A_n
\end{aligned}
\tag{9.13}
$$

where the factors A_c and A_n are readily seen to be complicated functions of the various parameters indicating the extent to which the primary reaction is augmented (or diminished) by averaging in the other possible reactions. For example, if our plasma system is one in which we inject a fraction f of deuterium ions and (1-f) of tritium ions, then we have

$$n_1 = nf$$

$$n_{2i} = n(1\text{-}f)$$

so that equation (9.10) now becomes

$$P_c = \tfrac{1}{4} n^2 \langle\sigma v\rangle_{12i} W_{12\alpha} \left\{ 4f(1\text{-}f) + 2f^2 \frac{\langle\sigma v\rangle_{11p}}{\langle\sigma v\rangle_{12i}} \pi_{21} + 2f^2 \frac{\langle\sigma v\rangle_{11n}}{\langle\sigma v\rangle_{12i}} \frac{W_{13}}{W_{12\alpha}} \pi_{31} + 2f^2 \frac{\langle\sigma v\rangle_{11n}}{\langle\sigma v\rangle_{12i}} \frac{W_{113}}{W_{12\alpha}} + 2f^2 \frac{\langle\sigma v\rangle_{11p}}{\langle\sigma v\rangle_{12i}} \frac{(W_{11p} + W_{112})}{W_{12\alpha}} \right\} \quad (9.14)$$

By the same token equation (9.11) also becomes

$$P_n = \tfrac{1}{4} n^2 \langle\sigma v\rangle_{12i} W_{12n} \left\{ 4f(1\text{-}f)\lambda_{12} + 2f^2 \frac{\langle\sigma v\rangle_{11p} \pi_{21} \lambda_{12}}{\langle\sigma v\rangle_{12i}} + 2f^2 \frac{\langle\sigma v\rangle_{11n}}{\langle\sigma v\rangle_{12i}} \frac{W_{11n}}{W_{12n}} \lambda_{11} \right\} \quad (9.15)$$

If the curly brackets in the above equation are denoted by A_{2c} and A_{2n}, respectively, then it is clear that we can write

$$P_c = \tfrac{1}{4} n^2 \langle\sigma v\rangle_{12i} W_{12\alpha} A_{2c}$$

$$P_n = \tfrac{1}{4} n^2 \langle\sigma v\rangle_{12i} W_{12n} A_{2n} \quad (9.16)$$

In the above expressions we have designated protons, neutrons, and $H_e{}^4$ ions by P, n and α, respectively, and the subscript i on a $\langle\sigma v\rangle$ indicates that the average is to be taken over the distribution of the injected ions neglecting the reaction product ions of that species.

If we similarly consider a system in which we inject a fraction f of deuterium and a fraction (1-f) of $H_e{}^3$ ions, we have

$$P_c = \tfrac{1}{4} n^2 \langle\sigma v\rangle_{13i} W_{13} A_{3c}$$

$$P_n = \tfrac{1}{4} n^2 \langle\sigma v\rangle_{11n} W_{11n} A_{3n} \quad (9.17)$$

where

$$A_{3c} = 4f(1\text{-}f) + 2f^2 \left[\left(\pi_{31} + \frac{W_{113}}{W_{13}} \right) \frac{\langle\sigma v\rangle_{11n}}{\langle\sigma v\rangle_{13i}} + \left(\frac{W_{11p} + W_{112} + \pi_{21} + W_{12\alpha}}{W_{13}} \right) \frac{\langle\sigma v\rangle_{11p}}{\langle\sigma v\rangle_{13i}} \right] \qquad (9.18)$$

$$A_{3n} = 2f^2 \left[\lambda_{11} + \pi_{21}\,\lambda_{12}\, \frac{W_{12n}\,\langle\sigma v\rangle_{11p}}{W_{11n}\,\langle\sigma v\rangle_{11n}} \right] \qquad (9.19)$$

From these formulas for P_c and P_n we obtain the ratio

$$\frac{P_c}{P_n} = \frac{\langle\sigma v\rangle_c\, W_c A_c}{\langle\sigma v\rangle_n\, W_n A_n} \qquad (9.20)$$

To calculate Q we also need to know the injected energy $\xi_i P_i$. If the ions are injected with an average energy E_o, and if the mean lifetime of an injected ion in the plasma (before either escaping or undergoing fusion) is τ, then

$$\xi_i P_i = \frac{nE_o}{\tau} \qquad (9.21)$$

The quantity τ must be averaged over the distribution of injected ions in the plasma, taking into account scattering collisions and fusion events, if possible, with each species present in the plasma (averaging suitably over their distribution).

In a mirror machine the end losses so predominate that the fusion events are normally neglected in calculating τ. Thus we have

$$Q = \frac{\left\{ \langle\sigma v\rangle_c\, A_c W_c + \langle\sigma v\rangle_n\, A_n W_n \right\} n\tau}{4E_o} \qquad (9.22)$$

With these results we shall examine four systems: two D-T and two $D\text{-}H_e{}^3$. All calculations will be for a mirror ratio, R of 3.3, although significant increases in Q and thus in ϵ due to improved containment would require a fairly large R. This is so because both τ and Q scale roughly as $\log_{10}$ R. For the D-T systems we shall assume an injection energy E_o of 100 keV for which the results of Futch *et al.* (1972) indicate that Q should be close to its maximum value. For the D-H_e^3 systems we use E_o = 800 keV which is also within the range considered by Futch *et al.* whose results show Q increasing with E_o at least as high as 1 MeV. Further, we shall make use of two blanket systems for each type of plasma. In the first blanket all neutrons are captured in Li^6, thus producing an extra 4.8 MeV per neutron (also breeding tritium, although this factor is ignored here). In the second blanket which we take from Post (1969) we have 60% Be for neutron multiplication plus 35% Na which gives 12.5 MeV per captured neutron. There is also 5% Nb for structural purposes plus enough Li^6 to maintain a 1.0 breeding ratio on D-T neutrons. While neither blanket, possibly, is quite appropriate for D-H_e^3 system we shall stick with them in these cases to provide easier comparison between the D-T and the D-H_e^3 plasmas. It is assumed that the fraction f of deuterium ions in the D-T plasma is maintained at f = 0.5 which roughly maximizes the power output. For the D-H_e^3 plasma the fraction f of deuterium ions is set (at f = 0.82) by the requirement that as much H_e^3 be generated in D-D reactions as is burned in D-H_e^3 reactions. The viability of the D-H_e^3 systems might be considerably improved if it were possible (for example by utilizing tritium decay as a further H_e^3 source) to increase the fraction of H_e^3 in the plasma to a value of 40% or better. Thus the four systems of interest are:

A) A D-T plasma, with R = 3.3, E_o = 100 keV, f = 0.5 and Li^6 blanket which gives λ_{11} = 3.0, λ_{12} = 1.34

B) A D-T plasma with R = 3.3, E_o = 100 keV, f = 0.5, and a Na-Be blanket which gives (according to Post) λ_{11} = 4.49, λ_{12} = 2.2

C) A D-H_e^3 plasma with R = 3.3, E_o = 800 keV, f = 0.812 and a Li^6 blanket which gives λ_{11} = 3.0, λ_{12} = 1.34

D) A D-H_e^3 plasma with R = 3.3, E_o = 800 keV, f = 0.812 and a Na-Be blanket which gives λ_{11} = 4.49, λ_{12} = 2.2

For systems A and B we will use the ion lifetime given by Futch *et al.*

$$n\tau = 2.67 \times 10^{10} E_o^{3/2} \log_{10} R$$

where E_o is given in keV. We also obtain from this source:

$$\langle\sigma v\rangle_{12i} = 9.4 \times 10^{-16} \text{ cm}^3/\text{sec}$$

$$\langle\sigma v\rangle_{11n} = 0.16 \times 10^{-16} \text{ cm}^3/\text{sec}$$

$$\langle\sigma v\rangle_{11p} = 0.13 \times 10^{-16} \text{ cm}^3/\text{sec}$$

assuming that the mean ion energy is about 100 keV. Substituting the appropriate numbers in equations (9.14) and (9.15) we obtain for f = 0.5

$$A_{2c} = 1.00277 + 0.0069\ \pi_{21} + 0.0443\ \pi_{31}$$

$$A_{2n} = (1.00 + 0.0069\ \pi_{21})\ \lambda_{12} + 0.00148\ \lambda_{11}$$

We see that even if the probabilities π_{21} and π_{31} approached one, they would have very little effect on the values of A_{2c} and A_{2n}. Therefore we have

$$A_{2c} \simeq 1.0$$

$$A_{2n} \simeq \lambda_{12} + 0.00148\ \lambda_{11}$$

In system A, then, with $\lambda_{11} = 3.00$ and $\lambda_{12} = 1.34$, we get

$$P_c = 8.27 \times 10^{-13}\ n^2 \text{ keV}$$

$$P_n = 4.44 \times 10^{-12}\ n^2 \text{ keV}$$

$$Q = 0.70$$

$$P_c/P_n = 0.186$$

With a thermal converter whose efficiency is $\xi_t = 0.50$ we finally get

$$\epsilon = 0.422 + 1.586\ \xi_{DC} - \frac{1.429}{\xi_i}$$

Figure 60 shows ϵ plotted versus ξ_{DC} for various values of ξ_i. If thermal converters are able to operate at 50% efficiency, both conventional and nuclear fusion power plants should have overall thermal efficiencies fairly close to that value. We also note from Figure 60 that this restriction limits the injector and the direct converter to efficiencies greater than 90%, in fact, near 95% unless

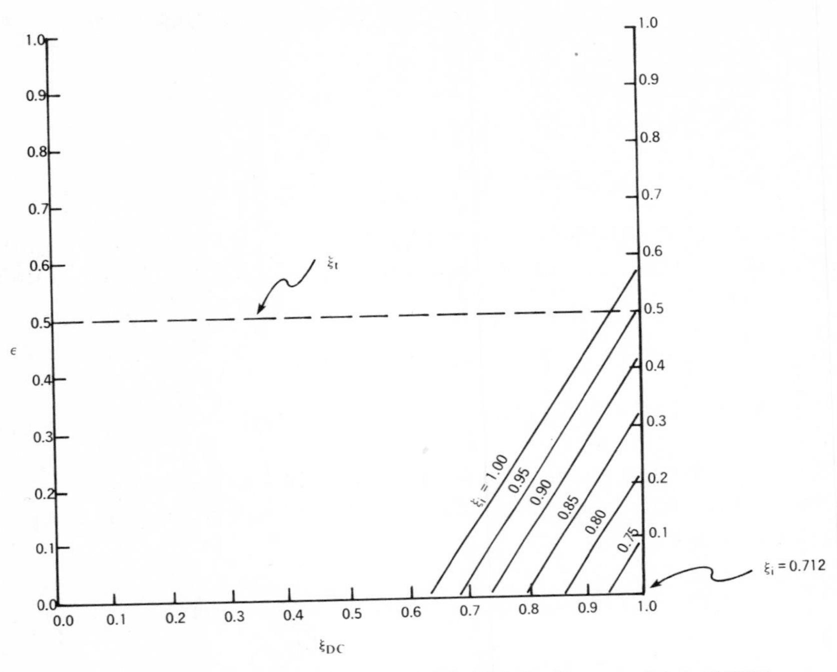

Figure 60. System A–D-T fusion, 100 keV injection neutron energy multiplication due to Li^6. $\xi_t = 0.50$, $Q = 0.700$, $P_c/P_n = 0.186$

one of them can be made with an almost 100% operating efficiency. While one might think at first glance that a highly efficient direct converter offers a promise of plant thermal efficiencies much greater than obtained with other power sources, Figure 60 shows that this is not so for a D-T system; even if both injector and direct converters were 100% efficient, the overall plant efficiency would be only 57.9%, a rather modest increase over the 50% thermal value. This is due, of course, to the fact that most of the energy from the D-T fusion is carried by the neutrons and must go through the thermal conversion cycle. With an injection power $\xi_i P_i$ about 1.4 times as great as the sum of P_c and P_n the losses involved in recirculating this power must be kept very small; thus ξ_i and ξ_{DC} must be very near unity, simply to make the overall system feasible.

In system B, with $\lambda_{11} = 4.49$ and $\lambda_{12} = 2.2$, we get

$P_c = 8.27 \times 10^{-13}\ n^2$ keV (as in system A)
$P_n = 7.32 \times 10^{-12}\ n^2$ keV
$Q = 1.09$
$P_c/P_n = 0.113$

Therefore with $\xi_t = 0.50$ we get

$$\epsilon = 0.45 + 1.019\ \xi_{DC} - \frac{0.918}{\xi_i}$$

Figure 61 shows ϵ plotted versus ξ_{DC} for various values of ξ_i. Comparing with Figure 60 we see that while the maximum possible ϵ (for $\xi_i = \xi_{DC} = 1$) has dropped slightly to 0.551, the curves in this figure lie above those of Figure 60 for all ϵ values less than 0.5 and are lower only in that small region for which ϵ is greater than 0.5. If the anticipated injector and direct converter efficiencies of $\xi_i = 0.95$ and $\xi_{DC} = 0.90$ are actually reachable, system B

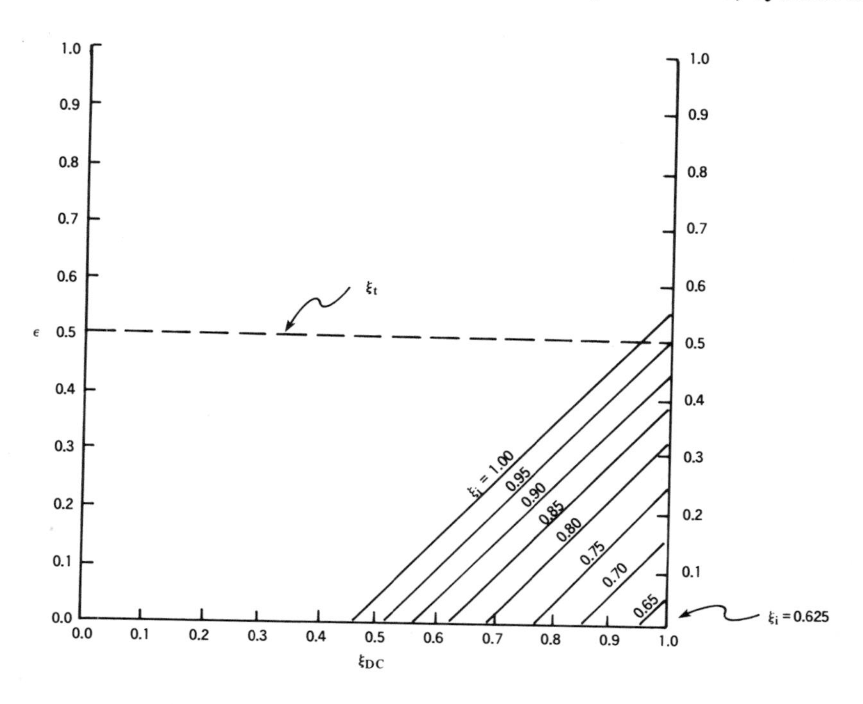

Figure 61. System B–D-T fusion, 100 keV injection, blanket Na+Be. $\xi_t = 0.50$, $Q = 1.09$, $P_c/P_n = 0.113$

would have a plant thermal efficiency of $\xi = 0.40$ which is just barely within our viability limits.

For an 800 keV injection energy D-$H_e{}^3$ system, as in our systems C and D, Futch *et al.* calculate Q = 0.216 for mirror ratio R = 10. Assuming that this scales as $\log_{10} R$, we would get Q = 0.108 for R = 3.3 However, this value is apparently for no energy multiplication in the blanket. Using the $\langle\sigma v\rangle$ averages from this reference we have

$$\langle\sigma v\rangle_{13i} = 2.17 \times 10^{-16}\ \text{cm}^3/\text{sec}$$

$$\langle\sigma v\rangle_{11n} = 1.0 \times 10^{-16}\ \text{cm}^3/\text{sec}$$

$$\langle\sigma v\rangle_{11p} = 0.86 \times 10^{-16}\ \text{cm}^3/\text{sec}$$

If the $H_e{}^3$ inventory is required to be constant, with no input from outside the plant, then

$$\frac{1-f}{f} = \frac{\langle\sigma v\rangle_{11n}}{2\langle\sigma v\rangle_{13i}}\ (1-\pi_{31}) \qquad (9.23)$$

Assuming that π_{31} can be neglected, this gives us, as noted earlier, f = 0.812. Using this value of f we obtain

$$A_{3c} = 0.753 + 0.101\ \pi_{21} + 0.608\ \pi_{31}$$

$$A_{3n} = 1.319\ \lambda_{11} + 6.53\ \pi_{21}\ \lambda_{12}$$

Since Futch *et al.* give us a value for

$$Q_c = \frac{P_c}{\xi_i P_i}$$

as well as Q, we can obtain their ratio P_c/P_n, and from that their ratio for A_{3c}/A_{3n}. Setting $\lambda_{11} = \lambda_{12} = 1.0$, if we assume that

$$\frac{\pi_{31}}{\pi_{21}} \simeq \frac{1}{4}\frac{\langle\sigma v\rangle_{13}}{\langle\sigma v\rangle_{12}}$$

we can solve for π_{21} and π_{31}, obtaining

$$\pi_{21} = 0.107$$

$$\pi_{31} = 0.0158$$

Utilizing these values we now get

$$A_{3c} = 0.774$$

$$A_{3n} = 1.319\ \lambda_{11} + 0.699\ \lambda_{12}$$

In system C, with $\lambda_{11} = 3.0$, and $\lambda_{12} = 1.34$ we now obtain

$$P_c = 0.769 \times 10^{-12}\ n^2\ \text{keV}$$

$$P_n = 0.300 \times 10^{12}\ n^2\ \text{keV}$$

and with $n\tau = 1 \times 10^{14}$ (from the same reference) we find

$$Q = 0.134$$

$$P_c/P_n = 2.56$$

Using $\xi_t = 0.50$ we finally get for this system

$$\epsilon = 0.141 + 8.18\ \xi_{DC} - \frac{7.461}{\xi_i}$$

The numerical coefficients obtained here do not appear to be very sensitive to the values of π_{21} and π_{31} used. Figure 62 shows ϵ vs ξ_{DC} for this system. We note that very high values of ξ_i and ξ_{DC} are required here if we are to have any net output power. In fact, for ξ_i less than 0.93, no value of ξ_{DC} will make ϵ greater than zero, while for ξ_{DC} less than 0.895, no value of ξ_i will make it greater than zero. Nevertheless, if somehow both ξ_i and ξ_{DC} can be made almost one, it is possible to obtain a plant thermal efficiency as high as 0.860. This large value is due to the fact that most of the fusion power (72%) appears in the form of charged particles and is processed by the direct converter.

Lastly, for system D, with $\lambda_{11} = 4.49$ and $\lambda_{12} = 2.2$ we get

$$P_c = 0.769 \times 10^{-12}\ n^2\ \text{keV (as in system C)}$$

$$P_n = 0.467 \times 10^{-12}\ n^2\ \text{keV}$$

$$Q = 0.156$$

$$P_c/P_n = 1.61$$

and thus, with $\xi_t = 0.50$ we get

$$\epsilon = 0.191 + 7.027\ \xi_{DC} - \frac{6.41}{\xi_i}$$

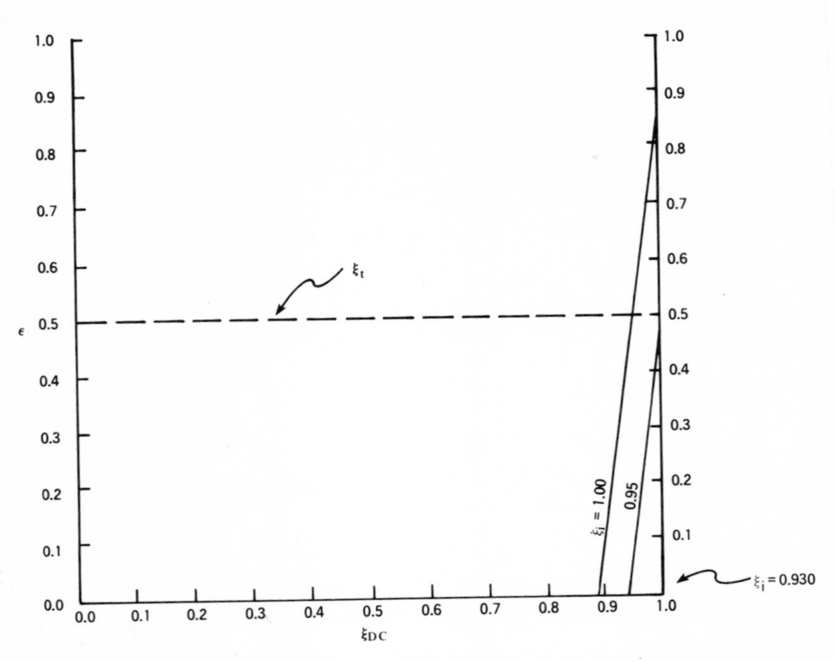

Figure 62. System C–D-He^3 fusion, 800 keV injection, blanket multiplication due to Li^6. $\xi_t = 0.50$, $Q = 0.134$, $P_c/P_n = 2.56$

Figure 63 shows that the plant thermal efficiency ϵ is increased slightly over that for system C in the range where $\epsilon < 0.50$, although owing to the lower ratio of P_c/P_n the maximum possible efficiency drops to 0.808.

In Figure 64 we attempt to show the degree of improvement in confinement necessary to make these fusion systems truly competitive. For this figure we assume that Post's estimates of $\xi_t = 0.50$, $\xi_i = 0.95$ and $\xi_{DC} = 0.90$ can be achieved, and assume that the ratio P_c/P_n found for each of the four systems will continue to hold. This last assumption precludes increasing Q by using greater power multiplication in the blanket. By plotting the plant efficiency ϵ versus Q, we can see how much improvement is necessary to reach a specified degree of competitiveness.

On the basis of this graph, we find reasonable agreement with other calculations when similar values of Q are taken. The reasonably satisfactory overall efficiencies obtained by Sweetman

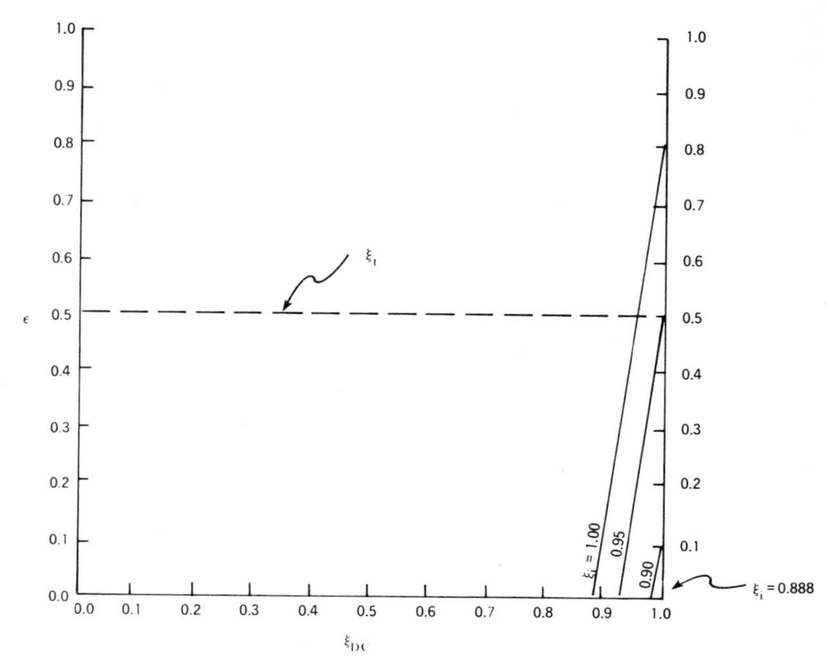

Figure 63. System D–He_e^3 fusion, 800 keV injection, blanket multiplication due to Na and Be. ξ_t = 0.50, Q = 0.156, P_c/P_n = 1.610.

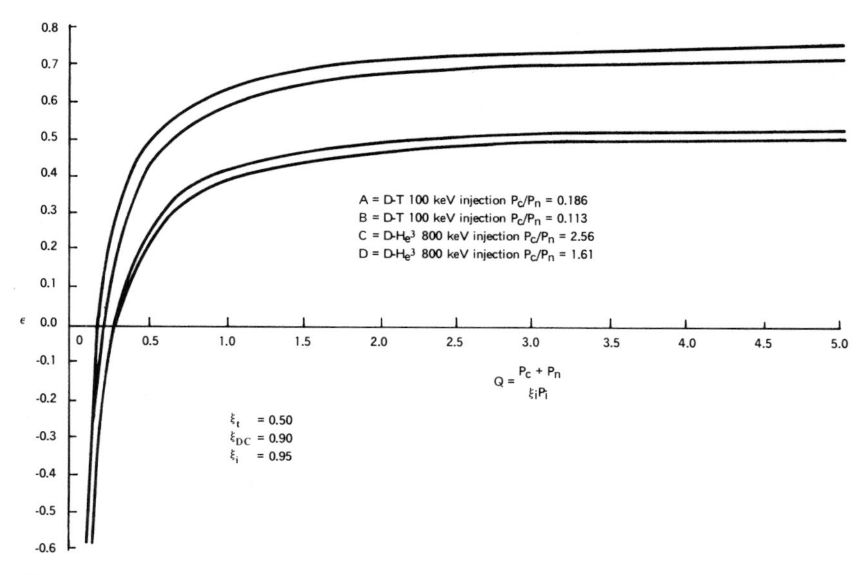

Figure 64. ε versus Q.

(1969), for example, appear to be the result of Q values which are generally two to four times as large as those employed in the above calculations. Figure 64 illustrates both the promise and the problem with $D\text{-}H_e{}^3$ fusion. While system A with Q = 0.70 has an efficiency $\epsilon = 0.355$ and system B with Q = 1.09 actually reaches $\epsilon = 0.40$, system C with Q = 0.134 and D with Q = 0.156 produce no net power; their efficiencies are $\epsilon = -0.351$ and $\epsilon = -0.231$, respectively.

Yet the improvement possible with a given percentage increase in Q is much more dramatic for the two $D\text{-}H_e{}^3$ systems than for the two D-T systems. For example, despite the fact that $D\text{-}H_e{}^3$ systems with their present Q values produce no net power, if we could increase the present Q value of each system by a factor of 5, system A would have an $\epsilon = 0.519$, system B an $\epsilon = 0.512$, system C an $\epsilon = 0.560$, and system D an $\epsilon = 0.551$. These numbers sound promising, but it is not all apparent that they are achievable. Increasing Q five times by improving the mirror ratio would require that the value of R be increased by a factor of $3.3^5 = 391$, which is almost certainly out of the question. Whether there is some other method by which the end losses, and thus $\xi_i P_i$, might be drastically reduced is a question which at present has no answer.

Thus on the basis of the calculations which are presently available it appears that a D-T mirror reactor may be marginally competitive with non-fusion power plants provided that the estimates of the achievable direct conversion and injection efficiencies prove realistic. The $D\text{-}H_e{}^3$ mirror reactor, however, appears less promising unless these efficiencies can be improved or the required injection power greatly reduced.

It might be enlightening and interesting to conclude this section by extending the above analysis to a toroidal reactor and compare the thermal efficiencies of the two systems. The detailed comparison will be left to Chapter 16, but for the present purpose we will simply note that no direct conversion (in the manner described above) will be utilized in the toroidal system and we can replace that portion of the power balance by the straight thermal conversion. Thus if we set $\xi_{DC} = \xi_t$ then we can write

$$P_{out} = \xi_t (P_n + P_c + \xi_i P_i)$$

$$P_{in} = P_i + P_a$$

and the net power from the reactor becomes

$$P_{net} = \xi_t (P_n + P_c) - (1 - \xi_i \xi_t) P_i - P_a$$

Again, if we ignore P_a on the assumption that it is small then the thermal efficiency ϵ as given by equation (9.4) becomes for the toroidal reactor

$$\epsilon = \frac{P_{net}}{P_n + P_c} = \xi_t - (1 - \xi_i \xi_t) \frac{P_i}{P_n + P_c}$$

and upon introducing Q it further becomes

$$\epsilon = \xi_t - \frac{(1 - \xi_i \xi_t)}{\xi_i Q}$$

We note from this expression that the thermal efficiency of such a system approaches ξ_t asymptotically, *i.e.*, as $Q \rightarrow \infty$. In other words the maximum possible efficiency of this reactor is that of the thermal converter, and the maximum is attained when the injected power is zero. If $\xi_t = 0.5$ and $\xi_i = 0.95$, then a reactor with $Q = 5$ will have a thermal efficiency ϵ of 39%, whereas for $Q = 1$, $\epsilon \cong 0$. These figures focus attention on the need to keep the Q value of the reactor as high as possible; this can be (ideally) achieved by operating the reactor with little or no injected power.

9.2. THE PRINCIPLE OF DIRECT CONVERSION IN MIRROR SYSTEMS

One of the major conclusions that can be drawn from the above discussion is that without direct conversion mirror devices might be questionable as serious contenders for fusion reactors. Moreover mirror machines utilizing the D-$H_e{}^3$ (*e.g.*, system C above) fuel cycle can have a very impressive plant thermal efficiency if both the injection and direct conversion efficiencies are maintained close to unity. This latter system has the special feature of having most of the fusion energy in charged particle reaction products rather than in the neutrons and hence may be viewed as a "cleaner" system. It is cleaner because the neutron-induced radioactivity may be less severe. In any case, none of these systems, it appears at present, can survive as a fusion reactor without direct conversion. For this reason we shall describe briefly the underlying principle of such a device.

The basic concept is the direct conversion to electricity of the energy of the plasma electrons and ions escaping from the mirror. This process can be thought of as proceeding through a sequence of four steps: 1) expansion, 2) charge separation, 3) deceleration and collection, and 4) conversion of the collected currents to a common potential.

Expansion

We recall from Chapter 1 that the underlying principle of mirror confinement is the adiabaticity of the particle magnetic moment, μ. An adiabatic expansion in a static magnetic field of the escaping particles also constitutes the first step of direct conversion. This expansion has two functions: 1) it converts the quasi-random motion of the escaping particles into energy of motion parallel to the field lines; and 2) it reduces the density of the emerging particle flux. This is accomplished by magnetically guiding the particle flux emerging from the mirrors through a region of weakening static magnetic field located outside the main confining fields. In so doing, the particle total energy (rotational plus translational) is transformed to energy of motion along the field lines. This proceeds in accordance with the two conservation equations:

$$\mu = \frac{W_\perp}{B} = \text{constant}$$

$$W(z) = W_\perp + W_\parallel = \text{constant}$$

where W is the total kinetic energy. If B(z) is 10^{-2} B(0) where B(0) is the field at the mirror, then expansion in such a field would result in transforming more than 99% of the particle kinetic energy into parallel energy available for conversion. The field lines in the expansion region will be in the shape of a fan at the end of which the plasma density will have fallen to a very small fraction (like $\leqslant 10^{-8}$) of the contained density. At such densities collective effects such as space charge effects ($\propto$ to ω_{pe}) will not be dominant and will not seriously jeopardize the direct conversion process.

Charge Separation

Although not essential, separation of the charges at the end of the expansion process is desirable especially since mirror-contained ions are much more energetic than the electrons. The separation process is accomplished by rapidly diverting the field lines at the end of the expansion region in a direction transverse to the fan as illustrated in Figure 65.

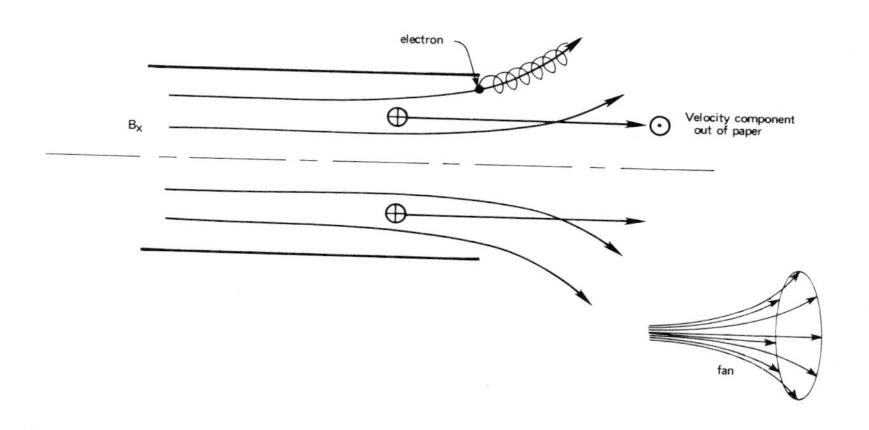

Figure 65. Expansion region in the direct conversion scheme.

In such a case the energetic ions will not behave adiabatically but will escape radially across the field lines. The electrons, on the other hand, will behave adiabatically and will follow the field lines away from the ions.

Deceleration and Charge Collection

To recover the energy of the ions emerging from the expander they are passed through an electrostatic field within which, in addition to being progressively decelerated, they are collected at potentials equal as close as possible to their kinetic energy. For high conversion efficiency the above process can be accomplished by passing the stream of ions through decelerating electrode structures of graded potential. As the energy of each particle of the stream approaches zero, that particle is diverted and deposited on an appropriate collector plate as shown in Figure 66. It has been shown that high overall collector efficiencies (> 90%) should be achievable with a reasonable number (10 to 20) of electrodes.

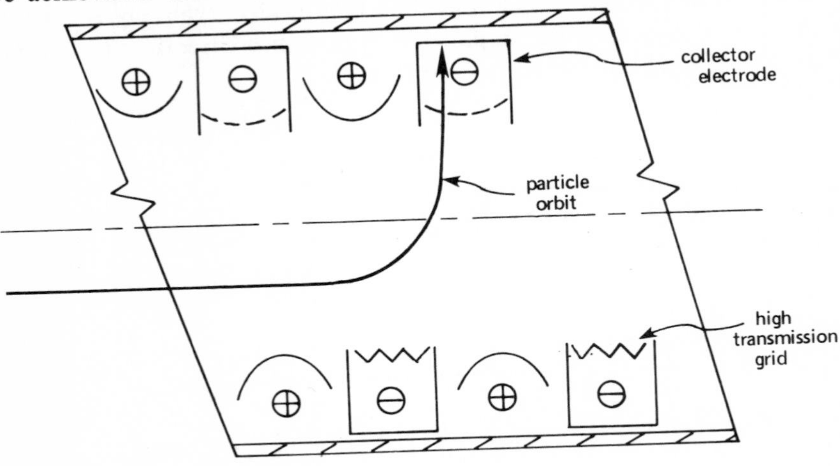

Figure 66. Collector electrode configuration.

Reduction to Common Potential

Since current is delivered to the several electrodes which are at different potential, a method must be found to transform the electrical energy thus recovered to a common potential for delivery to transmission line. This is accomplished efficiently by the arrangement shown in Figure 67 where the common potential is given by $\overline{V}$. Potentials above $\overline{V}$ are reduced through the use of DC to AC inverters while those below $\overline{V}$ are increased through the use of AC to DC rectifiers. If $\overline{V}$ is properly chosen the AC power output from the inverters will be just sufficient to supply the power needed by the rectifiers; thus proper choice of $\overline{V}$ makes the net power requirement zero.

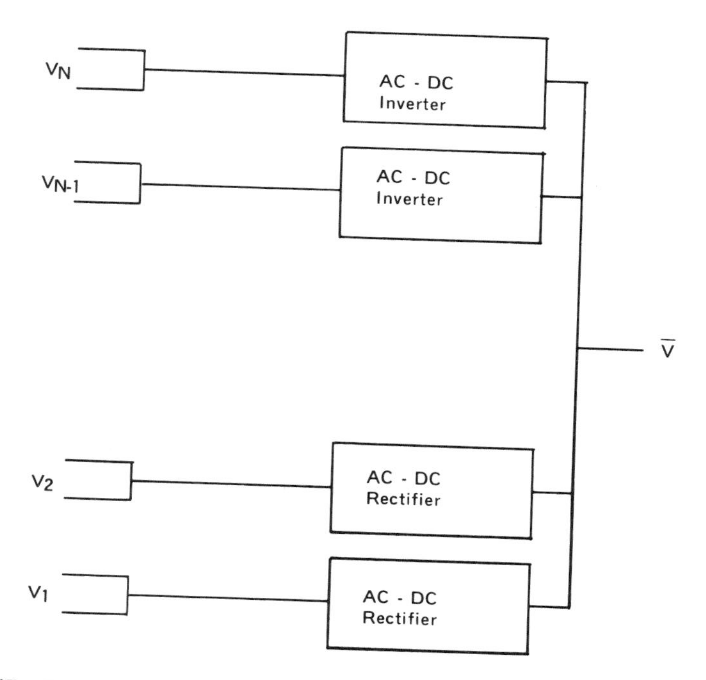

Figure 67. Reduction to common potential.

REFERENCES

1. Futch, A. H., J. P. Holdren, J. Killien, and A. A. Mirin. *Plasma Physics, 14,* 224 (1972).
2. Galbraith, D. L. and T. Kammash. *Nuclear Fusion, 11,* 575 (1971).
3. Moir, R. W. and W. L. Barr. *Nuclear Fusion, 13,* 35 (1973).
4. Post, R. F. BNES Nuclear Fusion Reactors (Proc. Int. Conf. Culham) 88 (1969).
5. Post, R. F. "Direct Conversion of Fusion Energy to Electricity," Las Vegas, 1970, p. 1-19.
6. Sweetman, D. R. BNES Nuclear Fusion Reactors (Proc. Int. Conf. Culham) 112 (1969).

CHAPTER 10

FISSION-FUSION HYBRID SYSTEMS

It has been often suggested in recent years that power from fusion reactors can be enhanced by making their blankets subcritical fission reactors. Since the neutrons from D-T fusion reactors are 14 MeV neutrons, the blanket might well be a fast breeder. In any case the resulting hybrid system illustrated in Figure 68 has certain advantages and disadvantages which must be weighed when compared individually to either fission or fusion systems.

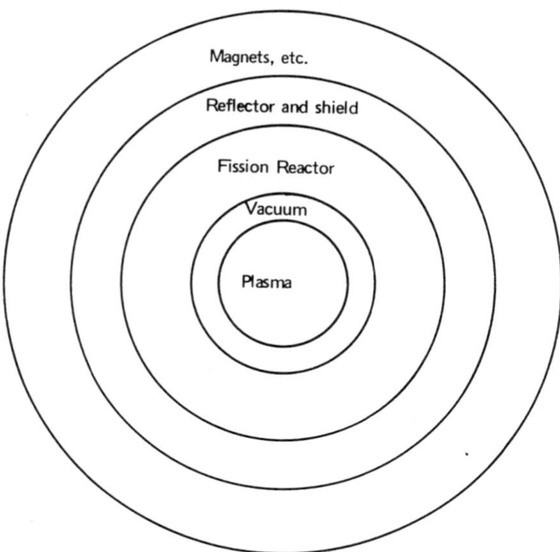

Figure 68. Hybrid system.

Chief among the disadvantages is that the hybrid system combines the hazards of the two systems–it combines the complexity and tritium handling problems of the fusion reactor with the radioactive waste problems of the fission reactor. Also the coolant and geometry problems may be more severe than for either "pure" system. The advantages most often cited are (1) greater power production than the equivalent fusion reactor, (2) faster fissile fuel breeding rate than is possible with pure fission, (3) safer than the fast breeder by itself, and (4) possibly faster development of fusion reactor design. These points are controversial and will not be debated or discussed here; rather the underlying technical principle will be examined.

The blanket reactions in the hybrid system are

$$Li^6 + n \longrightarrow T^3 + H_e^4 + 4.8 \text{ MeV}$$

$$Li^7 + n + 2.8 \text{ MeV} \longrightarrow T^3 + H_e^4 + n'$$

which are those associated with fusion, in addition to the fission reactions involving either uranium or plutonium, *i.e.*,

$$U^{238} + n \longrightarrow U^{239} + \gamma, \ \sigma = 36 \times 10^{-4} \text{ barns}$$

$$\longrightarrow U^{237} + 2n', \ \sigma = 0.55 \text{ b}$$

$$\longrightarrow U^{236} + 3n', \ \sigma = 0.80 \text{ b}$$

$$\longrightarrow \text{fission} \quad , \ \sigma = 1.20 \text{ b}$$

$$p^{239} + n \longrightarrow \text{fission}$$

The variation of the cross section vs neutron energy for these reactions is shown in Figure 69.

A number of calculations (though still very sketchy) have indicated that as much as 1100 MeV fission energy per 14 MeV fusion neutron may be possible in a blanket with a multiplication factor of k = 0.9. In a two-zone blanket a fairly detailed calculation has given the following results. For Zone 1, with 30-cm thickness consisting of 95% by volume of lithium (4% Li^6) and 5% Nb, and Zone 2, with 70-cm thickness and 65% U (0.4% U^{235}), 30% Li (4% Li^6) and 5% Nb, these calculations give:

energy generation = 103 MeV per 14.1 MeV neutron
tritium breeding = 0.986
U^{238} (n,γ) U^{239} breeding = 1.68 per 14.1 MeV neutron

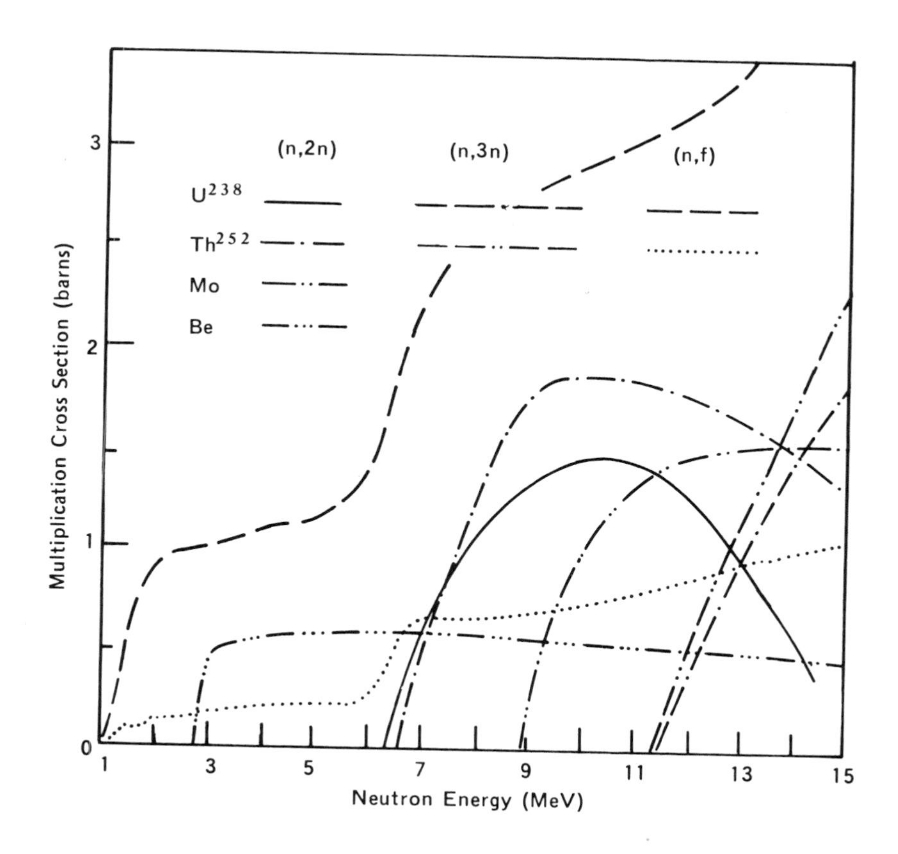

Figure 69. Cross sections for various neutron-induced reactions.

We turn now to the particle balance in a hybrid system and for that we consider two power-producing reactors which are coupled to one another through the production of fuel for one by the other. This is slightly different from the system considered by Lidsky (1969) in which the coupling only involves the production of fissile nuclei by D-T reaction neutrons. For simplicity we shall use the subscript 1 to denote quantities associated with the fusion reactor and subscript 2 for those identified with the fission reactor. The system of interest is illustrated in Figure 70 in which we have defined

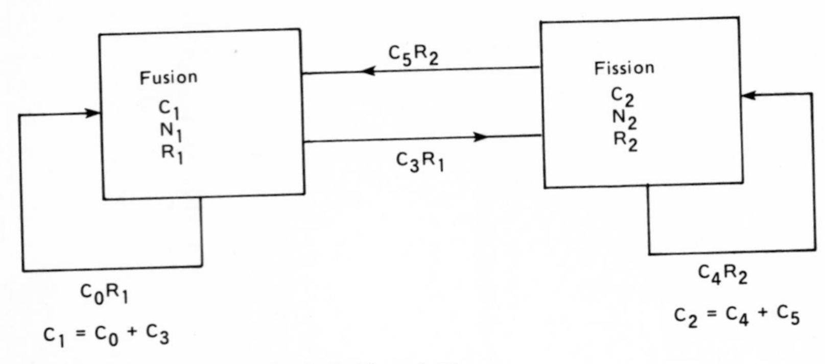

Figure 70. Schematic of a hybrid system.

N_1 = the number of tritium ions
N_2 = the number of fissile nuclei
R_1 = the number of fusions per sec
R_2 = the number of fissions per sec
C_0 = tritium produced per fusion event
C_3 = fissile atoms produced per fusion event
C_4 = fissile atoms produced per fission
C_5 = tritium produced per fission

The dynamics of the fuel inventories are described by the equations

$$\frac{dN_1}{dt} = R_1 (C_0 - 1) + R_2C_5 \tag{10.1}$$

$$\frac{dN_2}{dt} = R_2 (C_4 - 1) + R_1C_3 \tag{10.2}$$

We now define the "specific inventory" $n_j = N_j/R_j$ and the "time constant" $\tau_j = n_j/(C_j - 1)$, and further note that n_j has the dimension of time, (inventory)/(reactions/sec), and thus is inversely related to specific power. The time constant given by

$$\tau_j = \frac{n_j}{C_j - 1} = \frac{N_j}{R_j (C_j - 1)} \tag{10.3}$$

is the exponential time constant for fuel production in an isolated system with instantaneous fuel reinvestment. If we assume that the ratio of fission to fusion is constant, *i.e.*,

$$\frac{1}{N_1} \frac{dN_1}{dt} = \frac{1}{N_2} \frac{dN_2}{dt} = \frac{1}{\tau} \tag{10.4}$$

then from equations (10.1) and (10.2) we get

$$\frac{R_1 (C_1 - 1 - C_3) + R_2 C_5}{N_1} = \frac{R_2 (C_2 - 1 - C_5) + R_1 C_3}{N_2} = \frac{1}{\tau}$$

which also yields upon utilitization of equation (10.3)

$$\frac{1}{\tau} = \frac{1}{\tau_2} + \frac{R_1}{R_2} \frac{(C_1 - 1)(1 - \tau_1/\tau)}{(C_2 - 1)\tau_2} \tag{10.5}$$

This result can be transformed to yield R_1/R_2, *i.e.*,

$$\frac{R_1}{R_2} = \frac{(1 - C_2)(1 - \tau_2/\tau)}{(C_1 - 1)(1 - \tau_1/\tau)} = \frac{(1 - C_2) + n_2/\tau}{(C_1 - 1) - n_1/\tau} \tag{10.6}$$

If for the fusion D-T reactor we use $C_1 = 1.4$ and $\tau_1 = 0.113$ years, then the fusion-to-fission reaction ratio given by the above equation can be plotted against the system period τ for various fission reactors as shown in Figure 71. The various fission reactors

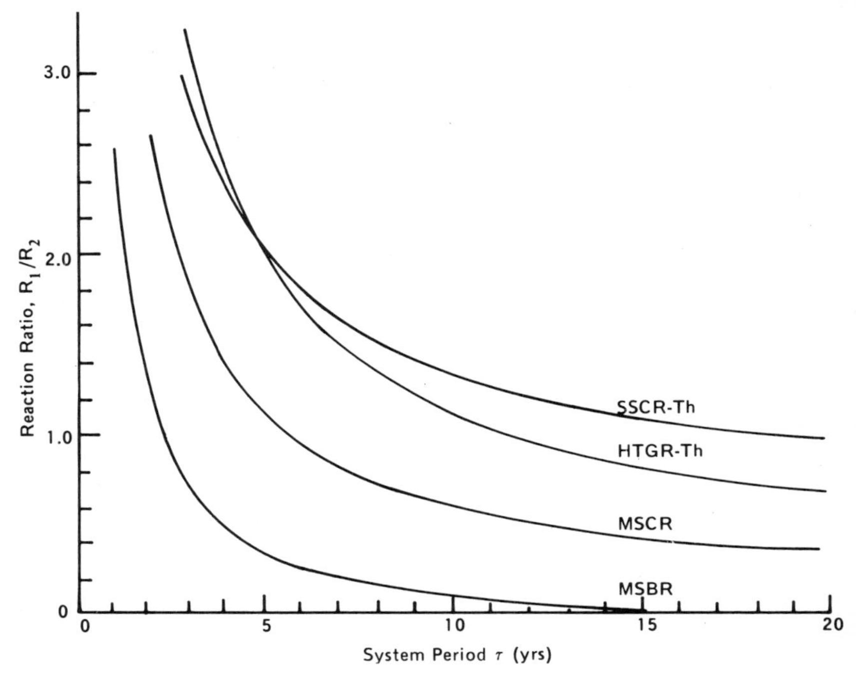

Figure 71. Fusion/fission ratio in a hybrid power plant as a function of system period (Lidsky, 1969).

used in the calculation as well as their individual C_2's and τ_2's are shown in Table 20. We note from Figure 71 that we can get reasonable periods for R_1/R_2 of the order of unity for all cases.

Table 20

Reactor Type	C_2	τ_2 *(yr)*
Spectral shift reactor: SSCR-Th	0.750	-10.7
High temperature gas: HTGR-Th	0.900	-33.4
Molten salt converter: MSCR	0.960	-42.8
Molten salt breeder: MSBR	1.049	+18.4

The power balance in a hybrid system can be calculated by first defining:

- U_1 = energy per fusion reaction including reactions in the blanket (~ 22.4 MeV)
- U_2 = energy per fission event (~ 200 MeV)
- W_L = power supplied to the plasma (injection and heating)
- W_n = the net power output of the system
- η = thermal converter efficiency
- η_i = injection and heating system efficiency

so that the relation of interest is

$$W_n = \eta\,[R_1 U_1 + R_2 U_2 + W_L] - \frac{W_L}{\eta_i}$$

$$W_n = \eta\, R_2 U_2 \left\{ 1 + \frac{R_1 U_1}{R_2 U_2} \left[1 - \left(\frac{1}{\eta\eta_i} - 1 \right) \frac{W_L}{R_1 U_1} \right] \right\} \qquad (10.7)$$

We recall from Chapter 9 that the "Q" value of the fusion system can be written as

$$Q = \frac{R_1 U_1}{W_L} \qquad (10.8)$$

and with this the net power output becomes

$$W_n = \eta\, R_2 U_2 \left\{ 1 + \frac{R_1 U_1}{R_2 U_2} \left[1 - \left(\frac{1 - \eta\eta_i}{\eta\eta_i} \right) \frac{1}{Q} \right] \right\} \qquad (10.9)$$

If we now use the typical values $Q = 1$, $\eta = 0.4$, $\eta_i = 0.8$, and $R_1/R_2 = 1$, then from the above equation we get

$$W_n = \eta\, R_2 U_2 \quad \{1 - 0.125\} \tag{10.10}$$

$$= 0.875\, \eta\, R_2 U_2$$

and we note that the power output for $R_1/R_2 \simeq 1$, and $Q \simeq 1$ depends only on the fission reactor portion of the hybrid system.

REFERENCES

1. Haight, R. C. and J. D. Lee. *Proceedings of ANS First Topical Meeting on The Technology of Controlled Nuclear Fusion,* San Diego, Calif. (April 1974).
2. Hansborough, L. D. *Proceedings of ANS First Topical Meeting on The Technology of Controlled Nuclear Fusion,* San Diego, Calif. (April 1974).
3. Lee, J. D. *Proceedings of ANS First Topical Meeting on The Technology of Controlled Nuclear Fusion,* San Diego, Calif. (April 1974).
4. Leonard, B. R., Jr. "A Review of Fusion–Fission (Hybrid) Concepts," *Nuclear Technology, 20,* 161 (1973).
5. Lidsky, L. M. *BNES Nuclear Fusion Reactors (Proc. Int. Conf. Culham),* 41 (1969).
6. Lidsky, L. M. *Proceedings of ANS First Topical Meeting on The Technology of Controlled Nuclear Fusion,* San Diego, Calif. (April 1974).
7. Lontai, L. N., *MIT Research Laboratory of Electronics Report TR-436* (1965).
8. Moir, R. W. *Proceedings of ANS First Topical Meeting on The Technology of Controlled Nuclear Fusion,* San Diego, Calif. (April 1974).
9. Wolkenhauer, W. C., C. W. Stewart, R. W. Werner, and J. D. Lee. *Proceedings of ANS First Topical Meeting on The Technology of Controlled Nuclear Fusion,* San Diego, Calif. (April 1974).
10. Wolkenhauer, W. C., W. R. Moir, B. R. Leonard, and R. W. Werner. *Proceedings of ANS First Topical Meeting on The Technology of Controlled Nuclear Fusion,* San Diego, Calif. (April 1974).

CHAPTER 11

INERTIAL-CONFINEMENT FUSION SYSTEMS

In all our discussions thus far we have been looking at controlled fusion systems in which the plasma is magnetically confined. We recall from the first chapter that (according to the Lawson criterion) in order for these systems to produce net power, $n\tau$ must exceed 10^{14} at 10 keV temperature with D-T fuel. At meaningful densities the confinement time τ must be on the order of a second; and though many of these devices are stable against MHD modes, they continue to be beset by microinstabilities. These instabilities tend to destroy confinement in times much shorter than the desired confinement time, and the Bohm diffusion time τ_B has been associated with the decay of a system due to microinstabilities. For a cylindrical plasma with radius r, we recall from equation (8.44) that

$$\tau_B = C \frac{r^2 B}{T} \simeq 10^{-14} \frac{r^2 B}{T} \tag{11.1}$$

and it is convenient to judge the quality of confinement in any system in terms of the factor τ/τ_B. In the same chapter we also noted that $\tau/\tau_B \sim 1000$ (and not classical confinement) was desirable for a low beta toroidal reactor since it leads to appropriate exhaust mechanism compatible with steady state operation. Nevertheless, stable confinement in accordance with the Lawson requirements on τ must first be accomplished in order to bring the reactor to ignition. Microinstabilities must somehow be circumvented at least initially.

If we wish to avoid the problem of suppressing these micro-instabilities we can attempt to operate the system at $\tau/\tau_B < 1$. In order to appreciate the characteristics of such a plasma, *i.e.*, its density (n) and dimensions (r), let's consider the following example:

$$\frac{\tau}{\tau_B} = 1 \qquad n\tau = 10^{14} \qquad T = 10^8\ ^\circ K \tag{11.2}$$

Combining these with equation (11.1) we obtain

$$n\,r^2\,B = 10^{26} \tag{11.3}$$

If we further assume that this plasma is in equilibrium with the confining magnetic field then

$$2nkT = B^2/8\pi \tag{11.4}$$

which upon combining with equation (11.3) yields

$$n\,r^{4/3} = 2 \times 10^{19} \tag{11.5}$$

The values of n and B as functions of the plasma radius r are shown in Table 21 where also shown are the plasma energy per unit length W and the quantity Wr which is of the order of the total plasma energy (see Linhart, 1970).

Table 21

Plasma Parameters for Different Radii

r (cm)	*0.1*	*1*	*10*	*100*	*1000*
n (cm^{-3})	4.3×10^{20}	2×10^{19}	0.93×10^{18}	4.3×10^{16}	2×10^{15}
B (MG)	20	5	1	0.2	0.05
W (MJ/cm)	0.054	0.25	1.15	5.4	25
Wr (MJ)	0.0054	0.25	11.5	540	25,000

We note from Table 21 that for densities less than, say, 10^{17} large stored energies are required, and for larger densities the pressures might be so large that most structural materials may not be able to withstand. It seems, therefore, that in order to avoid problems associated with long confinement ($\tau/\tau_B > 1$), one must be prepared to solve the problems of (a) creating high energy densities, *e.g.*, more than 1 KJ/cm^3, and (b) sustaining local pressures in excess of 10^4 atm. Moreover, one could also dispense with the problems of MHD stability if higher energy densities than 1 KJ/cm^3 could be obtained. In this case the lifetime of the plasma will have to be of the order of

$$\tau_o \simeq r/v_s \qquad (11.6)$$

where v_s is the velocity of expansion of the hot plasma in the absence of effective confinement; it is the sonic speed in the plasma, *i.e.*, $v_s \sim \sqrt{2kT/M}$. If we now return to the earlier example and, instead of taking $\tau = \tau_B$, we now use τ as given by equation (11.6) we find that

$$nr = 4 \times 10^{22} \qquad (11.7)$$

or

$$\rho r \sim 1 \text{ gm/cm}^2 \text{ (for D-T)}$$

For this case Table 21 will be replaced by Table 22 shown below.

Table 22

Plasma Density and Energy for Different Radii

r (cm)	*0.02*	*0.1*	*0.5*	*1*	*10*
n (cm^{-3})	2×10^{24}	4×10^{23}	0.8×10^{23}	4×10^{22}	4×10^{21}
W (MJ/cm)	0.1	0.5	2.5	5	50
Wr (MJ)	0.002	0.05	1.25	5	500

Once again we note the high energy densities required for plasma densities less than 10^{22}; in fact, if we restrict ourselves to $\tau \leqslant \tau_o$ and moderate energy densities, we find that we must work with plasma densities of the order of solid state densities, *i.e.*, $n \sim 10^{22}$. For such a system the Lawson criterion will be replaced by the "burn condition" given by equation (11.7), *i.e.*, the product of the material density, and the radius must be of the order of one gram per cm^2. It is clear that this system will have to be a pulsed and is perhaps an explosive system.

A great deal of enthusiasm has recently been generated for "Laser-Pellet" fusion in which fusion may be triggered and sustained by irradiating small D-T pellets with radiation from high-powered lasers.

This concept will be explored in some detail later in this chapter. Because of the relative infancy and possible military applications of this approach, some relevant material might be absent from the open literature making it difficult to corroborate the ideas and elementary discussion presented here. We shall therefore limit ourselves to a semi-intuitive analysis of inertial confinement systems and remind the reader that the numerical results cited are strictly illustrative and should be viewed as such.

11.1 TECHNICAL CONSIDERATIONS OF A MICROEXPLOSION

We consider an infinite D-T medium with density n_o and temperature $T \ll 10^8$ °K in which a finite amount of energy W_o is suddenly deposited in a small spherical volume of the medium heating it to a temperature much larger than 10^8. As result of this concentrated energy release, a spherical shock wave is propagated outward; as long as the energy losses (radiation to the surrounding cold medium) as well as the energy gain from fusion are small compared to W_o, the behavior of this shock wave is described by Taylor's similarity laws (Taylor, 1950). These relations give the radius r_o of the shock wave as a function of the time t elapsed since the explosion as

$$r_o = S(\gamma)\, t^{2/5}\, W_o^{1/5}\, \rho_o^{-1/5} \tag{11.8}$$

where γ is the ratio of the specific heats (taken as equal to 5/3), ρ_o is the mass density of the medium, and $S(\gamma)$ is a calculated function of γ. The appropriate similarity assumptions of an expanding blast wave of constant total energy are

$$\text{pressure,}\quad P/P_o = y = r_o^{-3}\,(t)\, f_1(y) \tag{11.9}$$

$$\text{density,}\quad \rho/\rho_o = \psi(y) \tag{11.10}$$

$$\text{radial velocity,}\quad u = r_o^{-3/2}\, \varphi_1(y) \tag{11.11}$$

where r_o is the radius of the shock wave forming at the outer edge of the disturbance; P_o and ρ_o are the pressure and density of the undisturbed medium, and f, ψ and φ are functions of the radial distance $y = r/r_o$. It is found that these assumptions are consistent with the equations of hydrodynamics, *i.e.*, the equations of motion, continuity and state.

The conditions at the shock front, y = 1, are given by the Hugoniot relations which may be expressed in the form

$$\frac{\rho_1}{\rho_0} = \frac{(\gamma - 1) + (\gamma + 1)\, y_1}{(\gamma + 1) + (\gamma - 1)\, y_1} \tag{11.12}$$

$$\frac{U^2}{a^2} = \frac{1}{2\gamma} \left\{ (\gamma - 1) + (\gamma + 1)\, y_1 \right\} \tag{11.13}$$

$$\frac{u_1}{U} = \frac{2\,(y_1 - 1)}{(\gamma - 1) + (\gamma + 1)\, y_1} \tag{11.14}$$

where ρ_1, u_1 and y_1 represent the values of ρ, u and y immediately behind the shock wave, and $U = dr_o/dt$ is the radial velocity of the shock wave. In equation (11.13) $a^2 = \gamma p_o/\rho_o$ where a is the velocity of sound in the undisturbed medium. The above conditions cannot be satisfied consistently with the similarity assumptions given by equations (11.9-11.11). On the other hand when y_1 is large so that the pressure is high compared to p_o, then the shock conditions (11.12-11.14) assume the approximate asymptotic form

$$\frac{\rho_1}{\rho_0} = \frac{\gamma + 1}{\gamma - 1} \tag{11.15}$$

$$\frac{U^2}{a^2} = \frac{\gamma + 1}{2\gamma}\, y_1 \tag{11.16}$$

$$\frac{u_1}{U} = \frac{2}{\gamma + 1} \tag{11.17}$$

These approximate boundary conditions are consistent with equations (11.9 - 11.11) and the equation for the radial velocity of the shock, *i.e.*,

$$U = \frac{dr_o}{dt} = A\, r_o^{-3/2} \tag{11.18}$$

where A is a constant. In fact equations (11.15 - 11.17) yield

$$\begin{aligned} \psi &= \frac{\gamma + 1}{\gamma - 1} \\ f &= \frac{\gamma + 1}{2\gamma} = \frac{f_1 a^2}{A^2} \\ \varphi &= \frac{2}{\gamma + 1} = \frac{\varphi_1}{A} \end{aligned} \tag{11.19}$$

The total energy W_o of the disturbance may be regarded as consisting of two parts:

$$\text{K.E.} = 4\pi \int_0^{r_o} \tfrac{1}{2} \rho\, u^2\, r^2\, dr \tag{11.20}$$

and the heat (internal) energy

$$\text{I.E.} = 4\pi \int_0^{r_o} \frac{pr^2}{\gamma - 1}\, dr \tag{11.21}$$

This last equation comes from the familiar form of internal energy of a perfect gas, *i.e.*, $\int C_v\, dT$ where C_v is the specific heat at constant volume. In terms of the functions given by equation (11.19) we can write

$$W_o = 4\pi A^2 \left\{ \tfrac{1}{2} \rho_o \int_0^1 \psi \varphi^2\, \eta^2\, d\eta + \left(\frac{p_o}{a^2(\gamma - 1)} \int_0^1 f \eta^2\, d\eta \right) \right\}$$

or since $p_o = a^2 \rho_o / \gamma$, the total energy might be put in the form

$$W_o = C \rho_o A^2 \tag{11.22}$$

where

$$C = 2\pi \int_0^1 \psi \varphi^2\, \eta^2\, d\eta + \frac{4\pi}{\gamma(\gamma - 1)} \int_0^1 f \eta^2\, d\eta \tag{11.23}$$

Since the two integrals in the above equation are functions of γ only, it seems that for a given value of γ, A^2 is simply proportional to W_o / ρ_o. For $\gamma = 1.4$ the integrals in equation (11.23) have been numerically evaluated, and the results of the component energies and the total energy become

$$W_o = \text{K.E.} + \text{I.E.} = C\,\rho_o\,A^2$$

$$= 1.164\,\rho_o\,A^2 + 4.196\,\rho_o\,A^2$$

$$= 5.36\,\rho_o\,A^2 \tag{11.24}$$

From this we note that the thermal energy comprises the major portion or

$$W_{Th} = \frac{4.196}{5.36}\,W_o = 0.78\,W_o$$

If a $\gamma = 5/3$ (as suggested by Linhart) is used, the above ratio changes slightly; it becomes

$$W_{Th} \simeq \frac{9}{13}\,W_o \tag{11.25}$$

We turn now to the velocity of the radial expansion of the shock, *i.e.*, equation (11.18). Substituting for A from (11.24) we can write

$$\frac{dr_o}{dt} = r_o^{-3/2}\,(C\,\rho_o)^{-1/2}\,W_o^{1/2} \tag{11.26}$$

The temperature T at any point is related to the pressure and density by the relation

$$\frac{T}{T_o} = \frac{P\rho_o}{p_o\rho} = \frac{1.33\,W_o\,r_o^{-3}}{p_o\,\psi}\,f \quad (\gamma = 1.4) \tag{11.27}$$

Since f (according to Taylor) tends to a uniform value of 0.436 in the central region ($r < r_o/2$), and ψ tends to the value $\psi = 1.76\,\eta^{7.5}$, then T tends to the value

$$\frac{T}{T_o} = \frac{W_o\,r_o^{-3}}{p_o}\,\eta^{-7.5} \quad (0.033)$$

At the shock front, $\eta = 1$, and the above expression becomes

$$\left(\frac{T}{T_o}\right)^{1/2} = \sqrt{0.033\,W_o/r_o^3\,p_o} \tag{11.28}$$

We now substitute this result in equation (11.26), and if we let $p_o = n_o kT_o$ and $\rho_o = Mn_o$, where M is the ion mass in the plasma, we find that

$$\frac{dr_o}{dt} = \frac{(0.033)^{-1/2}}{\sqrt{5.36}} \sqrt{\frac{kT}{M}}$$

or

$$\frac{dr_o}{dt} = \sqrt{6 \frac{kT}{M}} \tag{11.29}$$

This result differs slightly from that given by Linhart, *i.e.*, $r_o = (11kT/M)^{1/2}$, and this is partly due to the use of a different γ. The temperature in equation (11.29) is that immediately behind the shock front and it decreases with expansion—partly because of the expansion of the hot medium, and partly due to the shock heating of new material.

If we return to equation (11.27) we can rewrite it in the form

$$\frac{T}{T_o} = \frac{p\rho_o}{p_o\rho} = \frac{f}{\psi} \frac{A^2}{a^2} r_o^{-3} = \frac{\dot{r}_o^{\,2}}{a^2} \frac{f}{\psi}$$

and upon differentiation with respect to time it becomes

$$\frac{\dot{T}}{T_o} = \frac{2\dot{r}_o}{a^2} \frac{f}{\psi} \ddot{r}_o$$

but from equation (11.18)

$$\ddot{r}_o = -\frac{3}{2} A^2 r_o^{-4}$$

so that we may now write using the above expression for T/T_o,

$$\frac{\dot{T}}{T_o} = -3 \frac{\dot{r}_o}{r_o} \left(\frac{T}{T_o}\right)$$

or simply

$$\langle \dot{T} \rangle = -3 \frac{\dot{r}_o}{r_o} T \tag{11.30}$$

which gives the mean temperature drop as result of the shock wave expansion. We now turn to the calculation of the mean rise in temperature due to fusion in a 50-50 D-T system. Since only 9/13 of this energy is converted into thermal energy [according to equation (11.25)] we find that we have

$$\langle \dot{T}_F \rangle = 0.7 \times 10^{10}\ n \langle \sigma v \rangle_{D\text{-}T} \tag{11.31}$$

Initially we expect that $\langle \dot{T} \rangle$ be higher than $\langle \dot{T}_F \rangle$, but at some r_o they will be equal; since T decreases with increasing r_o, and T_F is independent of r_o, it is evident that after the moment when

$$\langle \dot{T} \rangle + \langle \dot{T}_F \rangle = 0 \tag{11.32}$$

T can only start growing. If this condition is not satisfied for any r_o, the temperature T will drop monotonically below the ignition temperature T_L of the D-T reaction ($T_L \sim 4 \times 10^7$ °K). We shall, therefore, take as a trigger-criterion

$$-\langle \dot{T} \rangle = \langle \dot{T}_F \rangle$$

or using equations (11.30) and (11.31)

$$n_o\, r_o = 6.5 \times 10^{-6}\ \frac{T_o^{3/2}}{\langle \sigma v \rangle_o} \tag{11.33}$$

where the subscript "0" indicates condition at the shock. The minimum of the function on the right-hand side of the above equation is about 0.55×10^{28} for $T \sim 1.5 \times 10^8$ °K. This gives

$$n_o\, r_o = 3.5 \times 10^{22} \tag{11.34}$$

The corresponding energy W_o as given by Linhart is

$$W_o = \frac{384\pi}{5}\, k n_o T_o r_o^3 \int_0^1 \frac{n}{4n_o} \frac{T}{T_o}\, \eta^2\, d\eta$$

and upon substituting the value of the integral, *i.e.*, 0.09 we get

$$W_o \simeq 700 \left(\frac{n_s}{n_o}\right)^2 \text{ megajoule} \tag{11.35}$$

and from equation (11.34) we also get

$$r_o \simeq 0.67\, (n_s/n_o) \tag{11.36}$$

where $n_s = \frac{1}{2} \times 10^{23}$ corresponding to solid state density.

In the case that a solid D-T medium is precompressed by a strong shock, we have $n_o = 4n_s$ and therefore $W_o = 43.7$ MJ and $r_o = 1.6$ mm. The above criterion may be considered "optimistic" since it ignores a number of physical processes which make triggering more difficult. These include: (1) energy loss due to heat conduction ahead of the shock; (2) alpha particle heating of the "cold" plasma, *i.e.*, plasma outside the fusion zone whose $T <$ 4×10^7 °K; (3) the difference in electron and ion temperature since the latter, T_i, is generally smaller than T_e; (4) the departure from Taylor's similarity solutions; (5) particle velocity distribution and deviation from the Maxwellian; and (6) radiation losses.

11.2 EFFECTS OF ELECTRON THERMAL CONDUCTION AND ALPHA HEATING

According to Linhart the above factors are listed in order of their relative importance to W_o. We shall therefore estimate the effects of the first two, namely the loss due to heat conduction and the effect of the range of the alpha particle on heat deposition.

From the familiar heat conduction equation we can write

$$Q = - K\nabla^2 T$$

where Q is the heat source per unit volume per unit time and K is the coefficient of thermal conductivity. If we now think of the electrons with temperature kT_o located in a volume with characteristic dimension δ as the heat source, then from the above equation we find

$$\frac{3/2\, n_o\, k\, T_o}{\tau_h} = K \frac{T_o}{\delta^2}$$

or

$$\delta \sim \sqrt{\frac{K\, \tau_h}{3/2\, n_o k}} \tag{11.37}$$

where τ_h represents the energy confinement time, which in this case is the time it takes the shock front to reach the heat source front. It is given by

$$\tau_h = \frac{\delta}{\dot{r}_o}$$

which upon using equation (11.29) and inserting in (11.37) yields

$$\delta \sim \frac{K}{n_o \sqrt{T_o}} \sqrt{\frac{2M}{3k^3}} \tag{11.38}$$

If we now assume that we can use Spitzer's formula for thermal conductivity (Spitzer, 1956), we have for a hydrogen plasma at thermonuclear temperature and solid state density

$$K \simeq 0.4 \times 10^{-5}\ T_o^{5/2}$$

which yields for δ the result

$$\delta \sim 0.4 \times 10^{7}\ T_o^{2}/n_o \tag{11.39}$$

At $T_o = 10^8$ °K, the above expression becomes

$$\frac{\delta}{r_o}\ (n_o r_o) \sim 4 \times 10^{22} \tag{11.40}$$

from which we see that δ is of the order of r_o. Since the electron thermal energy does not contribute directly to fusion it should be treated as another internal energy source, W_h, that can be added to W_o. Thus we have

$$W_h = \frac{3}{2} k n_o \langle T_e \rangle \frac{4}{3} \pi \left[(r_o + \delta)^3 - r_o^{\,3}\right] \tag{11.41}$$

and if we take as an example $\langle T_e \rangle = ½ T_o$, then

$$W_h = 0.17\ W_o \left[3 \frac{\delta}{r_o} + 3 \frac{\delta^2}{r_o^2} + \frac{\delta^3}{r_o^3}\right]$$

Substitution for the various quantities in this expression from equations (11.36 - 11.39) leads to

$$W_h \simeq 1.5\ W_o$$

The total triggering energy now becomes $1.5W_o + W_o$ or $2.5W_o$ which, when inserted into equation (11.35), yields

$$W_o' = 1750 \left(\frac{n_s}{n_o}\right)^2 \tag{11.42}$$

This result is probably an overestimate in view of the fact that many oversimplifications were employed (*e.g.*, plane skin depth, inexact K, etc.) to arrive at it. It should be regarded as an estimate and no more.

We turn now to the question of alpha particle heating. The slowing down of this particle in plasma can be reasonably accurately described by the "classical" version of the equation derived in Chapter 4, *i.e.*,

$$\frac{dE}{dt} = -\frac{n\pi(e_1 e_2)^2}{\mu V} \ln\left(\frac{1}{\beta^2}\right)\left[\frac{2m}{m+M}\,\frac{\Phi(R)}{R} - \frac{4}{\sqrt{\pi}}e^{-R^2}\right]$$

where

$$R = \left(\frac{M}{m}\,\frac{E}{kT}\right)^{½}$$

and the target ion has mass M and temperature T, while the test (alpha) particle has mass m and energy E (velocity V). It is reasonable as a first approximation to assume that only the plasma electrons are effective in slowing down the test particle, so that the range formula is simply (see Chapter 4)

$$\lambda = \int_0^{E_i} \frac{VdE}{dE/dt} = \frac{\mu}{\pi n(e_1 e_2)^2 \ln(1/\beta^2)} \int_0^{E_i} \frac{V^2\,dE}{\left[\frac{2m}{m+M}\,\frac{\Phi(R)}{R} - \frac{4}{\sqrt{\pi}}e^{-R^2}\right]} \tag{11.43}$$

where E_i is the initial energy of the test particle. For an alpha particle slowed down by electrons, $m/M_e = 7344$ so that the dimensionless quantity R is quite small unless E is much larger than kT. For $R \ll 1$, we have

$$\frac{2m}{m+M}\,\frac{\Phi(R)}{R} - \frac{4}{\sqrt{\pi}}e^{-R^2} \simeq \frac{4}{\sqrt{\pi}}e^{-R^2}\left[(1 + 2/3\ R^3 + \ldots) - 1\right]$$

$$= \frac{8R^3}{3\sqrt{\pi}}$$

Putting this into the range formula we obtain

$$\lambda_\alpha \simeq (8.75 \times 10^{20}) \frac{(kT_e)^{3/2} E_i^{1/2}}{n \ln (1/\beta^2)} \tag{11.44}$$

where kT_e and E_i are in keV, and n is in particles per cm^3. The range λ_α is in cm. The quantity $\ln(1/\beta^2)$ is normally set equal to 40 in most plasma problems, but for solid state densities it is only about half that large. Thus if we take $E_i = 3.5 \times 10^3$ keV, and $\ln(1/\beta^2) = 24$, we obtain

$$\lambda_\alpha = 2.16 \times 10^{21} \frac{(kT_e)^{3/2}}{n}$$

so that if we put in the value of k, *i.e.*, 8.6×10^{-8} keV/°K, we get the result

$$\lambda_\alpha = 5.46 \times 10^{10} \frac{T_e^{3/2}}{n} \tag{11.45}$$

where the temperature is expressed in degrees Kelvin. This result is slightly different from that given by Linhart, and that may be due to the difference in the values of $(1/\beta^2)$. We recall that the approximate formula (11.44) is valid so long as

$$R = \left(\frac{M}{m} \frac{E}{kT_e}\right)^{1/2} = \left(\frac{1}{7344} \frac{E}{kT_e}\right)^{1/2}$$

is much less than unity. Since E cannot exceed the initial energy, 3.5 MeV, the above condition requires that kT_e be of the order of 10 keV or $T_e \geqslant 10^8$ °K. Use of the formula for electron temperatures on the order of 1 keV or smaller would tend to underestimate the range of the alpha particle. Table 23 gives comparisons of the approximate ranges found from the formula mentioned above with the ranges found from numerical integration of the more exact formula, for plasma densities of 10^{22}.

Table 23

Alpha Range in 10^{22} cm^{-3} Density Plasma

Plasma Temperature (keV)	*Approx. Range (cm)*	*Exact Range (cm)*
10	5.00	5.41
1.0	0.158	0.337
0.1	0.005	0.0462
0.01	158×10^{-6}	702×10^{-4}

In the exact formula, the effects of plasma ions (50% D and 50% T) are included assuming that the ions are at the same temperature as the electrons.

If we now consider n and T_e as corresponding to the plasma at r_o, we have from equation (11.45) at $T_e = T_o = 1.5 \times 10^8$.

$$\lambda_\alpha n_o \sim 9 \times 10^{22}$$

and using equation (11.34) we get

$$\lambda_\alpha / r_o = 9/3.5 \simeq 2.6 \tag{11.46}$$

This shows that the alpha particles are stopped outside the shock, *i.e.*, in the cold plasma. This result which differs from that given by Linhart—although the two range formulas are similar—points to the need for extreme care in calculating important parameters and the assumptions invoked in obtaining them. Only when the electron temperature is low (~ 1 keV) will the range of the alphas be comparable or less than r_o. Since, from equation (11.44), the range varies directly with $\sqrt{E}$, then an alpha particle with about $(1/2.6)^2 E_i \simeq 1/5\ E_i$ energy will have a range equal to r_o (or δ). This indicates that only 1/5 of the alpha particle energy goes to heating the electrons within δ. If we therefore require that the alphas produced at $r < r_o$ be able to supply the heating to the advancing heat-precursor (electron heat conduction) as well as the shock, then we can write approximately that

$$\dot{W}_F = \dot{W}_o + 4\pi r_o^2 \left(\frac{\delta}{r_o} + \frac{1}{2.6} \frac{\delta^2}{r_o^2} \right) \dot{r}_o \frac{3}{2} kT_o$$

and upon substituting for W_F from equation (11.31), and from (11.35) and (11.36) we get

$$n_o r_o \sim 35 \times 10^{21} \left[\frac{\delta}{r_o} \left(1 + \frac{1}{2} \frac{\delta}{r_o} \right) + 1 \right]$$

The value of δ / r_o can now be put in from equation (11.40) to give

$$n_o r_o \sim 6.3 \times 10^{22} \tag{11.47}$$

which in turn gives for the trigger energy

$$W_o = 2500 \left(\frac{n_s}{n_o} \right)^2 \tag{11.48}$$

and the corresponding heat-precursor energy of $W_h \sim 0.6W_o$. Therefore the total trigger criterion is

$$W_o' \sim 4000 \left(\frac{n_s}{n_o}\right)^2$$

This is also a pessimistic result for the same reasons mentioned earlier and also due to the shortening of the alpha range because of collective effects. Some more accurate calculations have shown that

$$W_o = 1000 \left(\frac{n_s}{n_o}\right)^2 \text{ MJ}$$

$$r_o = 0.75 \left(\frac{n_s}{n_o}\right) \text{ cm} \qquad (11.49)$$

which is somewhat larger than the optimistic result cited earlier.

11.3 THE CONCEPT OF LASER-FUSION

In the previous section we examined in a very elementary way an inertially confined fusion system based on a "microexplosion" resulting from the sudden deposition of a great amount of energy in the center of a spherical fusing medium. The laser-fusion concept as it is currently envisaged is an "implosion" system which consists of a tiny spherical pellet of deuterium-tritium surrounded by a low-density atmosphere extending to several pellet radii located in a vacuum chamber and a laser capable of generating an optimally time-tailored pulse of electromagnetic energy. Before the main pulse is applied the atmosphere may be produced by ablating the pellet surface with a prepulse. The pellet is subsequently compressed to a very high density, and if compression occurs at the time of energy arrival a thermonuclear "burn" is initiated in the central region. A thermonuclear burn front propagates radially outward from the central region igniting the dense fuel. This sequence of events is schematically illustrated in Figure 72.

As will be noted in the next section, the interaction of laser radiation with plasma—especially the all important absorption

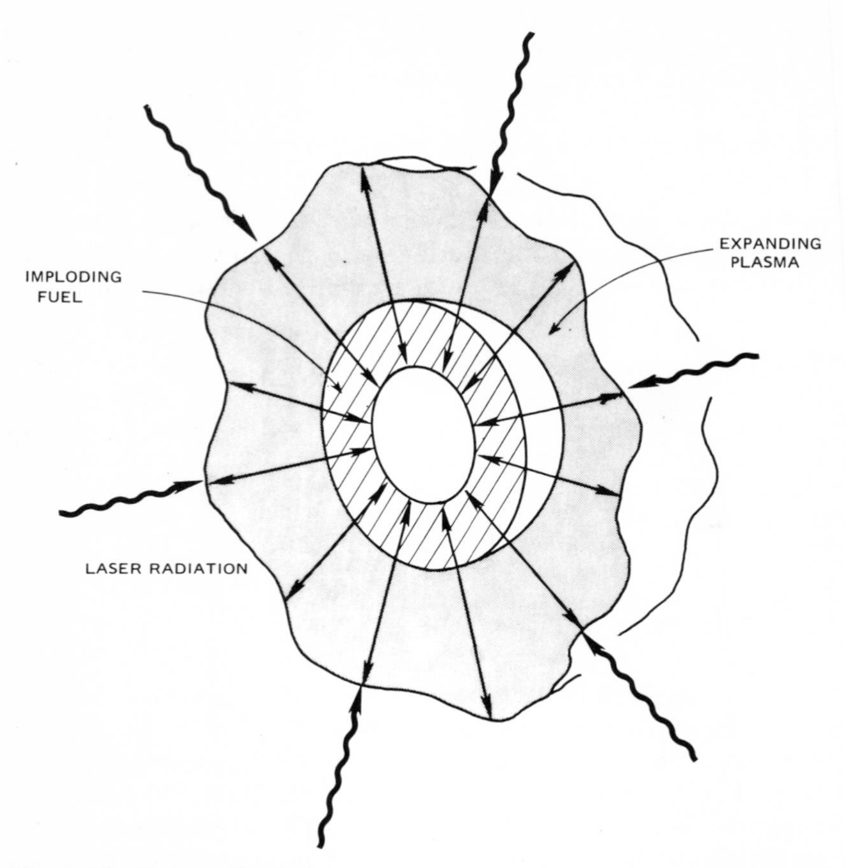

Figure 72. Laser-pellet fusion.

processes—is a very complicated phenomenon that has not yet been thoroughly understood or properly tested in the laboratory. Nevertheless, we can (as we did in the previous section) assess the potential of this approach to fusion by examining in some detail the energy break-even conditions assuming that energy from the laser does couple to the plasma at some efficiency.

We shall again restrict our discussion to a D-T pellet where we recall that most of the 17.6 MeV energy released in fusion is carried by the neutron. Following Brueckner (1973) we focus our attention on the multiplication of the laser energy by fusion reactions for which we define an energy multiplication factor given by

$$M = \frac{E_{fusion}}{E_{laser}\ (\text{output})} \tag{11.50}$$

and an efficiency for coupling of laser energy to the D-T plasma denoted by

$$\epsilon_L = \frac{E_{plasma}\ (\text{Thermal})}{E_{laser}\ (\text{output})} \tag{11.51}$$

The first factor which underscores the energy gain in a practical system depends on the characteristics of the laser and on the system for converting the fusion energy into laser input energy. The second is determined by the laser-plasma interaction and the mechanisms of energy transfer from the laser to the medium covering the sequence of processes leading to the production of fusion energy. If we let W be the energy produced per fusion reaction, then the rate of fusion energy production by a sphere of radius r is

$$\begin{aligned}\left(\frac{dE}{dt}\right)_f &= \left(\frac{4}{3}\pi r^3\right)\left(n_D n_T\right) \langle\sigma v\rangle (\theta_i)\ W \\ &= \frac{4}{3}\pi r^3\ \frac{n^2}{4} \langle\sigma v\rangle (\theta_i)\ W\end{aligned} \tag{11.52}$$

where we have used $n_D = n_T = n/2$, *i.e.*, 50-50 mixture of deuterium and tritium. The reaction rate $\langle\sigma v\rangle$ is a function of the ion temperature $\theta_i = kT_i$, and the initial plasma thermal energy is simply the sum of the electron and ion thermal energies or

$$E_{Th} = (4/3\ \pi r^3)\ 3/2\ n\left[\theta_i(o) + \theta_e(o)\right] \tag{11.53}$$

Because of their light mass the electrons tend to be accelerated more readily than the ions by the electric field of the laser radiation, and as a result the laser energy is deposited almost entirely into the electrons. Through coulomb collisions the electrons transfer energy to the ions and from equation (4.104) we recall that the thermal equilibrium between two species of Maxwellian distributions is described by the equation

$$\frac{d\theta_i}{dt} = \frac{\theta_e - \theta_i}{\tau_{Th}} \tag{11.54}$$

where the (classical) thermalization time τ_{Th} is given by

$$\tau_{Th} = \frac{3\, m_e\, m_i}{8\sqrt{2\pi}\; ne^4\, \ln \Lambda} \left(\frac{\theta_e}{m_e}\right)^{3/2} \tag{11.55}$$

For plasmas of interest to magnetic confinement we recall that $\ln\Lambda \simeq 20$, but for densities and temperatures in the range of interest to laser fusion $\ln\Lambda \simeq 5$; and if we let m_i be the average D and T ion masses, then τ_{Th} assumes the form

$$\tau_{Th} = 5 \times 10^{12}\, \theta_e^{3/2}/n \text{ sec cm}^{-3},\ \theta_e = \text{keV} \tag{11.56}$$

We would expect a net energy production from this system if the fusion reaction time is substantially larger than τ_{Th}, *i.e.*, if the following condition is satisfied

$$nt \gg 5 \times 10^{12}\, \theta_e^{3/2} \tag{11.57}$$

and for $\theta_e = 10$ keV it becomes $nt \gg 5 \times 10^{14}$ sec/cm^3.

If we now assume that the above condition is satisfied and we set $\theta_e(o) = \theta_i(o) = \theta_o$ in equations (11.52) and (11.53) we get

$$E_f = 1/3\, \pi r^3\, n^2\, \langle\sigma v\rangle\, (\theta_o)\, Wt \tag{11.58}$$

$$E_{Th} = 4\pi r^3\, n\, \theta_o$$

It should be noted at this point that both E_f and E_{Th} are functions of time since the temperatures vary with time; but if we assume that θ_e and θ_i do not increase appreciably during the reaction, then the constant can be evaluated at the initial temperature and the result is such that E_{Th} remains constant while E_f increases only linearly with time. From equations (11.50) and (11.51) we can relate E_f to E_{Th}, *i.e.*,

$$E_f = \frac{M}{\epsilon_L}\, E_{Th} \tag{11.59}$$

and if we substitute from (11.58) we further obtain

$$nt = \frac{M}{\epsilon_L}\; \frac{12\, \theta_o}{W\langle\sigma v\rangle\, (\theta_o)} \tag{11.60}$$

For this relation to yield a meaningful energy requirement of the system, some sort of a plasma life time must first be put in for "t." As in the previous case, this time is given by the time it takes a rarefaction (or sound) wave to propagate from the center to the surface of the pellet or $r = V_s t$. Using the isothermal sonic speed $V_s = \sqrt{kT/M}$ or simply $V = V_o \theta^{1/2}$ where $V_o = 3.5 \times 10^7$ cm/sec and θ is in keV, then the required laser energy can be readily calculated from equation (11.51) by utilizing equations (11.52) and (11.60). The result is

$$E_L = \frac{1}{\epsilon_L} E_{Th}$$

$$= \frac{1}{\epsilon_L} (4\pi r^3 n_o \theta_o) \qquad (11.61)$$

$$= \frac{1}{\epsilon_L} (4\pi) (V^3 t^3) \frac{n_o^3}{n_o^2} \theta_o$$

$$= 4\pi \frac{M^3}{\epsilon_L^4 n_o^2} \theta_o \left[\frac{12 \theta_o^{3/2} V_o}{W\langle\sigma v\rangle (\theta_o)}\right]^{-3}$$

The above quantity has a broad minimum at θ_o of 10 keV which when put into equation (11.60) yields

$$nt = 5.7 \, M/\epsilon_L \, 10^{13} \text{ sec cm}^{-3} \qquad (11.62)$$

At this temperature E_L takes on the value

$$E_L = \frac{M^3}{\epsilon_L^4} \left(\frac{n_s}{n}\right)^2 (1.6)\text{megajoules} \qquad (11.63)$$

where $n_s = 4.5 \times 10^{22}/cm^3$ is solid state density. We note in equation (11.62) that if $M/\epsilon_L = 3$ it reduces to the familiar Lawson condition for energy break-even. Moreover we observe that for this value of M/ϵ_L, equation (11.57) is not valid and hence the derivation of equation (11.60) is equally not valid. For laser-fusion applications, however, the laser coupling efficiency ϵ_L is much smaller than unity while the efficiency M can be substantially larger than one. If for example we assume that the thermal energy of the system can be converted to electrical energy at 40% efficiency and that the laser has a 25% efficiency, then we note from equation (11.50) that M = 10. It has been suggested that

M/ϵ_L can be as high as 200, in which case the conditions of equation (11.57) can be easily met. Such a large energy multiplication, however, invalidates the assumption used earlier that the temperature remains constant during the reaction. Nevertheless it is possible for the fuel to "ignite" and burn at high temperatures if self-heating occurs which would significantly increase energy production due to the sharp increase of the reaction rate at high temperatures. Insofar as the requirements on the laser energy are concerned, we note from equation (11.63) that it depends heavily on the coupling efficiency ϵ_L and on the fuel compression ratio (n_s/n).

The coupling efficiency represents the combined efficiency of the laser energy deposition into the plasma and the subsequent transfer of energy from the deposition region to the dense reacting region of the pellet. Although an efficient deposition of energy can take place, it does not necessarily follow that an efficient transfer of the energy to the dense region will also occur. When a laser beam is made to impinge on a pellet, the initial penetration of the laser energy results in the ionization of the pellet surface and the formation of a plasma at solid state density. As will be noted in the next section the plasma soon reaches a density (for which the plasma frequency is equal to the laser frequency) which renders it opaque to the incident radiation. The laser flux is then reflected at the density discontinuity with a relatively low absorption taking place as a result of penetration of the laser radiation into the over-dense region a distance of the order of the laser wave length. The absorbed energy appears as the thermal energy of the absorbing layer, as the kinetic energy of the expanding plasma, and as the thermal and kinetic energy of the over-dense plasma into which energy is transferred primarily by electron conduction. The sequence of events is illustrated in Figure 73.

Following the thermal conduction and energy penetration into the over-dense region, a rarefaction wave moving outward is set up and the energy corresponding to the outward motion and expansion of the heated plasma accounts for the marked depletion of the energy deposited by the laser. An estimate of the energy partition among the processes described above can be made that would enable us to estimate the laser coupling efficiency referred to earlier. The pressure driving the implosion arises from

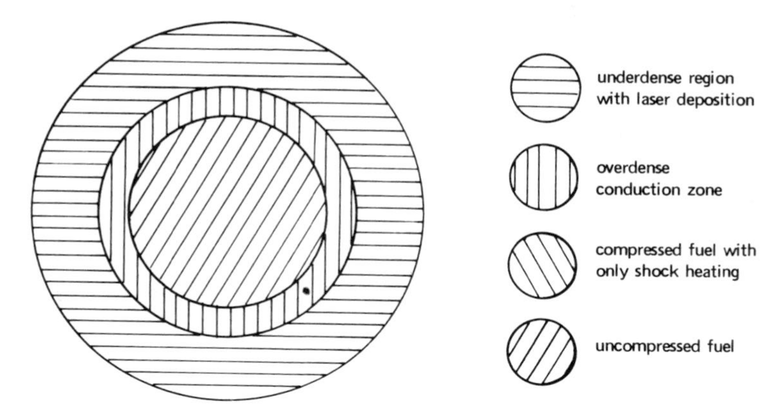

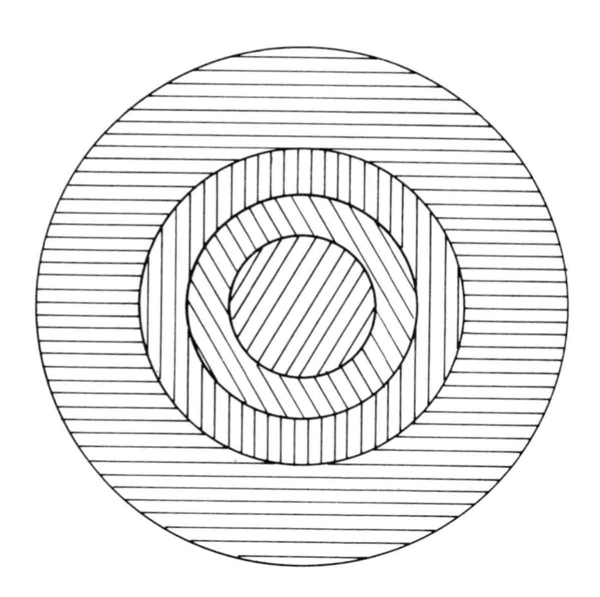

Figure 73. Early stages of laser pellet interaction (Brueckner, 1973).

the removal of material from the surface as result of initial deposition of laser energy, and energy balance requires that the incident laser flux be equal to the energy of the ablated material. In order to allow for efficient energy transfer by electron conduction, it will be assumed that the region of energy flow is isothermal even though the density drops rapidly outward from the dense heated surface as result of outward acceleration of ablated material. With this isothermal condition and in one dimension, the equations of motion and continuity assuming equal ion and electron temperatures are

$$m_i\, n \left(\frac{\partial}{\partial t} + v \frac{\partial}{\partial z} \right) v = - 2\, \theta\, \partial n/\partial z \tag{11.64}$$

$$\frac{\partial n}{\partial t} + \frac{\partial nv}{\partial z} = 0 \tag{11.65}$$

In these equations z represents the distance from the surface and t is the time. The solutions can be readily shown to be

$$n = n_o \exp\,(- z/C_T\, t) \tag{11.66}$$

$$v = v_s + z/t \tag{11.67}$$

where $v_s = \sqrt{2\theta/m_i}$ is the isothermal sound speed. The total energy per unit area in the plasma is the sum of the kinetic and thermal energies or

$$E = \int_o^\infty dz\, [\tfrac{1}{2}\, m_i\, n\, (v_s + z/t)^2 + 3n\theta]$$

$$= 8\, n_o\, \theta\, v_s\, t \tag{11.68}$$

and the energy balance in plane geometry requires that the above energy be equal to the laser flux per unit area or

$$\phi_L = E/t = 8\, n_o\, \theta\, v_s \tag{11.69}$$

The mass flow into the rarefaction zone is given by

$$\frac{dm}{dt} = m_i n_o v_s$$

and the ablation pressure $P_A = n_D\theta_D + n_T\theta_T = 2\, n_o\, \theta$.

We now use these equations to determine the acceleration of the surface. If we let m_o be the initial mass per unit area of the accelerating high density layer then the rate of mass loss is given by

$$m(t) = m_o - \frac{dm}{dt} t = m_o - \dot{m}\, t \qquad (11.70)$$

and the equation of motion becomes

$$(m_o - \dot{m}t)\, \ddot{r} = P_A \qquad (11.71)$$

The above equation can be further written as

$$\frac{dv}{dt} = \frac{P_A}{m_o - \dot{m}t}$$

which, upon integration subject to the initial v = o at t = o, becomes

$$v = \frac{P_A}{\dot{m}} \ln \frac{m_o}{m_o - \dot{m}t}$$

If we now substitute for P_A and $\dot{m}$ their equivalents, we find that $P_A/\dot{m} = v_s$. The equation for the velocity becomes

$$v(t) = v_s \ln \frac{m_o}{m_o - \dot{m}t} \qquad (11.72)$$

This result enables us to write the kinetic energy of the accelerated layer as

$$\tfrac{1}{2}m(t)\, v^2(t) = \tfrac{1}{2}m(t)\, v_s^2 \left(\ln \frac{m_o}{m_o - \dot{m}t}\right)^2 \qquad (11.73)$$

The laser energy as given by equation (11.69) can be further written, if we substitute for t from (11.70), as

$$E_L = \phi_L t = 8\, n_o\, \theta\, v_s \left(\frac{m_o - m(t)}{\dot{m}}\right)$$

$$= 4\, v_s^2 \left[m_o - m(t)\right] \qquad (11.74)$$

where use has been made of the expression for the sound speed. The ratio of the kinetic energy of the accelerated layer to the incident laser energy therefore is

$$\frac{\frac{1}{2}m(t)\,v^2(t)}{E_L} = \frac{1}{8}\,\frac{m(t)}{m_o - m(t)}\left(\ln\frac{m_o}{m(t)}\right)^2$$

This energy transfer maximizes at $m_o/m(t) \simeq 5$ thereby giving

$$\left|\frac{\frac{1}{2}m(t)\,v^2(t)}{E_L}\right|_{max} = 0.081 \qquad (11.75)$$

which suggests that the laser coupling efficiency ϵ_L is about 8%. A number of simplifying assumptions have gone into the calculation of the above result including the spherical nature of the motion. More detailed calculations have shown that typically $\epsilon_L \simeq 0.05$, and if we use M = 10 for the multiplication factor we find from equation (11.63) that the laser energy requirement is

$$E_L = 2.6 \times 10^8\,(n_s/n)^2 \text{ megajoules} \qquad (11.76)$$

The above requirement can be quite prohibitive unless a very large compression ratio is attained. Some of the recent numerical results on strong shocks in spherical geometry have shown that a compression ratio of 33 can be achieved by the passage of a single shock. For a gas with $\gamma = 5/3$, equation (11.15) shows that the passage of the first converging shock gives a density increase of $\gamma + 1/\gamma\text{-}1 = 4$. The ensuing adiabatic compression leads to further density ratio of 15 and upon reflection from the center the returning shock gives a further compression to a maximum density ratio of 33. According to equation (11.76) this compression leads to a laser energy requirement of about 2.6×10^5 megajoules which is still beyond the reach of any present day lasers. If on the other hand a succession of shocks of increasing strengths can be initiated in the pellet in such a way as to not cancel each other or overtake each other at the center of convergence, then very high compression ratios and hence lower laser energy requirements can in principle be achieved. Such multiple shock results are illustrated in the computer experiments shown in Figure 74.

The above results must be viewed as estimates, and until better understanding of the processes associated with the evaluation of ϵ_L is gained they should remain so. As will be noted in

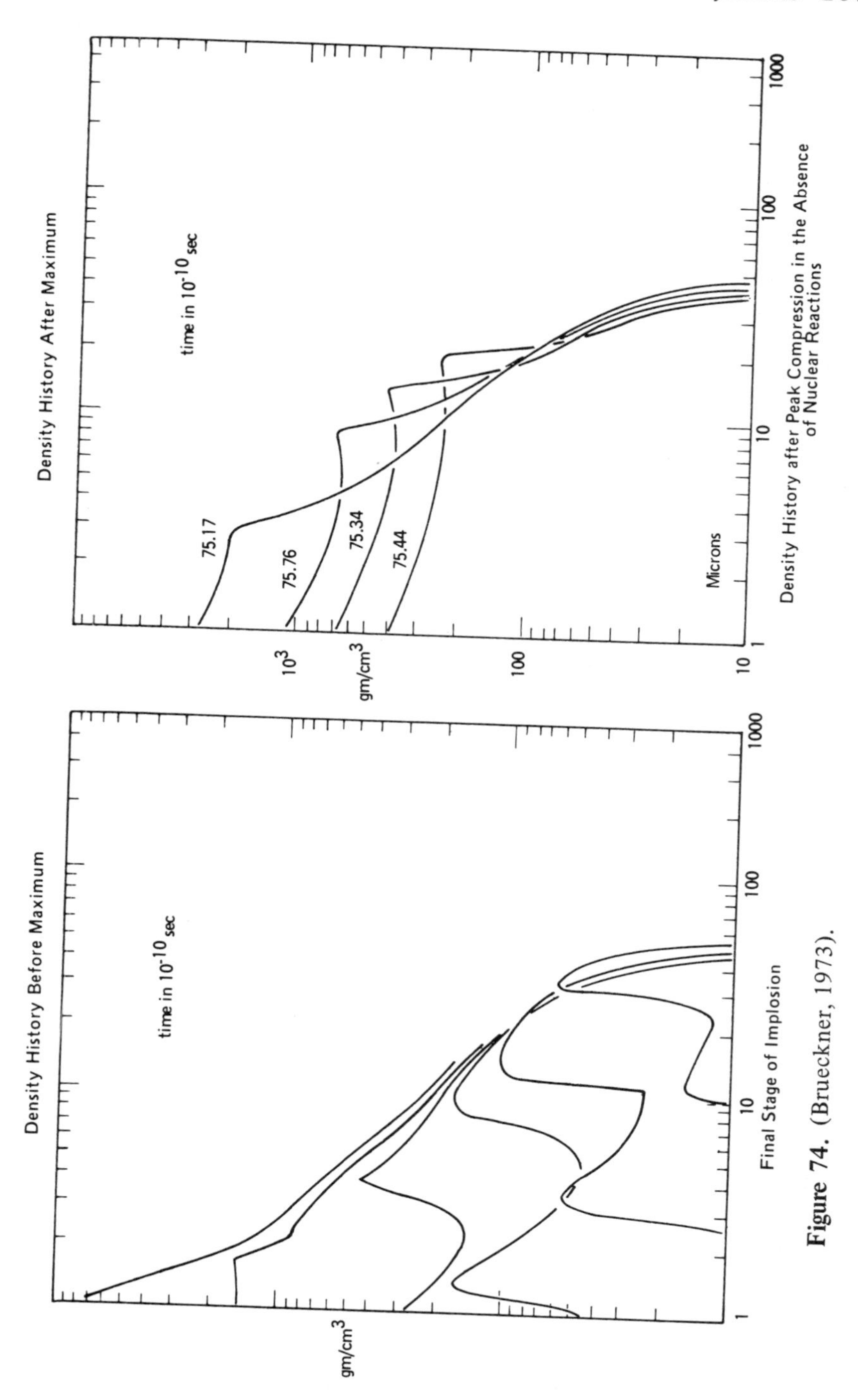

Figure 74. (Brueckner, 1973).

the next section the interaction of laser radiation with plasma is indeed a complex phenomena that has thus far eluded exact and deterministic assessment. Nevertheless a number of conceptual designs of fusion reactors utilizing laser-driven fusion have been advanced—in one case culminating in the feasibility study of a central station for commercial electric power generation.

In this investigation, L. A. Booth and his colleagues at the Los Alamos Scientific Laboratory proposed the reactor concept illustrated in Figure 75 in which liquid lithium is used to form a protective layer on the inner-cavity wall. This layer protects the wall from the damaging effects of the blast and thus prevents vaporization of the wall material by thermal radiation and erosion by high-velocity material, both of which emanate from the fuel pellet. By contrast to another concept called "the Blascon" by A. P. Fraas of the Oak Ridge National Laboratory in which the cavity is defined by the vortex of a swirling liquid blanket, the Wetted-Wall concept has the analytically attractive feature in that the cavity is well defined by a solid wall. Moreover, it possesses another important advantage in that it provides a passage by mechanical means for exhausting the hot gases in the cavity prior to the next shot. It is a pulsed system in which D-T pellets are injected, as shown, and upon arriving at the center of the cavity they are zapped by a laser pulse producing 200 megajoules per pulse. The main features of the blast-containing design evolve around a system of three walls: an innermost porous wall which is thin and allows the passage of lithium to form a protective coating on the inside surface; a main pressure-vessel wall thick enough to restrain the internal pressures in the cavity and blanket so that it does not seriously affect the breeding ratio; and an inner structural wall located between the former two walls which is thick enough to restrain the inward motion and thin enough so that it still permits an adequate breeding ratio. The power plant envisaged in the design consists of ten modules, each with the characteristics shown in Figure 75, generating a total thermal power of 2025 megawatts at a net overall plant efficiency of about 41%.

A cavity with a "dry wall" is the basis of yet another design of a laser-fusion reactor by a group at the Lawrence Livermore Laboratory. The characteristics of this design along with those of the two mentioned earlier are summarized in Table 24.

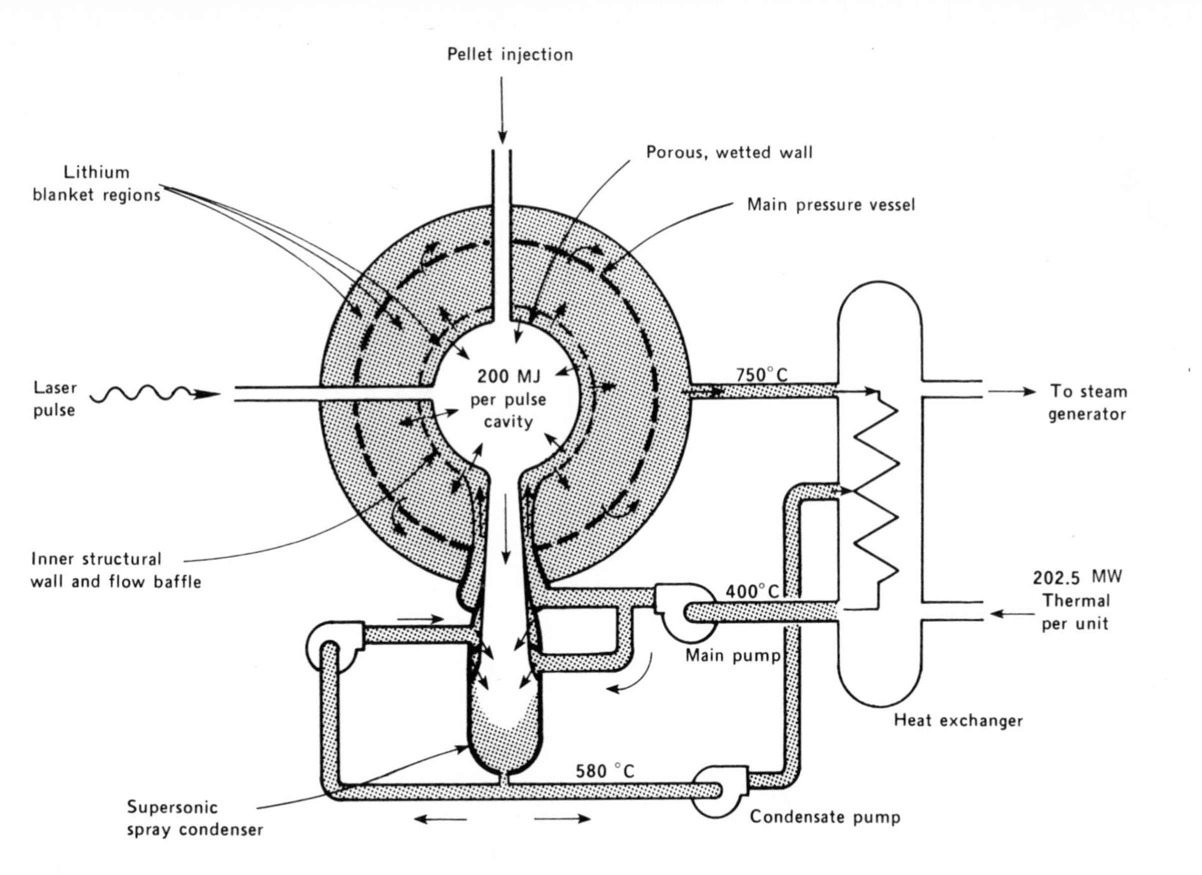

Figure 75. Wetted wall ICTR concept.

Table 24

Parameters of Various Proposed Systems

System	*Blascon* *Oak Ridge*	*Wetted-Wall* *Los Alamos*	*Dry Wall* *Livermore*
Thermonuclear Energy/pulse	1500 MJ	200 MJ	10-80 MJ
Repetition rate of laser	0.1 Hz	1 Hz	~100 Hz
Mean power	150 MW(Th)	200 MW(Th)	1000-8000 MW(Th)
Inner Wall	Lithium vortex (with bubbles)	Porous wall wetted by 2mm Li	10 chambers
Inner wall diameter	4.5 m	2 m	-

11.4 ECONOMIC CONSIDERATIONS

It would be interesting to assess the economic feasibility of such systems, especially that examined in section 2. To do this let us first recall the reactor criterion as expressed by the Lawson criterion given in Chapter 2. In the current notation it can be written as

$$\epsilon_c \epsilon_p (\dot{W}_F \tau + \dot{W}_r \tau + W_P) \geqslant W_P + \dot{W}_r \tau \qquad (11.77)$$

where ϵ_c denotes the efficiency of conversion of heat into the energy of the source, ϵ_p is the efficiency of conversion of the source energy into the energy W_p of the D-T plasma, and $\dot{W}_r$ is the rate of energy loss due to bremsstrahlung. In the familiar Lawson criterion, $\epsilon_p \epsilon_c$ is taken as 1/3 and the reactor condition becomes

$$n\tau = 10^{14} \qquad T = 10^8 \ {}^\circ K$$

In the case of pulsed reactors it is likely that $\epsilon_p \ll 1$, and $\dot{W}_r \ll \dot{W}_F$, so that with $\epsilon_c = 1/3$ equation (11.77) becomes

$$1/3 \ \epsilon_p \dot{W}_F \tau \geqslant W_p$$

which yields for $T = 10^8$ the approximate relation

$$n\tau \geqslant 10^{14}\ \epsilon_p^{-1} \tag{11.78}$$

When inertial confinement is used, $\tau = K\tau_o$, where $\tau_o = \sqrt{M/2kT}\ r$ so that the above condition becomes

$$nKr \geqslant 10^{22}\ \epsilon_p^{-1} \tag{11.79}$$

Since $n \sim \epsilon_p^{-1}$, then in all cases the stored energy must be

$$W_s = \frac{W_p}{\epsilon_p} \sim \epsilon_p^{-2} \tag{11.80}$$

For inertial confinement in cylindrical geometry Linhart has shown that

$$W_s \geqslant 87 \frac{r_o}{K\epsilon_p^{\,2}\epsilon_c}$$

which may be viewed as the reactor criterion. The economic criterion for pulsed reactors used for electrical generation can be written as

$$C < \epsilon_e \dot{W}_F \tau f \tag{11.81}$$

where C is the cost of fuel and apparatus per shot, ϵ_e is the efficiency of conversion of W_F to electricity, and f is the cost of one joule of electrical energy delivered by the reactor. If all costs are expressed in dollars, and $\dot{W}_F\tau = Y$ in megajoules, and if the energy corresponding to f is put in kwh, then for $\epsilon_e = ½$ we find

$$\frac{C}{Y} < 0.014\ f \tag{11.82}$$

Moreover, if Y is expressed in kilotons of TNT then we can write

$$\frac{C}{Y} < 700\ f$$

A typical value of f for coal-fired plants is 3 mills per kilowatt-hour. If we suppose that the storage system has a lifetime of N shots, then equation (11.81) is replaced by

$$\frac{C}{NY} < 1/5\ \epsilon_c\ f \tag{11.83}$$

when C is the total cost of the energy storage system destroyed after the release by the reactor of NY megajoules of energy. If the stored energy is Y_o, it is clear that for a reactor the energy gain $\alpha = Y/Y_o > \epsilon_c^{-1}$; we have

$$C_o < 1/5\ \epsilon_c f \alpha N \tag{11.84}$$

where now C_o is the cost of energy storage per megajoule. Once more taking $\epsilon_c = ½$, $\alpha = 9$, and f = (1/3) x 10^{-2} \$/kwh, we get

$$\frac{N}{C_o} > (1/3) \times 10^3 \tag{11.85}$$

If the storage system is condensers then the best available ratio N/C_o for fast condensers is currently of the order of 10. It is clear that such a system is not economically feasible, and an improvement factor of about 30 is needed for a break-even operation.

11.5 LASER ABSORPTION

All the calculations presented thus far contain the implicit assumption that a certain amount of energy can be deposited (and absorbed) in the target plasma to trigger the fusion reaction. We have seen also that the triggering condition depends on the efficiency with which the incident energy is absorbed by the plasma. In the case of laser energy there are a variety of mechanisms and different physical processes of importance depending on the laser intensity, frequency and pulse duration as well as on the target materials and configuration. The efficiency of getting the laser energy into the dense fuel is limited, perhaps to 10% or less. In the remainder of this section we shall briefly discuss the various processes (believed to be) responsible for laser energy absorption.

When intense laser light falls on a solid it will ionize it and, if the electron density in the resulting plasma is sufficiently high, the light will be reflected. Reflection occurs when the plasma frequency ω_p equals the laser frequency, ω, *i.e.*,

$$\omega_p = \sqrt{\frac{4\pi n e^2}{m_e}} = \frac{2\pi c}{\lambda} = \omega \tag{11.86}$$

This defines the critical density for a particular light frequency and it corresponds to 0.15 μ-light (ultraviolet) at solid D-T densities. Longer wavelengths will be reflected; but if a hydrodynamic expansion of the plasma takes place, an underdense region where $\omega_p < \omega$ will develop in which there will be some "collisional" absorption or absorption by "inverse bremsstrahlung." In addition there are many mechanisms which enhance this absorption, and in what follows we shall discuss each mechanism briefly.

The collisional absorption of energy is a process exactly opposite to the process of radiation by electrons discussed in Chapter 2, *i.e.*, bremsstrahlung. In the inverse process the electron absorbs radiation, and in order for this to take place the electron must be moving in the field of an ion. As a result of this absorption the electrons' energy and hence their temperature increases and they transfer some of it to the ions through coulomb collisions. From Chapter 4 we recall that such energy transfer is characterized by a relaxation time which introduces another critical time scale. If, for example, this relaxation time is longer than the fusion reaction time, and there are no other mechanisms to enhance absorption, then the whole idea of laser-fusion might be in jeopardy. The absorbed energy does, however, give rise to hydrodynamic motion and subsequent ion heating under the proper conditions. Collisional absorption decreases with increasing electron temperature and increasing laser energy because the electron collision time gets longer with increasing electron energy. At laser intensities where collisional absorption decreases, collective effects referred to as "anomolous" absorption become important. These are not too well understood at this time.

Another process by which laser energy gets absorbed occurs when the coherent energy is incident at an oblique angle to the critical surface so that a component of the electric field of the wave is normal to the critical surface where it is "resonant" with plasma oscillations. If the electron ion collision frequency is small compared to the wave frequency, these oscillations build up to large amplitudes until they break up in phase (spatial and velocity) space and very hot electrons are produced.

If on the other hand the collision frequency is much larger than the wave frequency, collisions will limit the amplitude of the oscillations and the electron distribution will be thermal.

Maximum absorption seems to occur at angles of incidence centered about 10° to the normal; as much as 50% of the incident energy can be absorbed by this process in favorable cases.

As pointed out earlier there are anomolous processes of absorption which start to compete with inverse bremmstrahlung at intensities of 10^{12} watts/cm^2 at wavelengths of 1 μm; this threshold varies inversely with the square of the wavelength. By contrast to the inverse bremsstrahlung, these processes are collisionless and arise from a large number of instabilities which are predicted by theory and numerical simulations. They are generally divided into three types: electrostatic instabilities, electromagnetic instabilities, and relativistic mechanisms.

When a laser beam is incident on a "critical" surface, the oscillating current flow caused by the electric field of the incident wave can drive electrostatic instabilities in the plasma. These are longitudinal oscillations that arise from charge separation, and they have a threshold of 10^{12}-10^{14} W/cm^2 at 1 μm in a hydrogen plasma. Their growth rates decrease with increasing electron temperature. The most important among these instabilities are the streaming instabilities and the parametric decay instability (Boyer, 1972). The effect of these instabilities is to increase the plasma resistivity which can result in a very strong absorption. They can also lead to the production of a small component of high energy electrons in addition to the thermal distribution. If the fraction of these energetic electrons is not small, they can carry a lot of energy away from the region of interest because their collision mean free path is long. This is not desirable because presumably in laser-pellet fusion one would like to compress the material to solid state densities and deliver the heat at the instant of compression in order to initiate "burn."

Electromagnetic instabilities in which the oscillation has both an electric and a magnetic vector (as opposed to just an electric vector in the case of electrostatic instabilities) occur when the incident power density is sufficiently large that the v x B term in the Lorentz equation (6.10) can no longer be neglected. For 1μm light the threshold is 10^{14} watts/cm^2, and the two main cases named in analogy to molecules and solid state processes are the stimulated Raman and stimulated Brillouin instabilities. In the first case the incident and reflected waves interact with plasma oscillations through the v x B force. Although some electron heating occurs through the breaking of the wave in phase space, this process is not expected to be a very important one. In the stimulated Brillouin instability the interaction occurs between the incident and reflected waves on the one hand and an ion accoustic wave (which is another longitudinal oscillation whose frequency is $\sim \sqrt{T_e/m_i}$ where m_i is the ion mass) on the other hand (*i.e.*, three-wave interaction) with the result of heating electrons and ions equally.

At very high intensities, above 10^{17} W/cm^2, new phenomena associated with relativistic electrons and radiation pressure come into play. At these intensities the electrons are driven to velocities near the speed of light and the resulting current ($n_e ec$) could indirectly shield the plasma from the incident wave preventing it from further penetration. These nonlinear interactions of the wave and the matter could also produce strong forces which affect the absorption mechanisms described above. Many of these phenomena are still under intensive study and it may take many years before the whole picture concerning laser energy absorption by plasma is clarified.

REFERENCES

1. Bickerton, R. J. *Nuclear Fusion, 13,* 457 (1973).
2. Booth, L. A. "Central Station Power Generation by Laser-Driven Fusion," Los Alamos Scientific Laboratory Report LA-4858-MS, Vol. 1 (1972).
3. Boyer, K. *Bull. Am. Phys. Soc. II, 17,* 1019 (1972).
4. Boyer, K. "Review of Laser-Induced Fusion," presented at the 1972 Annual Meeting of the Division of Plasma Physics, Monterey, Calif., (November 1972).
5. Brueckner, K. A. *IEEE Trans. on Plasma Science, 1,* 13 (1973).
6. Floux, F. *Nuclear Fusion, 11,* 635 (1971).
7. Fraas, A. P. USAEC Rep. ORNL - TM 3231 (1971).
8. Hancox, R. and I. J. Spalding. *Nuclear Fusion, 13,* 385 (1973).
9. Linhart, J. G. *Nuclear Fusion, 10,* 211 (1970).
10. Lubin, M. J. and A. P. Fraas. *Sci. American, 224,* 21 (1971).
11. Nuckolls, J., J. Emmett, and L. Wood. *Physics Today,* 46 (August 1973).
12. Rosenbluth, M. N. and R. Z. Sagdeev. *Comments on Plasma Physics and Controlled Fusion, 1,* 129 (1972).
13. Spitzer, L., Jr. *Physics of Fully Ionized Gases.* (New York: Interscience, 1956).
14. Taylor, G. I. *Proc. Roy. Soc. A, 201,* 159 (1950).

15. Williams, J. M. "Laser CTR Systems Studies," Los Alamos Scientific Laboratory Report.
16. Williams, J. M. in W. C. Gough, Ed. *CTR Engineering Systems Study Review Meeting,* USAEC Report Wash. 1278 (1973).

CHAPTER 12

RADIOLOGICAL ASPECTS OF FUSION REACTORS

We have seen in Chapter 1 that much of the enthusiasm for controlled fusion as an energy source is based on an expectation that the radiological burden associated with fusion power will present fewer potential difficulties than that associated with fission power. In this chapter we will assess the radiological aspects of a fusion reactor using a D-T fuel cycle and whenever possible we will compare the results with those of a comparable fission reactor. Since the neutrons from the D-T reaction are the main source of both penetrating radiation and radioactive contamination due to activation, one may argue readily that these radiological problems can be easily circumvented by using non-neutron-producing fusion nuclear reactions. Those most often mentioned are reactions involving lithium. They are shown in Table 25 along with the maximum cross section and the corresponding energy.

Table 25

Non-Neutron-Producing Fusion Reactions

Reaction	*Cross Section (mb)*	*Energy (MeV)*
${}_3Li^6\ (p, He^3)_2He^4$	200	1.8
${}_3Li^6\ (He^3, p)_4Be^8(2\alpha)$	30	~ 5
${}_3Li^6\ (He^3, p)_4Be^8(\gamma, 2\alpha)$	60	> 5
* ${}_3Li^7\ (p, He^4)_2He^4$	65	~ 3
${}_3Li^6\ (D, \alpha)_2He^4$	30	~ 3.7

*Has a sharp (12 keV) resonance at 440 keV, $\sigma_{res} = 7$ mb.

The last reaction in Table 25 should be viewed with caution since it involves deuterons; some D-D and D-T reactions are bound to take place in such a system giving rise to neutron-associated radiological problems although not to the same extent as in a system utilizing the D-T fuel cycle. The striking feature of almost all the reactions given in the above table is that they all exhibit much more demanding requirements for ignition than the D-T reaction. Of the more likely reactions for the first generation fusion reactors, *i.e.*, reaction (1.4) in Chapter 1, the D-T reaction is the least demanding ignition-wise, but it involves the largest inventories of radioactive materials and hence poses the most demanding requirements for radiological impact. We shall, therefore, focus our attention on a D-T reactor; in order to estimate the magnitude of the associated radiological problems we choose a reference reactor with a power output of 2000 megawatts.

The tritium-breeding blanket of this system is assumed to be composed of 75% lithium (4% Li^6), 5% niobium structure, and 20% void (by volume) and is 1.5 meters thick (Lee, 1970). The neutrons and the tritium of the D-T (and D-D) reaction are the primary causes of the radiological problems. Severe as they may be, these D-T reactors have two primary advantages over a comparable fission system: 1) the basic deuterium-tritium reaction produces stable helium and hydrogen while fission produces a multitude of highly radioactive elements; 2) fusion reactors are incapable of a nuclear runaway. The fusioning plasma is so tenuous that there is never enough fuel present at any one time to support a nuclear excursion. We have seen examples of that in our discussion of reactor dynamics where various fuel removal mechanisms were examined. Fission reactors on the other hand must contain a critical mass of fissionable material containing large amounts of potential energy.

The radiological problems of a D-T fusion reactor can be summarized as follows:

1. shielding primary and secondary neutrons and gammas,
2. containing radioactive materials during normal operation,
3. containing radioactive materials in case of an accident, and
4. storage of radioactive waste products.

12.1 RELEVANT NUCLEAR REACTIONS

A neutronics calculation using a Monte Carlo code was performed on the reference reactor (E. Plachaty, 1968) and the results obtained are shown in Table 26.

Table 26

Blanket Neutronics

Reactions	*per 14.1 MeV Neutron*
Energy	17.1
Tritium breeding (total)	1.31
a) $Li^7(n, n'\alpha)T$ reactions	0.704
b) $Li^6(n,\alpha)T$ reactions	0.602
Nb^{93} (n,γ) Nb^{94} reactions	0.232
Li^7 (n,γ) Li^8 reactions	1.79×10^{-4}
Nb^{93} $(n,2n)$ Nb^{92} reactions	0.125

The excess tritium breeding of 0.31 is just an example. It will normally be dictated by the need for tritium such as for starting up new reactors or to fuel special-purpose reactors that have no tritium breeding blankets. In any case the excess can be varied by changing the ratio of Li^6 to Li^7 in the blanket (see Chapter 3). The blanket structural material Nb^{93} undergoes a number of reactions when bombarded by neutrons–two of which are shown in the above table. The activation chains of interest for this isotope are shown in Figure 76.

Some of the interesting decay schemes in the above chains are shown below. The (n,γ) reaction of Nb^{93} produces two isomeric states of Nb^{94} with half-lives of 6.6 minutes and 2×10^4 years. These transitions are shown in Figure 77, and we note that the decay described by the above half-lives constitutes 99% of the total.

The decay schemes for the above radioisotopes are given by

$$Nb^{94} \xrightarrow[t_{1/2} = 6.3 \text{ m}]{} Nb^{94} + \beta^- (1.3 \text{ MeV}) + \gamma\ (0.8 \text{ MeV})$$

$$Nb^{94} \xrightarrow[2 \times 10^4 \text{ yr}]{} Mo^{94} + \beta^- (0.5 \text{ MeV}) + \gamma\ (0.7 \text{ MeV})$$

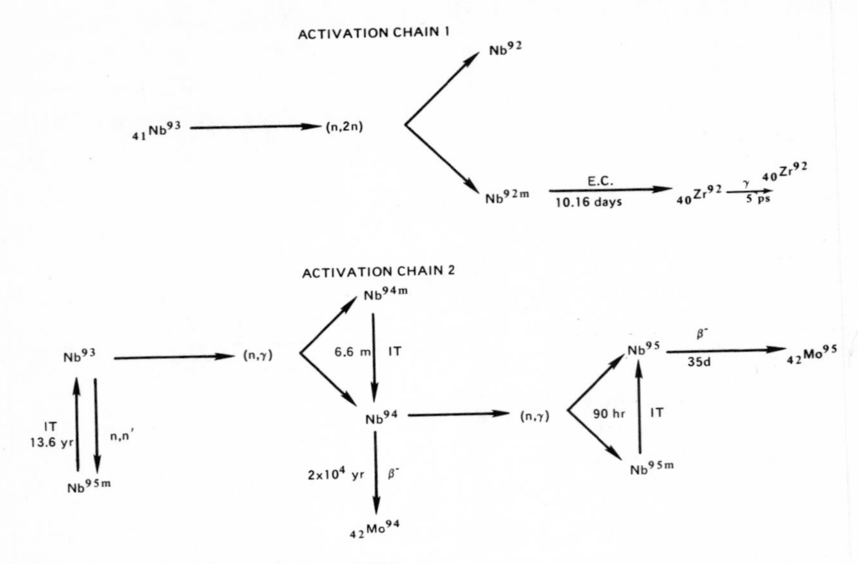

Figure 76. Activation chains of Nb.

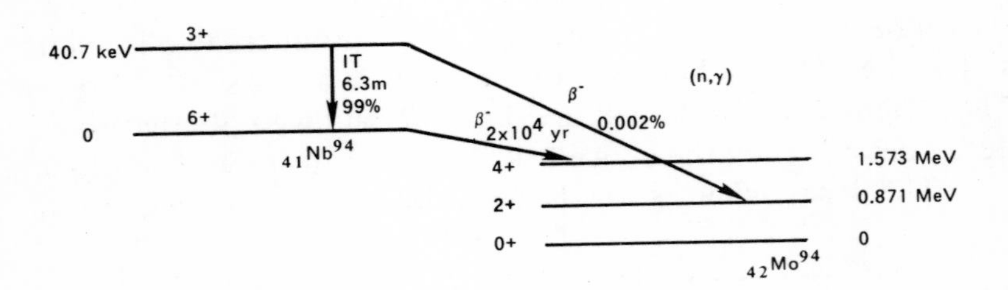

Figure 77. Energy level diagram.

and, though apparently a minor long-lived radioactivity problem, it is an important prompt gamma source. By contrast, the (n,n′) reaction (inelastic scattering) illustrated in Figure 78 may be regarded as a major source of long-lived radioactivity, although not much data is available on it.

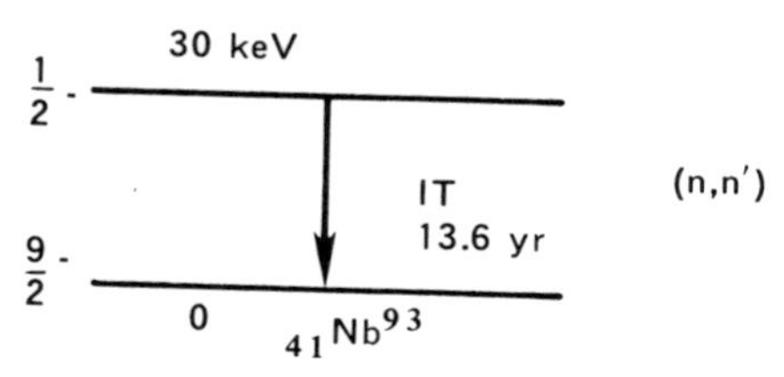

Figure 78. Isomeric transition of Nb^{93}.

There are other reactions involving Nb^{93} such as (n,p), (n,α), and (n,np) which may not be important from the radiological point of view, but they are important from the point of view of radiation damage due to the helium and hydrogen bubbles they introduce in the materials, and the introduction of impurities such as zirconium. At 14 MeV these reactions have relatively low cross sections; they are

$$\left.\begin{array}{l} \sigma(n,p) \sim 40 \text{ mb} \\ \sigma(n,\alpha) \sim 10 \text{ mb} \\ \sigma(n,np) \sim 336 \text{ mb} \end{array}\right\}$$

Another blanket structural material that has been given serious consideration because of its nuclear characteristics is vanadium ($_{23}V^{51}$). Though not a refractory its obvious advantage over niobium is that it has a much smaller charge number and is hence of lesser impact as an impurity (*e.g.*, in bremsstrahlung). If the vacuum wall is made of natural vanadium then it contains 99.75% of $_{23}V^{51}$, and 0.25% of $_{23}V^{50}$, and at 14 MeV it has a total cross section of about 2.3 barns. The two important activation chains involving this isotope are illustrated in Figure 79, and some of the neutron reactions are as shown following Figure 79.

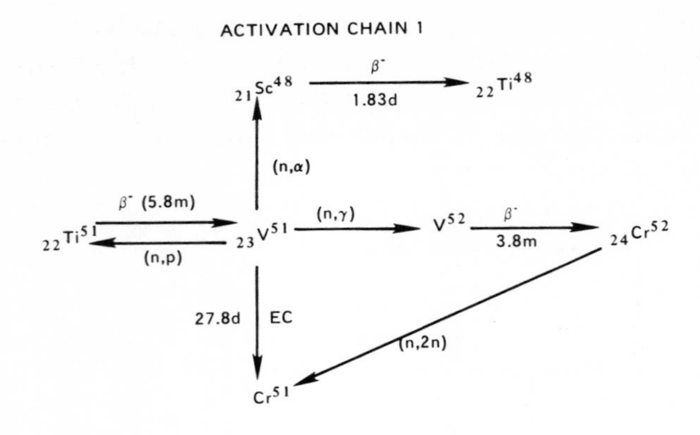

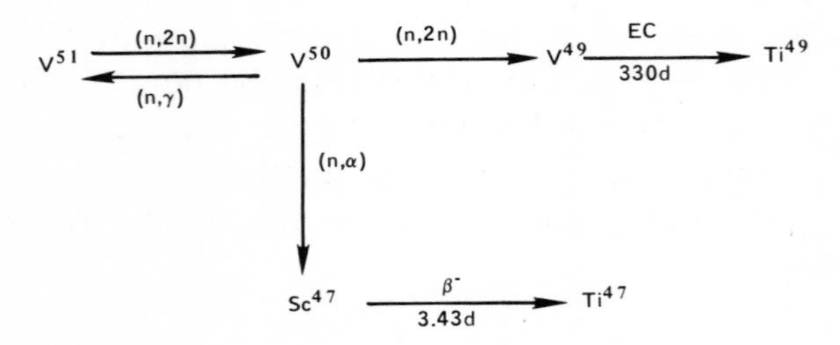

Figure 79. Activation chains of V.

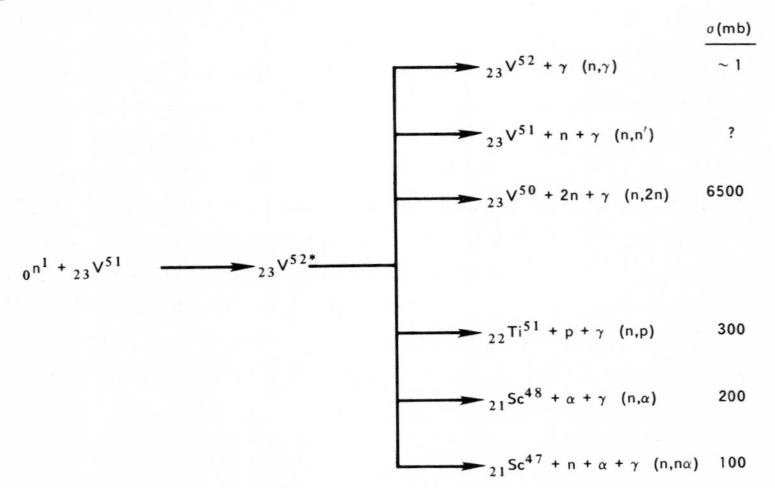

The (n,γ) branch of the first chain has a short half-life of 3.8 minutes and, as shown in the energy level diagram below, it results in a stable chromium isotope. This reaction does not therefore contribute to any measurable long-lived radioactivity.

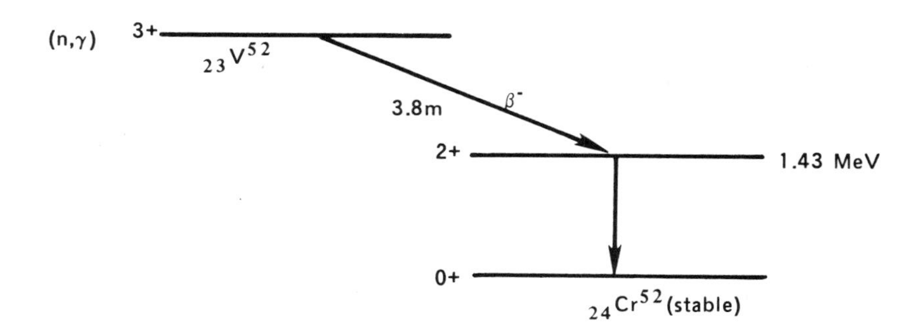

Since the isotope $_{23}V^{51}$ does not appear to have any significant isomeric states the inelastic scattering reaction (n,n′) can only be a source of prompt gammas. There are also no known isomeric states of V^{50} and, since the cross section at 14 MeV is about 6 barns, the (n,2n) reaction of V^{51} is of no consequence. The (n,p) reaction which constitutes one of the legs in chain 1 has a cross section of about 300 mb and gives rise to the titanium isotope which decays according to the scheme shown to V^{51} by beta emission.

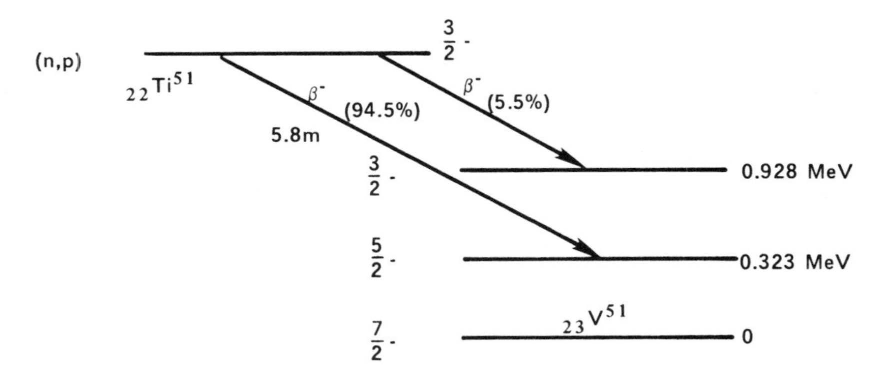

Again, this reaction is simply a source of prompt gammas and of no consequence to long-lived radioactivity. As a result of the (n,α) reaction with a cross section of about 200 mb, the isotope scandium is formed which decays by beta emission to Ti^{48} as shown in the level diagram.

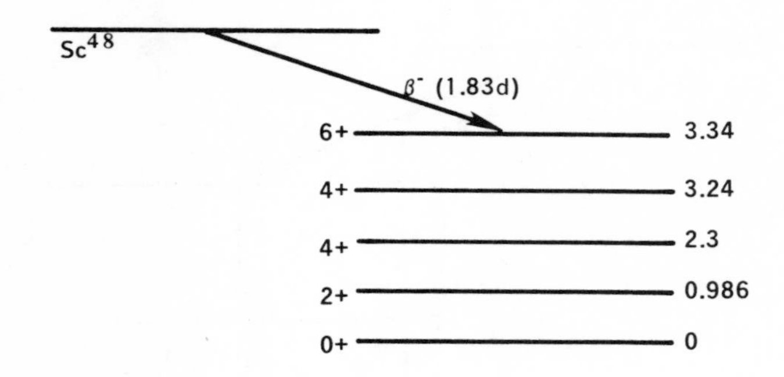

It also is of no apparent long-lived radioactivity but it is a source of some concern as an "afterheat" problem. The (n,α) reaction involving the isotope V^{50} leads to Sc^{47} which in turn decays by beta emission to Ti^{47} with a half-life of 3.4 days. This leg of the second activation chain shown below has a large cross section and constitutes another source of afterheat which is of no serious long-lived consequence. In summary, it appears that vanadium is a much more desirable blanket structural material from the nuclear standpoint because it is not associated with any serious long-lived radioactivity problems as niobium is.

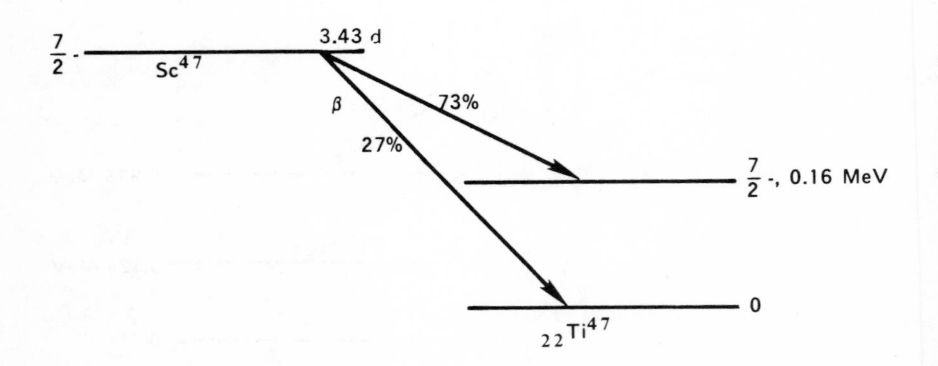

It is evident from the above brief discussion that the biological shielding requirements of a D-T fusion reactor will be greater than those required for a fission reactor. This follows from the simple fact that whereas 80% of the D-T fusion reaction energy appears in neutrons, only 10% of the fission reaction energy appears in the form of penetrating neutrons and gammas.

12.2 RADIOISOTOPE INVENTORY

Fortunately, tritium is the only radioactive isotope produced by the breeding reactions

$$Li^6 + n \longrightarrow T + He^4 + 4.8 \text{ MeV}$$

$$Li^7 + n \longrightarrow T + n' - 2.47 \text{ MeV}$$

since the helium by-product is stable. The isotope Li^7 does undergo a neutron capture reaction but its cross section is so low and the Li^8 half-life is so short (0.085 sec) that it is of no consequence. For a 2000 MW_t plant tritium consumption is about 260 grams per day. Tritium holdup in the blanket and other elements of the tritium loop will dictate the tritium inventory. Holdup is estimated to be about 1000 grams (9.48×10^6 curies) in a 2000 MW plant (Johnson, 1969). If we now combine the tritium inventory with that of niobium discussed in the previous section, we get the total radioisotope inventory shown in Table 27.

Table 27

Radioisotope Inventory of a 2000 MW
DT Fusion Reactor

Material	*Half-Life*	*Inventory (curies)*
Tritium	12.6 y	9.48×10^6
$Niobium^{94}$	2×10^4 y	1.13×10^4
$Niobium^{92}$	10.1d	2.05×10^9
$Niobium^{94*}$	6.6m	3.43×10^9

We have already seen the decay schemes of Nb^{94}. The schemes for the remaining isotopes making up the remainder of the inventory are

$$T \xrightarrow[12.6y]{} He^3 + \beta^- \ (0.018\ MeV)$$

$$Nb^{92} \xrightarrow[10.1d]{} Zr^{92} + \gamma \ (2.07\ MeV)$$

It is apparent from the above inventory calculations that D-T reactors must be housed in accident-proof containment structures for the same reasons fission reactors must. To get a feeling for a comparison between the hazard potentials of the radioisotope inventories of a D-T reactor and a fission reactor, the radioisotopes of each system are weighted by the International Commission on Radiological Protection (ICRP) maximum permissible continuous (40-hr week) concentrations (mpc's) of each in air with the following results using "A" as the activity in curies:

Fusion

$$\frac{A(H^3)}{mpc(H^3)} = \frac{9.48 \times 10^6}{5 \times 10^{-6}} = 1.9 \times 10^{12}$$

$$\frac{A(Nb^{92})}{mpc\ (Nb^{92})} = \frac{2.05 \times 10^9}{2.5 \times 10^{-5}} = 8.0 \times 10^{13}$$

$$\frac{A(Nb^{94})}{mpc(Nb^{94})} = \frac{1.13 \times 10^4}{10^{-8}} = 1.1 \times 10^{12}$$

Fusion total 8.3×10^{13}

Fission

$$\sum_{i=1}^{20} \frac{A(\text{fission product i})}{mpc\ (\text{fission product})} = 2.0 \times 10^{16}$$

$$\frac{A(U_{235})}{mpc(235)} + \frac{A(U_{238})}{mpc(238)} = \frac{4.75}{5 \times 10^{-10}} + \frac{493}{7 \times 10^{-11}} = 7 \times 10^{12}$$

$$\frac{A(Pu^{239})}{mpc(239)} = \frac{6.13 \times 10^4}{2 \times 10^{-12}} = 3 \times 10^{16}$$

Fission total $2\text{-}3 \times 10^{16}$

The fission products used in the above example are those from U^{235} fission after one year of operation and one day decay. The amounts of U and Pu are based on a boiling water reactor with a specific power density of 900 kW/kgram of U^{235} and a fast reactor with a specific power density of 2000 kW/kg of Pu (Etherington, 1958). For the fusion reactor the inventory calculations ignore the short half-life isotopes and assume that the long half-life Nb^{94} isomer has been building up for one year. It is significant that the fusion reactor inventories are a few orders of magnitude lower than those of fission reactors. From the containment point of view fusion reactors have two major advantages. First, the radioisotope inventory can be limited to a small number of chemical elements. In the above example, there are only two, tritium and niobium. Secondly, there is no fissile material that must be kept from reconstructing a critical mass. These advantages should make the fusion reactor easier to contain.

It is also evident from the above example that the major contamination problem during the normal operation of a D-T fusion power plant will be tritium leakage. It has been estimated (*Nuclear News*, June 1970) that the ICRP recommended annual dose limit of 0.5 rem/y would limit the acceptable tritium oxide discharge rate to 5×10^5 curies/year. This discharge rate corresponds to a leak of 0.3 grams per day. There is no reason to believe that tritium leakage from a fusion system cannot be limited to any acceptable level.

12.3 NUCLEAR AFTERHEAT

A problem which is closely associated with the hazard potential of the radioactive inventories presented above is the energy-release rate from the decay of the radioactive material. During shutdowns, either planned or emergency, the decay of activated materials will produce afterheat which might lead to vaporization and dispersal. The only significant source of afterheat in the fusion reactor is that associated with the structural material of the blanket, and in our example they are given in Table 28.

Table 28

Sources of Afterheat

Material	*Half-Life*	*Activity (curies/watt)*	*Q (MeV)*
Nb^{92}	10.1 d	1.0	2
Nb^{94*}	6.6 m	1.71	2

The afterheat power P_a can be expressed as

$$P_a = P_o \times 1.18 \times 10^{-12} \, (1.0 \, e^{-0.0286t} + 1.71 \, e^{-6.29t})$$

where P_o is the power level of the blanket during operation, and t is the time after shutdown in hours. Sample results of this equation are given below.

t (hrs)	P_a/P_o
0.1	0.0226
1.0	0.0118
10.0	0.0115
100.0	0.00886

and we note that the blanket will require cooling during both planned and emergency shutdown to keep from damaging the blanket and containment structures.

Steiner and Fraas (1972) have calculated as a function of time after shutdown (in seconds) the afterheat (as a fraction of the reactor thermal power) for fission-reactor fuel and for alternate structural materials (niobium and vanadium) in a reference fusion reactor with 1000 MW(t) power output. These results are shown in Figure 80, while the afterheat power densities associated with the niobium structure of the reference fusion reactor (RFR), the oxide fuels of a typical pressurized water reactor (PWR), and with a liquid-metal cooled fast breeder reactor (LMFBR) are given in Table 29.

We note that the afterheat power density in the fusion reactor is much less than that of the fission reactor fuels and that it is significantly less for vanadium than it is for niobium. Because the specific heats of niobium and the oxide fuels are comparable, the rate of rise of the niobium temperature will be much less than that of the fission fuels in the event of a loss-of-coolant accident.

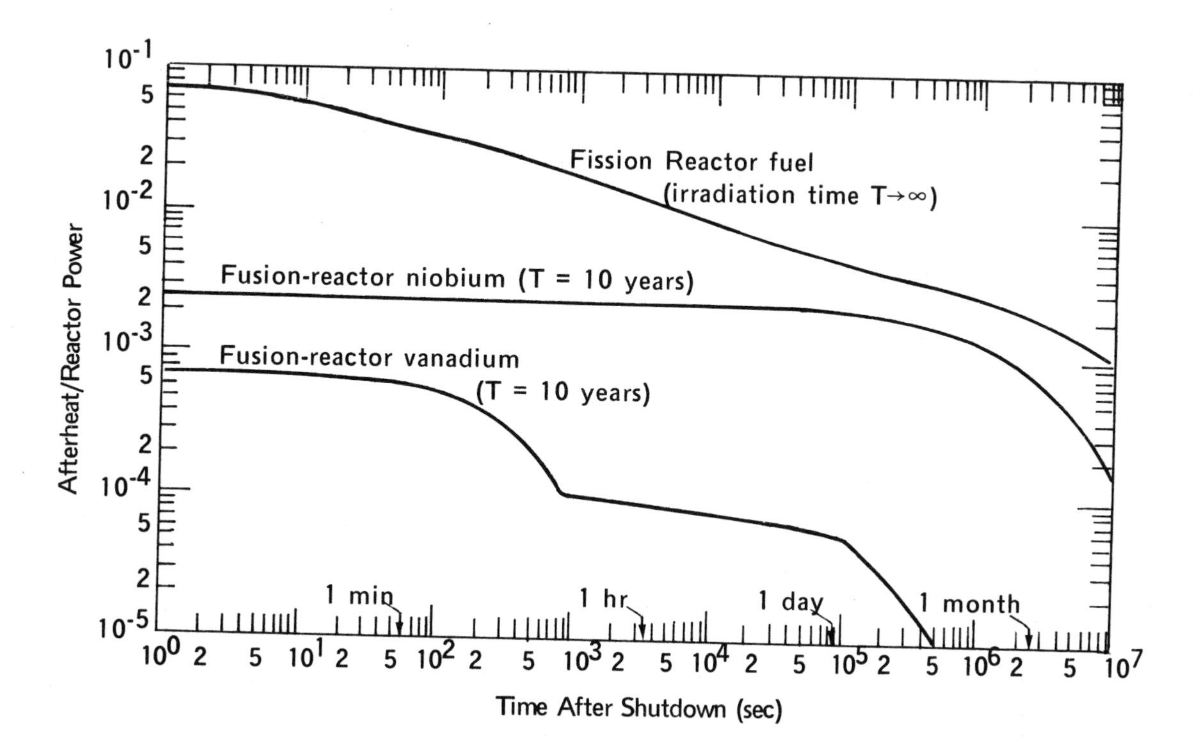

Figure 80. Afterheat versus time after shutdown.

Table 29

Afterheat Power Densities

	RFR	*PWR*	*LMFBR*
Average afterheat power density at shutdown, W/cm^3	0.15	19	105
Maximum-to-average afterheat power density	~ 2	~ 2	~ 2

At shutdown, for example, the adiabatic temperature rise associated with the average afterheat power density would be about 0.1°F/sec for the niobium, about 10°F/sec for the PWR fuel, and about 50°F/sec for the LMFBR fuel. The adiabatic temperature rise for vanadium is about one-third that of niobium. On the basis of these results it appears that the problem of afterheat removal is much less severe in the fusion reactor than it is in its fission counterpart. Nevertheless the per cent afterheat in fusion reactors will be significant (at shutdown ~ 0.25% for Nb and ~ 0.07% for V) and must be taken into account in the design considerations of these systems.

The radioactive waste problem associated with fusion reactors is also coupled to the radioactive material produced in the blanket, *i.e.*, the activated niobium. Storage of this isotope should be simpler than storage of the multitude of fission products, some of which are gaseous. Since the radioactive waste products are produced in the structural material and not in the basic D-T fusion reaction or the tritium breeding reaction it should be possible to minimize the problem by proper choice of the structural material. Vanadium is only one choice and there might well be better ones.

REFERENCES

1. Ajzenberg, F. and Lauretson. "Energy Levels of Light Nuclei, IV," *Rev. Mod. Physics, 24,* 321 (1952).
2. Alley, W. E. and R. M. Lessier. "Neutron Activation Cross Sections" *Nuclear Data Tables, 11,* (1973).
3. Etherington, H. *Nuclear Engineering Handbook.* (New York: McGraw-Hill, 1958).

4. Fraas, A. P. and H. Postma. "Preliminary Appraisal of the Hazards Problems of a D-T Fusion Reactor Power Plant," USAEC Report ORNL-TM-2822 (Rev., November 1970).
5. Johnson, E. F. BNES Nuclear Fusion Reactors (Proc. Int. Conf. Culham) (1969), p. 441.
6. Johnson, E. F. "Overall Tritium Balances in Fusion Reactors," (Princeton, N.J.: Princeton University, 1969).
7. Lederer, C. M., J. M. Hollander, and I. Perlman. *Table of Isotopes*, 6th edition (1967).
8. Lee, J. D. *Fusion Technology*, Energy 70 Las Vegas, Nevada (1970), p. 58.
9. *Neutron Cross Sections*, BNL-325 (updated).
10. *Nuclear News, 13(6)*, 31 (June 1970).
11. Plachaty, E. Lawrence Livermore Laboratory Report UCRL-50532 (1968).
12. Postma, H. "Engineering and Environmental Aspects of Fusion Power Reactors," *Nuclear News* (April 1971).
13. Steiner, D. *Nuclear Fusion, 11*, 305 (1971).
14. Steiner, D. *Nuclear Fusion, 14*, 33 (1974).
15. Steiner, D. and A. P. Fraas. *Nuclear Safety, 13*, 353 (1972).
16. Strehlow, R. A. and D. M. Richardson. "Chemistry of Tritium in Controlled Fusion Devices," USAEC Report, ORNL-3836 (1965).

CHAPTER 13

DESIGN CONSIDERATIONS OF FUSION REACTORS

Although most of the technical discussions presented in the previous chapters bear heavily on the design of various components in a fusion reactor, no attempt has been made to synthesize this information in terms of its relevance to the design of the reactor as a whole. In this chapter we shall utilize this information to estimate the major parameters needed to carry out a conceptual design of a fusion reactor. We shall first focus our attention on a reactor with toroidal geometry simply because much work has been done on this type of reactor; and except for the field geometry (or its absence as in the case of inertially-confined reactors) much of the analysis can be carried over to other geometries as will be seen in the subsequent sections. Because no actual fusion reactor has been built no attempt will be made to carry out detailed calculations of specific components; rather we shall assume that any component that enters the design consideration will function in accordance with the requirements placed on it and at an appropriate efficiency. At this point in time it is perhaps more important to emphasize (and hopefully translate) the role of the physics in the engineering considerations of fusion power systems.

As pointed out repeatedly the first-generation fusion reactors are expected to be D-T fueled because of their comparatively low demand on ignition. We shall therefore choose for our first conceptual fusion reactor a toroidal steady state system fueled by a 50/50 mixture of deuterium and tritium. We will assume adequate control of plasma equilibrium and stability. Although the source of the energy is the plasma, it is possible to establish the minimum dimensions of the system without any consideration of the plasma parameters. This arises because of two factors which dictate the dimensions of the structure external to the plasma. The first has to do with the efficient recovery of the fast neutron energy, and the second has to do with the desirability of using superconducting coils for the generation of magnetic fields. The design which will emerge from these and other considerations is shown in Figure 81 and follows closely the work of Carruthers *et al.* (1967).

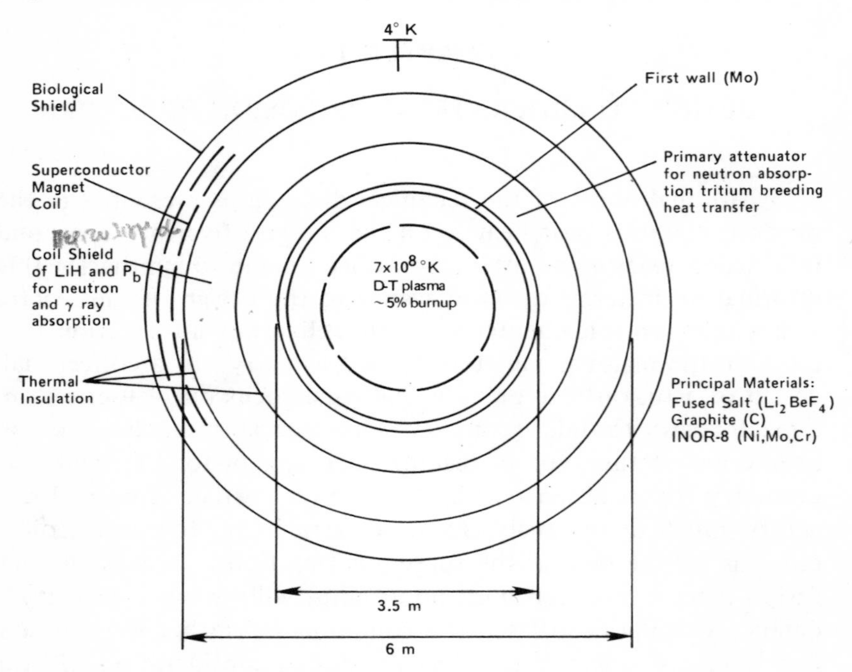

Figure 81. Schematic of a D-T fusion reactor.

13.1 WALL LOADING

One of the major (if not the most) critical factors in the design of a fusion reactor is the thermal loading on the "first" (or vacuum) wall which surrounds the plasma. If we take as an example, the model B reactor of Chapter 8 in which the confinement time is characterized by drift wave turbulence, we find that the reference reactor had a power of 3 gigawatts. Since the minor radius of the reactor is 2 meters, and the major radius is

10 meters, the wall loading (assuming these dimensions to be those of the vacuum chamber) can be readily seen to be

$$P_W = \frac{3 \times 10^9}{(2\pi a)\,(2\pi R)} = \frac{3 \times 10^9}{4\pi^2 a\ R} = 375 \text{ watts/cm}^2 \qquad (13.1)$$

This is of course a conservative estimate since it includes only the power of the neutrons from the D-T reaction. A number of estimates have been made that include the neutrons from Li^6(n,T) reactions in the blanket and a maximum wall loading of 750 W/cm^2 has been derived. On the other hand, Carruthers *et al.* point out that if molybdenum is selected for the wall material (it has the special advantages of low neutron capture and high neutron muiltiplication cross section) the wall loading can be increased to 1300 watts/cm^2 of wall from all energy sources in a D-T fusion reactor. Using this estimate for P_W we proceed to calculate the wall radius r_W shown in Figure 82.

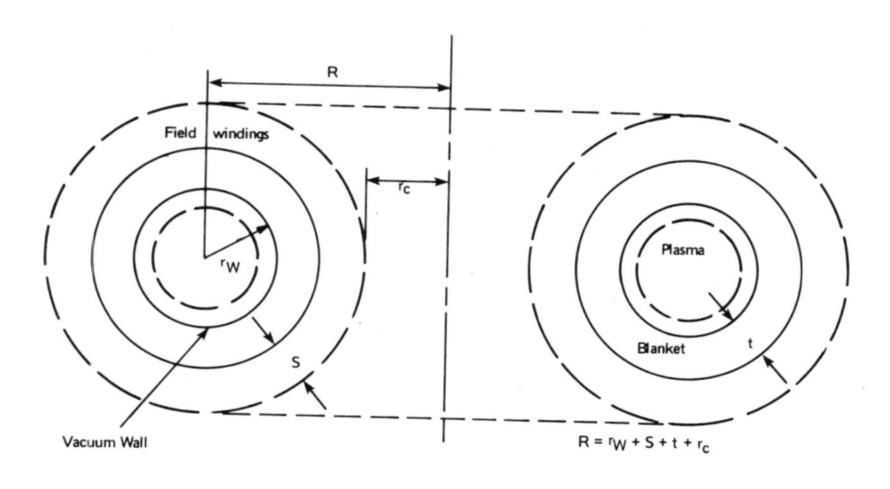

Figure 82. Toroidal fusion reactor cross section.

If we average the fusion power over the engineered volume of the reactor we obtain for the power density P_D per unit volume the result

$$P_D = \frac{2\pi\, r_W P_W}{\pi(r_W + t + S)^2} \qquad (13.2)$$

where t is the blanket thickness and S is the distance characterizing the thermal insulation of the magnet windings. These latter distances are independent of r_W, since they are determined by the nuclear properties of the blanket and those of the superconducting coils. Therefore equation (13.2) can be maximized with respect to the wall radius r_W. Letting $b = t + S$, differentiating equation (13.2) with respect to r_W, and setting the result equal to zero we get

$$\frac{d(P_D/P_W)}{dr_W} = \frac{2}{(r_W + b)^2} - \frac{4\, r_W}{(r_W + b)^3} = 0 \qquad (13.3)$$

or

$$r_W = b = t + S$$

Homeyer (1965) has given a value of 110 cm for the blanket thickness t, and allowing for additional nuclear shielding of the magnet coils to reduce refrigeration cost, Carruthers *et al.* have added an additional 15 cm to get t = 125. An estimate of 50 cm for S has also been given by these authors so that the value of r_W for optimum average power density becomes 175 cm.

If we now substitute these quantities in equation (13.2) and use $P_W = 1300\ \text{W/cm}^2$, we find that the average power density for the reactor under consideration is $P_D = 3.5\ \text{W/cm}^3$. For economic considerations it is interesting to compare this figure with that of a fast breeder reactor. A 1000 MW(e) liquid metal cooled fast breeder reactor might carry a power rating of 2 to 2.5 W/cm^3 in the core; we note that this is of the same order of magnitude as that of our fusion reactor.

The maximum rated power output of the reactor can be calculated using the maximum permissible wall loading given above, *i.e.*, $P_W = 1300\ \text{W/cm}^2$. If we call this thermal power P_T then we note that it is given by

$$P_T = (2\pi R)\,(2\pi r_W)\,P_W = 4\pi^2\, r_W\, R\, P_W \qquad (13.4)$$

The above quantity has a minimum at $r_c = 0$ where, as shown in Figure 82, r_c represents the clearance radius of the central winding. The minimum P_T is therefore

$$(P_T)_{min} = 1300\ [4\pi^2\ (175)(350)] = 3100\ \text{Mw(Th)} \qquad (13.5)$$

and if we assume a thermal conversion efficiency, η, of 0.4 then the electrical output of our fusion system is 1250 MW(e). Clearly, it is quite difficult and perhaps impossible to design a magnet system with $r_c = 0$; so if we choose the power output of our system to be 5000 MW(Th) [corresponding to 2000 MW(e)] then with S = 50 we find that $r_c = 200$ cm which in turn gives R = 5.5 meters. The aspect ratio for this system is about 3 which appears to be practicable for toroidal confinement; an aspect ratio of 5 was used in the model reactors examined in Chapter 8. A generating station of the above size, *i.e.*, 5000 MW(Th) and 2000 MW(e), appears to be the size predicted for electrical generation units in the United States and Western Europe for the mid-eighties.

13.2 MAGNETIC FIELD AND PLASMA DENSITY

The determination of the magnetic field depends ultimately on the requirements of the plasma physics and at this point we will simply estimate it in order to examine related quantities such as containment parameters and cost estimates. We recall from Chapter 8 that the required magnetic field is strongly related to the confinement law underlying plasma containment. For example, in the case of the reference reactor, a 60-KG magnetic field was inadequate for Bohm-type confinement while it was adequate for classical and type B anamolous confinement. In the latter case we recall that the reactor had a power output of 300 MW(Th) with a corresponding β (ratio of plasma pressure to magnetic field pressure) of 0.12, and an aspect ratio of 5. In this section we shall examine the problem from the opposite end by first writing the expression for the nuclear power per cm length in a cylindrical geometry. The reactor power is simply the product of the fusion reaction rate and the plasma volume. For a cylinder with length L and plasma radius r_p the power is

$$P = \pi r_p^2 L R_F \tag{13.6}$$

with

$$R_F = \tfrac{1}{4} n^2 \langle \sigma v \rangle Q_F$$

where Q_F is the energy release per fusion reaction. Since

$$\beta = \frac{nkT}{B^2/8\pi}$$

we solve for n in terms of β and substitute in the above equation; the result for the power per unit length, $P_L = P/L$, is

$$P_L \sim \beta^2 B^4 r_p^{\,2} \langle\sigma v\rangle/(kT)^2 \qquad (13.7)$$

The above result seems to indicate, because of the dependence on B^4, that the highest possible magnetic field is desirable. This is an erroneous conclusion because of the limitations pointed out earlier. The wall loading P_W is dictated by the wall material; P_T is determined by the blanket requirements and by the size of a commercially viable reactor. Therefore P_L is indirectly determined by the above considerations and if the B^4 dependence in equation (13.7) is to be utilized, the remaining parameters must be changed to keep P_L constant. An alternative form to (13.7) which depicts more clearly the restriction on the magnetic field can be obtained by noting that in the temperature range of interest (*e.g.*, 10-30 keV), the product of the plasma density and confinement time $n\tau = CkT \simeq A$ where C and A are constants. Moreover, if we express τ in terms of the Bohm diffusion time, *i.e.*, $\tau = \alpha\tau_B$, then we can write $A = n\,\alpha\,\tau_B$ and for β we get

$$\beta \simeq \frac{nkT}{B^2} = \frac{n^2\,\alpha\,\tau_B}{B^2}$$

Substituting for n^2 from the above expression into P_L of equation (13.7) and recalling that $\tau_B \sim B/kT$ [see equation (8.44)] we find that

$$P_L \sim \frac{\beta kT}{\alpha} B \qquad (13.8)$$

The deciding factor in the choice of B is clearly related to cost which in turn is based on the state of development of the technology of superconductors. From the various examples studied in Chapter 8 it appears that B = 100 KG is reasonable and should be within reach of magnet technology.

We turn now to determining the plasma density. This quantity is determined by the engineering parameters P_W and r_W

obtained earlier. If the energy per fusion reaction, Q_F, is expressed in keV then the nuclear power density in the plasma is

$$P_n = 1.6 \times 10^{-16} \frac{n^2 \langle \sigma v \rangle}{4} Q_F \text{ watts/cm}^3 \tag{13.9}$$

and the power per cm length is given by

$$\pi r_p^2 \, P_n = 2\pi \, r_W P_W \tag{13.10}$$

For the D-T fusion reaction and including the energy from the Li^6(n,T) reaction, $Q_F = 17.6 + 4.8 = 22.4$ MeV, which if substituted in equation (13.9) and combined with equation (13.10) gives

$$n = 1.49 \times 10^6 \left(\frac{P_W}{\langle \sigma v \rangle r_W \, y} \right)^{1/2} \tag{13.11}$$

where $y = r_p/r_W$. The quantity y must be less than unity and if we take $y \simeq 0.7$ with $r_W = 1.75$, we get $r_p = 1.25$ which yields a clearance of 0.5 m between the plasma and the vacuum wall. With y, and knowing P_W and $\langle \sigma v \rangle$, the plasma density can be readily calculated. A larger value of y might be used but a true determination of this parameter depends on the performance of the confinement system.

13.3 THE CONTAINMENT PARAMETER $n\tau$

For zero net useful power output the product $n\tau$ is calculated from the Lawson criterion which has been detailed in both Chapters 2 and 11. We recall, for example, from equation (11.77) that this criterion can be written as

$$3nkT + P_b\tau = \eta\epsilon \, [\tfrac{1}{4} n^2 \langle \sigma v \rangle Q_F\tau + P_b\tau + 3nkT] \tag{13.12}$$

where P_b is the rate of radiation loss by bremsstrahlung and Q_F is the fusion energy referred to earlier. In Lawson's simple treatment fuel is injected into the reaction space with energy in excess of the plasma thermal energy, 3nkT, to compensate for the radiation losses P_b during time τ, other losses neglected. The values of $n\tau$ are then derived for zero net energy output, all output energy being used to heat the fuel at the high overall efficiency of $\eta\epsilon$ =

1/3. In other words the thermonuclear reaction output, $\frac{1}{4}n^2 \langle\sigma v\rangle Q_F \tau$, and the injected energy are all assumed recovered with overall efficiency $\eta\epsilon$ where ϵ is the fraction of output used for preheating fuel at overall efficiency η. This overall plant efficiency η is the product of the thermal conversion efficiency η_T and the injection efficiency η_I, and as pointed out earlier $\epsilon = 1$ gives Lawson's simple criterion. From Chapter 2, equation (2.15) we recall that

$$P_b = 5.35 \times 10^{-31} \zeta n^2 (kT)^{1/2} \text{ w/cm}^3 \tag{13.13}$$

where ζ is a correction for electron-electron bremsstrahlung (Wandel *et al.*, 1959). If we now combine equations (13.12) and (13.13) we get

$$n\tau = \frac{12 kT}{\frac{\eta\epsilon}{1-\eta} Q_F \langle\sigma v\rangle - 1.34 \times 10^{-14} \zeta (kT)^{1/2}} \tag{13.14}$$

which reduces to the familiar criterion when $\eta\epsilon = 1/3$. For a 50/50 D-T fueled reactor with a lithium blanket, $Q_F = 22.4$ MeV and a plot of the above equation for various $\epsilon\eta$ is shown in Figure 83.

Equation (13.14) may now be viewed as that representing the plasma power balance for an economic injection-heated fusion reactor; for no useful power output ϵ must be less than unity. The reaction rate $\langle\sigma v\rangle$ and the ratio P_b/P_n of bremsstrahlung to fusion power as functions of the temperature for D-T and D-D fusion reactors with lithium blankets are shown in Figure 84. As an alternate to injection heated reactor we can also consider a reactor which is heated in the steady state by the charged particles of the fusion reaction. We have examined such reactors in Chapter 8, and in the case of D-T fuel the heating is achieved by the 3.52 MeV alpha particle. For such systems we can write a simple power balance condition comparable to Lawson's treatment of injection heating; it is

$$\frac{1}{4} n^2 \langle\sigma v\rangle Q_c\tau - P_b\tau = 3nkT \tag{13.15}$$

where Q_c in the case of a D-T reactor is 3.52 MeV. If we substitute for P_b from equation (13.13) the above balance relation reduces to

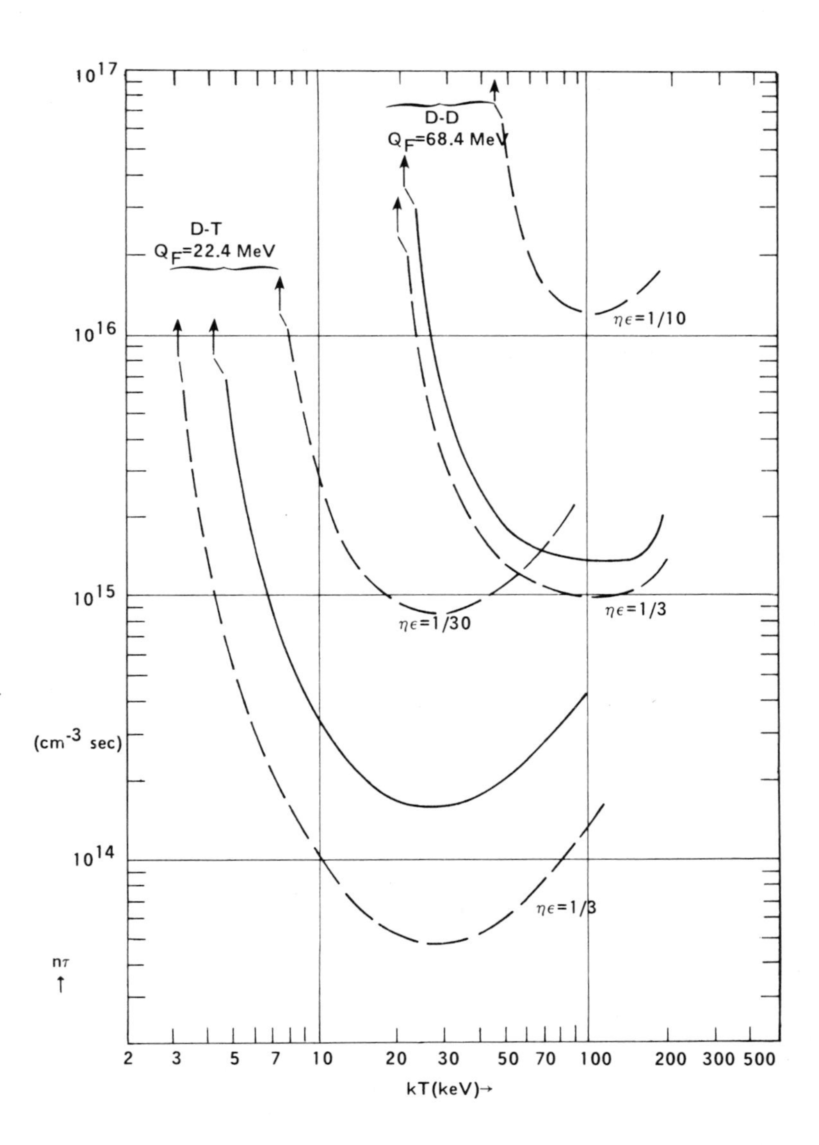

Figure 83. $n\tau$ versus temperature for injection heating (dashed) and charged particle heating (solid).

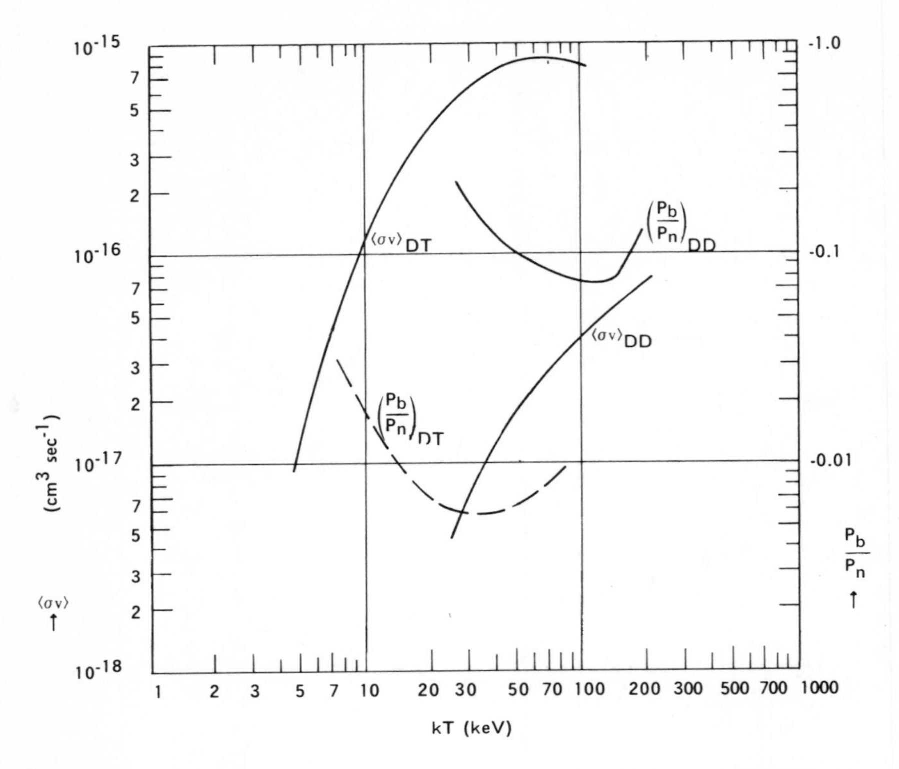

Figure 84. Reaction rate and the ratio of bremsstrahlung to fusion power as a function of temperature.

$$n\tau = \frac{12\ kT}{Q_c\ \langle\sigma v\rangle - 1.34 \times 10^{-14}\ \zeta(kT)^{1/2}} \tag{13.16}$$

which is plotted in Figure 85 using the values of $\langle\sigma v\rangle$ from Figure 84. The $n\tau$ from equation (13.16) is shown by curve A while that denoted by curve B takes into account in simplified terms radiation loss due to cyclotron radiation (Kofoed-Hansen, 1960).

Having calculated the required density for our reactor, *i.e.*, equation (13.11) we can now derive the "full power operation" containment time, $\tau = n\tau/n$. However, we must bear in mind that this value of τ is only an indication of the system requirement; at a reduced power, n might be lower and τ larger. In any case the design will have to be capable of supporting longer confinement times to permit control. We recall from Chapter 8 that

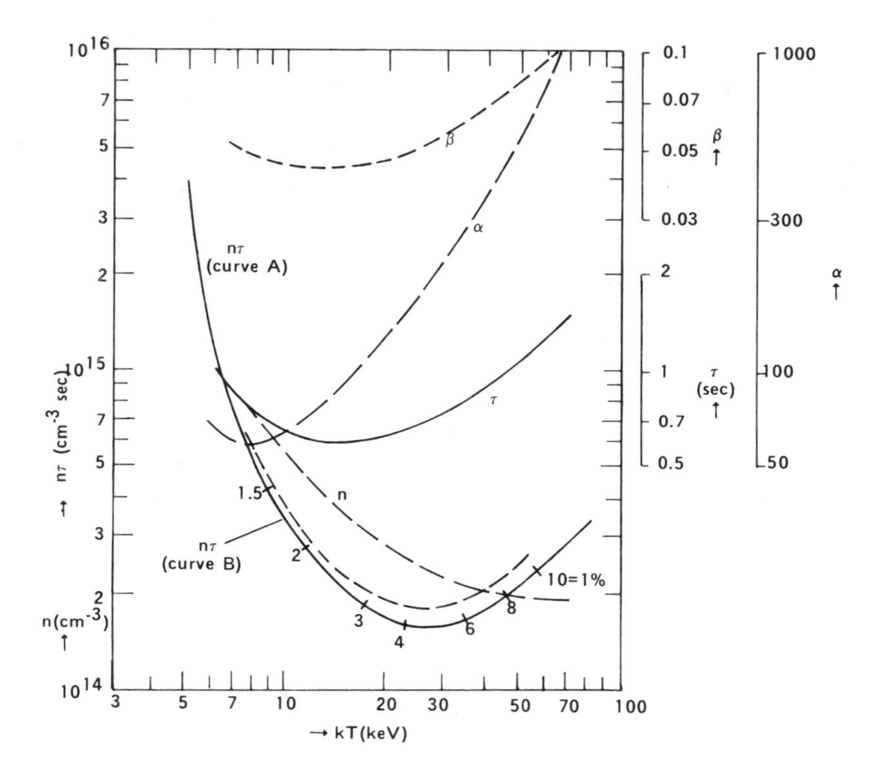

Figure 85. Plasma parameters for charged particle heated D-T reactor.

a charged particle heated D-T reactor operating on a "drift wave" confinement time is thermally stable; it is self-regulating in that the reactor always returns to its equilibrium operating conditions whenever an excursion in the plasma temperature (or density) takes place. This implies that the confinement time varies with the plasma temperature in such a way as to maintain stability, *i.e.*, the reactor is capable of operating over a wide range of τ values.

The remaining confinement parameters for our reactor can now be calculated for B = 100 KG. They are

$$\beta = \frac{2nkT}{B^2/8\pi}$$

the ratio of plasma to magnetic pressure, and $\alpha = \tau/\tau_B$, the ratio of τ to the Bohm diffusion time τ_B. In addition, the burn-up fraction f can be calculated from equation (2.38) of Chapter 2 and has the form f = ½ nτ $\langle\sigma v\rangle$ x 100. These results for a charged particle-heated reactor are shown in Figure 85 while those for an economic injection-heated reactor with $\eta\epsilon$ = 1/30 are shown in Figure 86.

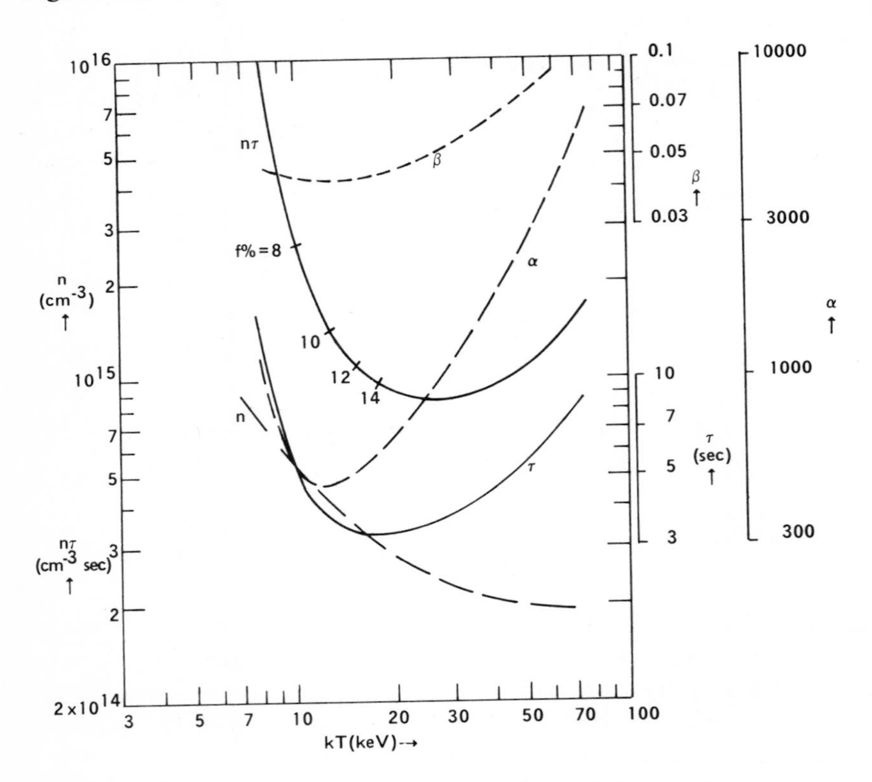

Figure 86. Plasma parameters for injection heated D-T reactor.

We note from Figure 85 that in order to operate a charged particle-heated D-T reactor at short containment times, the plasma temperature should be between 10 and 20 keV with preference for the higher temperature because of lower bremsstrahlung on the wall—as indicated in Figure 84. At 20 keV both n and nτ are relatively low, and the burn-up fraction is about 3.5%, a value not so low as to aggravate the tritium problem. Moreover at this temperature, β is close to its minimum value and the required containment time is about 120 Bohm diffusion times.

By contrast, Figure 86 for an injection-heated D-T reactor shows that at a temperature of 13 keV the burn-up fraction is more than 10% and α is about 470. The data for both heating systems are shown in Table 30 where the parameter β is corrected to include alpha particles. These results have been obtained on the basis of a simplified analysis in which the electron and ion temperatures are taken equal and the confinement time is the same for all ionic components. The containment parameters for the charged particle-heated reactors of Chapter 8 are more meaningful since distinction was made between all the charged particles—and between fuel and ash.

Table 30

Containment Parameters for D-T Reactor, 5000 MW(Th)

Plant efficiency $\eta = 0.42$, B = 100 KG
Plasma radius $r_p = 1.25$ m
Wall loading $P_W = 1300$ W/cm^2
Wall radius $r_W = 1.75$ m
$\epsilon = 7\%$

Item	*Units*	*Plasma Heating*	
		Charged Particle	*Injection*
$n\tau$	sec/cm^3	1.7×10^{14}	1.4×10^{15}
Temp	keV	20	13
n	cm^{-3}	2.8×10^{14}	4×10^{14}
τ	sec	0.6	3.5
τ_B	sec	4.9×10^{-3}	7.5×10^{-3}
α	τ/τ_B	120	470
β	$2mk\,T/B^2/8\pi$	0.075	0.043
f	percent	3.5	10

13.4 MIRROR REACTOR FEASIBILITY

It is interesting and useful to carry out a similar analysis on a mirror reactor and compare its feasibility with the toroidal (Tokamak) reactor discussed above. In order for the comparison to be meaningful it would be desirable to examine the experimentally obtained parameters for a mirror machine (2X-II device at the Lawrence Livermore Laboratory in California) and a Tokamak device (the T-3 machine at the Kurchatov Institute in the USSR) shown in Table 31.

Table 31

Experimentally Obtained Results

	Tokamak	*Mirror Machine*
Plasma density	5×10^{13} cm^{-3}	5×10^{13} cm^{-3}
Plasma temperature	0.5 keV	8 keV
Confinement time	25 m sec	0.4 m sec
Beta	0.03	0.05
Plasma volume	1 m^3	0.005 m^3

We note from these results that, even though the temperature and beta are larger for the mirror machine, the parameters of the Tokamak put it closer to the Lawson criterion. At any rate our objective here is to examine the feasibility of a mirror machine as a fusion reactor and compare it with Tokamak feasibility. The first comparison can be made by examining Figure 87 where the Lawson plot for a charged particle-heated reactor is shown in the upper part of the figure. This is the same curve drawn in Figure 85, moved here to show the fusion regions of the Tokamak and mirror. The plasma confinement in the Tokamak is in accordance with the neoclassical theory mentioned in Chapter 8, and it is shown here for various values of the "safety factor," q, and various values of beta. Complete stability is assumed here. The diagonal straight lines represent the best confinement conditions in mirrors (Kuo-Petrovic, *et al.*, 1970), and as we see they do not intersect the upper Lawson plot. This means that there is no region for energy production from charged particle-heated mirror

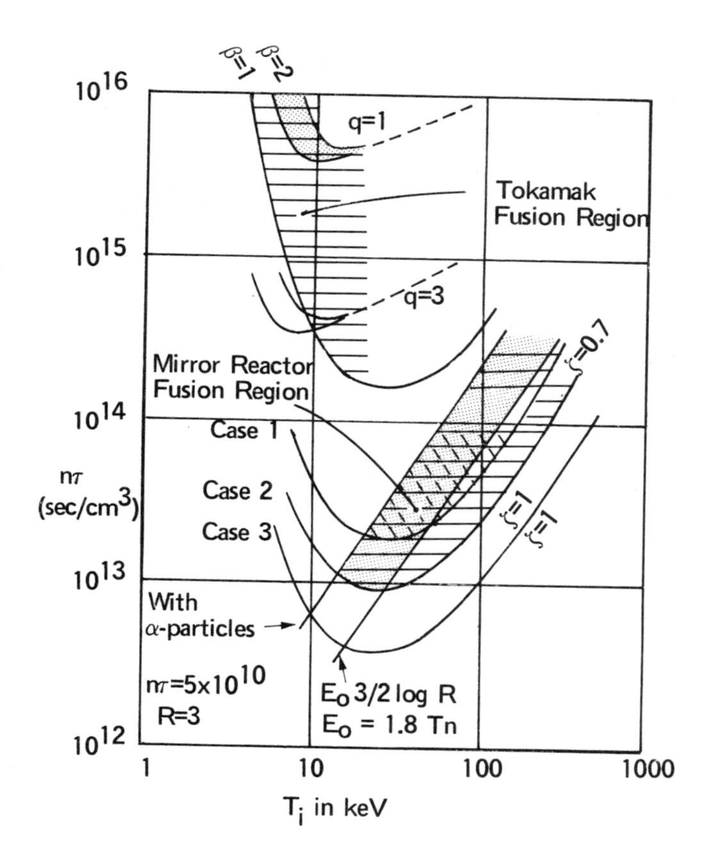

Figure 87. $n\tau$ versus temperature for mirror and Tokamak reactors (Golovin, 1970). Case 1: $\eta_1 = \eta_{DC} = 0.9$, $\eta_T = 0.4$, $E_T = 22.4$ MeV. Case 2: $\eta_1 = 0.9$, $\eta_{DC} = \eta_T = 0.6$, $E_T = 30$ MeV Case 3: $\eta_1 = \eta_{DC} = 0.9$, $\eta_T = 0.6$, $E_T = 30$ MeV.

reactors. This is due primarily to the loss of ions through the mirrors after the first collision and the corresponding decrease in the fusion reactions needed to maintain the plasma temperature. On the basis of this it seems clear that an economically feasible fusion reactor cannot be made from charged particle-heated mirror machine. On the other hand a steady state of a hot plasma can be maintained in mirrors with sufficient injected energy.

In order to assess the feasibility of such a mirror reactor we will deduce a Lawson criterion for such a system and we shall represent it by the flow diagram shown in Figure 88. Direct conversion (see Chapter 9) represented by the efficiency η_{DC} will be assumed as part of the system, and a portion ϵ of the total power output will be used for injection.

The energy balance relation for this system can be written as

$$\epsilon_o = \tfrac{1}{4} n^2 \langle \sigma v \rangle Q_c = \frac{3}{2} n \frac{T_e + T_i = T_t}{\tau} + P_b \tag{13.17}$$

where ϵ_o is the rate of energy injection and P_b is the power due to bremsstrahlung radiation. The rate of injection is

$$\epsilon_o = \eta_I P_I = \eta_I \epsilon P_e = \eta_I \epsilon \left(P_{e_1} + P_{e_2}\right) \tag{13.18}$$

but

$$P_{e_1} = \eta_{DC} \frac{3}{2} n \frac{T_t}{\tau}$$

$$P_{e_2} = \eta_T P_b + \eta_T \frac{n^2}{4} \langle \sigma v \rangle Q_N$$

so that equation (13.18) becomes

$$\epsilon_o = \eta_T \epsilon \left\{ \eta_{DC} \frac{3}{2} n \frac{T_t}{\tau} + \eta_T P_b + \eta_T \frac{n^2}{4} \langle \sigma v \rangle Q_N \right\} \tag{13.19}$$

If we now substitute in equation (13.17) we get

$$\frac{3}{2} \frac{nT_t}{\tau} \left\{ \eta_I \eta_{DC} \epsilon - 1 \right\} + P_b \left\{ \epsilon \eta_I \eta_T - 1 \right\} + \frac{n^2}{4} \langle \sigma v \rangle \left\{ \epsilon \eta_I \eta_T Q_N + Q_c \right\} = 0$$

which upon solving for $n\tau$ yields

$$n\tau = \frac{6(T_e + T_i)\,[1 - \epsilon \eta_I \eta_{DC}]}{\langle \sigma v \rangle\, [Q_c + \epsilon \eta_I \eta_T Q_N] - [1 - \epsilon \eta_I \eta_T]\, \frac{4P_b}{n^2}} \tag{13.20}$$

We readily note from this expression that an increase in ϵ results in lowering the $n\tau$ (or the Lawson) plot and vice versa. In fact, calculations have shown on the basis of this equation that at an ϵ less than 0.6 - 0.7 there is no intersection between the Lawson plot and the diagonal lines and hence no reliable fusion region. This may sound ominous for the mirror as a fusion reactor

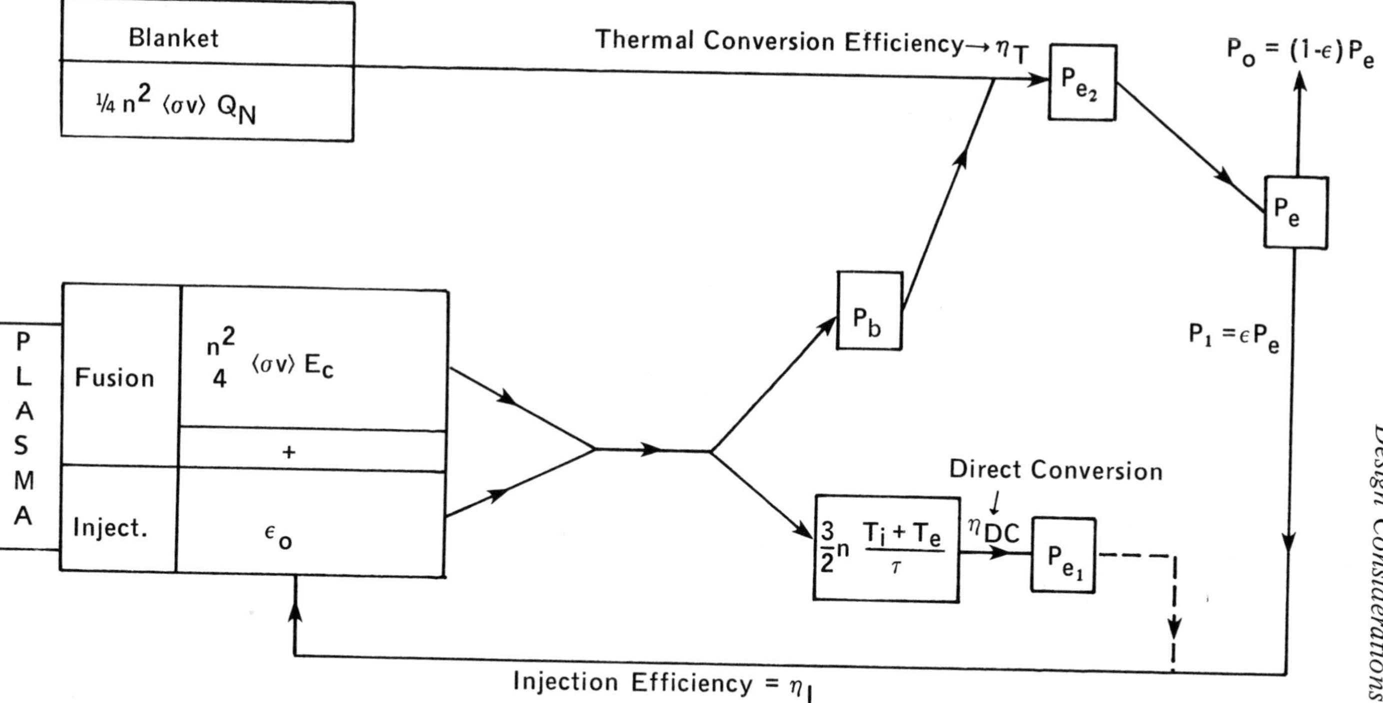

Figure 88. Power flow diagram for a mirror reactor.

since for modern power plants economic recirculation of power is taken at 10-12 per cent and not at 60-70 per cent. On the other hand injection energy could be fully supplied by the directly converted energy of particles lost through the mirrors (as indicated by the dotted line in Figure 88) provided the Q value of the plasma (see Chapter 9) has the value

$$Q > \frac{E_T}{E_o}\left(\frac{1}{\eta_I \eta_{DC}} - 1\right) \tag{13.21}$$

where E_T is the total energy release from fusion including the secondary reactions in the blanket (22.4 MeV for D-T-Li reactor) and E_0 is the injected energy. For the D-T fuel cycle with E_T = 22.4 MeV and $\eta_I = \eta_{DC} = 0.9$ it is necessary to have Q = 1.5. This is quite feasible and corresponds to $\epsilon = 0.7$ at $\eta_T = 0.4$. If the above condition is met the fusion is sustained by the direct conversion, and the total power obtained through the usual thermal cycle from the heat in the blanket and on the vacuum wall is the net power output except for a small part which is spent for the usual auxiliaries. It is interesting to note that the mirror power cycle as illustrated in Figure 88 is quite complicated compared to that of a Tokamak. In the latter, after the plasma has been heated (by some method) to the ignition temperature, it is possible (as we have seen in Chapter 8) to introduce the cold fuel into the plasma, to remove the heat from the vacuum wall and blanket and generate electricity through the usual thermal cycle (*e.g.*, heat exchanger-boiler-turbo-generator).

13.5 MAIN DESIGN PARAMETERS OF A MIRROR REACTOR

We recall from Chapter 1 that simple mirror geometry is susceptible to MHD instabilities and that mirrors with "magnetic wells" are stable against these modes. The latter type of geometry is quite complex, expensive and makes injection quite difficult. For the purposes of this discussion we will assume the simple axially symmetric fields and assume that plasma stability can be achieved by some other method (*e.g.*, feedback stabilization).

As in the case of Tokamak we let P_W be the reactor wall loading and note that in the case of mirrors the charged particles do not contribute significantly to this loading since they escape through the mirrors. In other words the wall is heated primarily by bremsstrahlung, cyclotron radiation and a fraction of the neutron energy. Nevertheless we shall take the maximum loading $(P_W)_{max} = 1300\ W/m^2$ as in the Tokamak because at this condition the radiation damage and hence the wall lifetime in both reactors will be the same.

The total thermal power of the reactor is given by

$$P_T = 2\pi r_W L P_W \tag{13.22}$$

where L is the distance between the mirrors, and the remaining parameters are as defined before. The distance L is proportional to the coil radius x and depends on the mirror ratio R (ratio of the magnetic field strength at the mirror to that in the center) through some function f(R), *i.e.*,

$$L = x f(R) \tag{13.23}$$

At R=3, f(R) = 3, and at R=5, f(R) = 3.8 provided the coil cross section is small enough as it would be with superconductors. We further assume that the coil radius is related to the vacuum wall radius by the relation

$$x = \frac{r_W + b}{1 - \frac{1}{2}\alpha} \tag{13.24}$$

where b is the blanket thickness, and $\alpha = t/x$ is the nondimensional coil thickness. If we now take the blanket thickness to be 120 cm, almost the same as that of the Tokamak, we can readily express r_W in terms of P_T and the results are shown in Figure 89.

The plasma density can be obtained by equating P_T in equation (13.22) to the fusion power from the plasma, *i.e.*,

$$P_T = \tfrac{1}{4} n^2 \langle\sigma v\rangle Q_F \pi r_W^2 y^2 L \tag{13.25}$$

where Q_F is the total energy release per fusion (22.4 MeV for D-T), and y is [see equation (13.11)] the ratio of the plasma radius to the wall radius. This result is shown in Figure 90 for

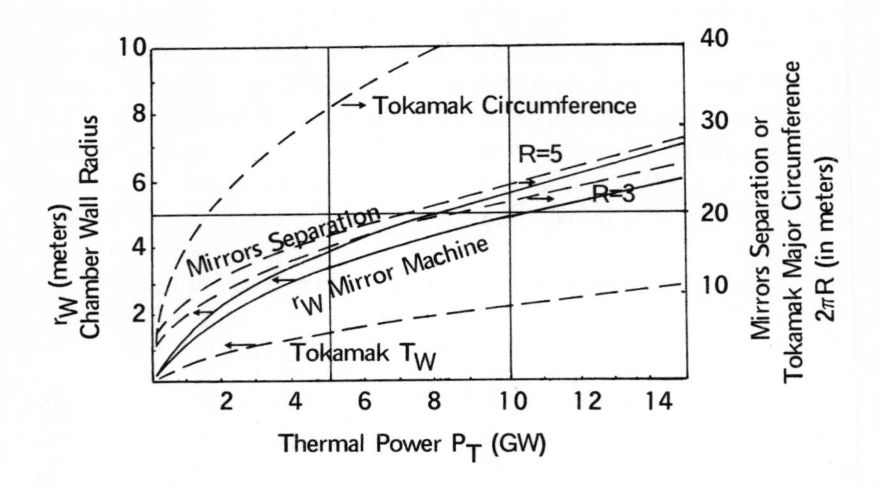

Figure 89. Wall radius versus thermal power.

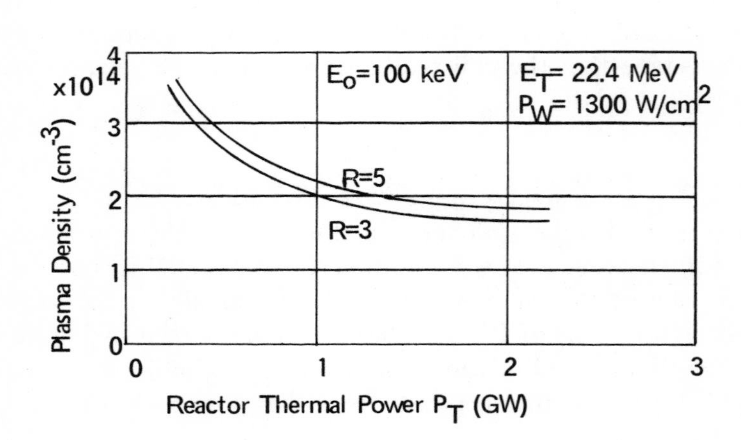

Figure 90. Plasma density versus thermal power.

different mirror ratios and at an injection energy of 100 keV. Using these values of the plasma density one can calculate the maximum power containment time by utilizing equation (13.20) for $n\tau$. Again, the discussion concerning this parameter carries over from that of Tokamak in that the system must be capable of supporting a range of values of $n\tau$ in order to provide for control.

The most important parameter in a mirror reactor is perhaps the magnetic field which, as in the case of Tokamaks, is also limited by the characteristics of superconductors. In terms of β in the center of the machine the magnetic field according to the calculations of Kuo-Petravic *et al.* (1969) can be put in the form

$$B_m = R\left[6.4\pi\,\frac{n E_o}{\beta}\right]^{1/2} \tag{13.26}$$

and in the presence of plasma the mirror ratio R_p differs from its vacuum value R by

$$R_p = \frac{R}{\sqrt{1-\beta}} \tag{13.27}$$

A plot of the magnetic field as a function of the reactor thermal power is thown in Figure 91 which also shows a comparable curve for Tokamak.

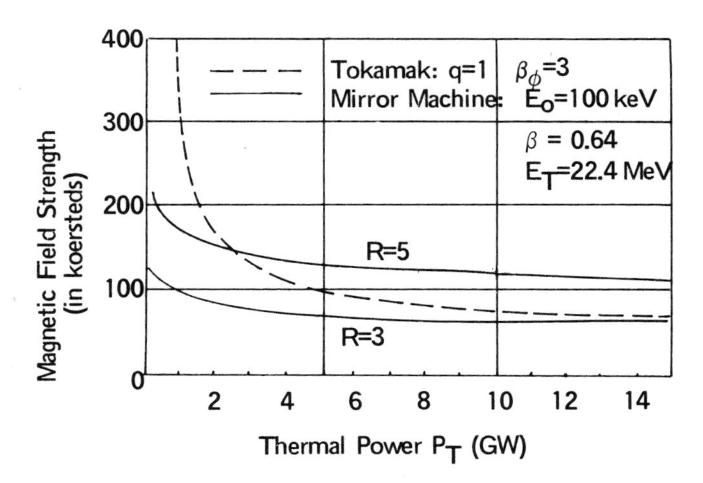

Figure 91. Magnetic field strength versus thermal power.

The injection requirement for maintaining a constant plasma density n in a mirror reactor can be estimated by the formula

$$I = \frac{nV}{\tau} = \frac{n^2 V}{n\tau}$$

where I is the injection current and V is the plasma volume. The commonly used expression for $n\tau$ in mirrors is

$$n\tau = 5 \times 10^3 \; E_o^{3/2} \log R_P$$

with E_o being the injection energy in keV. With this, the injection current becomes, utilizing equation (13.25)

$$I = \frac{4 \; P_T}{3.5 \times 10^{10} \; E_o^{3/2} \langle\sigma v\rangle \; Q_F} \tag{13.28}$$

and at E_o = 100 keV, and Q_F = 22.4 MeV it further becomes

$$I = 6 \times 10^{-6} \; P_T$$

For a reactor with a total power P = 600 MW(Th), we get I = 3600 equivalent amps, r_W = 1 meter, L = 7 meters, and n = $2.5 \times 10^{14}/cm^3$. Table 32 shows how close (or far) the present day 2X-II mirror machine and T-3 Tokamak are from reactor conditions.

Table 32

Plasma Parameters for Mirror and Tokamak Reactors

	Tokamak		*Mirror Reactor*	
	Necessary	*Obtained*	*Necessary*	*Obtained*
Plasma density	$3 \times 10^{14}/cm^3$	$5 \times 10^{13}/cm^3$	$2 \times 10^{14}/cm^3$	$5 \times 10^{13}/cm^3$
Plasma temperature	10 keV	0.5 keV	~ 100 keV	8 keV
Containment time	700 msec	25 msec	200 msec	0.4 msec
Beta	$\geqslant 0.3$	~ 0.03	$\geqslant 0.5$	0.05
Plasma volume	$\geqslant 200$ m^3	1 m^3	20 m^3	0.005 m^3

It might be pointed out that a disadvantage of Tokamak is that it is impossible to make a low power reactor out of it; the reactor we examined earlier with a power rating of 5000 MW(Th) or 2000 MW(e) is about the minimum feasible size. Mirror reactors with several hundred megawatts are possible and with the realization of highly efficient direct conversion, we recall from Chapter 9 that a D-He^3 mirror reactor can have as high as 90% plant thermal efficiency. The major advantage of Tokamak is its simplicity and this may be illusory because it is difficult to build closed-line magnetic configurations.

13.6 DESIGN CHARACTERISTICS OF A PULSED FUSION REACTOR

The reactors examined in the previous two sections are envisaged to operate in the steady-state mode and are generally characterized as low-beta systems. A system which operates in the pulsed mode and at high values of beta is expected to be based on the "theta-pinch" concept and this section will be devoted to the examination of the underlying principle of such a reactor. By contrast to the so-called Z-pinch where pinching of the plasma results from an azimuthal magnetic field generated by a current flowing along the (Z) axis of the cylinder, the magnetic field in the theta-pinch arises from currents flowing in the (θ) azimuthal direction. Because it is a pulsed system the necessary confinement time can be short and hence the plasma density and correspondingly the plasma beta can be substantially larger than those in steady-state systems.

The conceptual design of the Los Alamos Reference Theta-Pinch Reactor (RTPR) calls for a staged θ-pinch illustrated in Figure 92 in which two separate sources of energy are used. In the shock heating stage a magnetic field B_s having a rise time of a few hundred nano-seconds and a magnitude of a few tens of kilogauss drives the implosion of a fully ionized plasma whose initial density is of the order of 10^{15} cm^{-3}. The ion energy associated with the radially directed motion of the implosion subsequently thermalizes and the plasma assumes a temperature T_E characteristic of the equilibration of ions and electrons in accordance with equation (4.104). In Figure 92 the shaded areas represent magnetic fields perpendicular to the plane of the paper. After a few microseconds the adiabatic compression field—

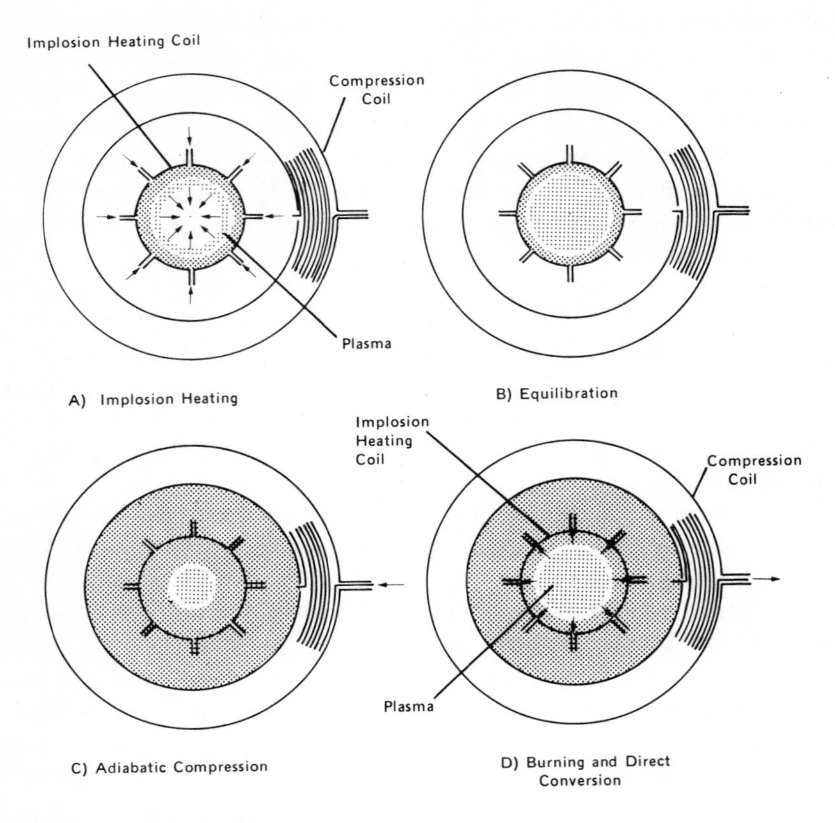

Figure 92. Staged theta pinch reactor.

characterized by a rise time of a few milliseconds and a maximum value B_o of 100-200 KG–is applied by energizing the compression coil shown in Figure 92. In that figure the arrow indicates the direction of magnetic energy flow into the system. In the final compression stage the plasma radius is substantially reduced and its temperature is raised to a value T_o = 10-20 keV. As the D-T plasma burns for several tens of milliseconds it produces 3.5 MeV alpha particles which partially thermalize with the D-T ions and the electrons as the burned fraction of the plasma increases to about 11%. As a result the plasma is further heated. If we assume that the ratio of plasma pressure to magnetic field pressure, β, is approximately unity, and that the magnetic field B remains constant with the value B_o, the plasma expands against the magnetic field, doing work ΔW which is approximately 7% of the fusion energy produced by the D-T reaction. This work produces an emf (electromotive force) which forces magnetic energy out of the compression coil (as illustrated in Figure 92d) and back into the compression magnetic energy store. In this sense there is direct energy conversion in the theta-pinch reactor as in the mirror system previously examined.

Thermonuclear burn in this system is considered to begin when the compression field reaches its final value B_o. If we include the energy produced in the blanket due to the $Li^6(n,\alpha)T$ reaction, *i.e.*, 4.8 MeV, then the energy produced by each neutron from the D-T fusion reaction is $Q_n = 18.9 \text{ MeV} = 2 \times 10^{-12}$ J. For a plasma of radius "a" and $n_D = n_T = n/2$ the energy produced per unit length of the blanket is

$$P_n = \frac{\pi}{4} a^2 n^2 Q_n \langle\sigma v\rangle \tag{13.29}$$

If we now express B_o in KG, then before appreciable burn-up occurs, pressure balance at $\beta = 1$ requires that

$$n_o kT_o = 1.24 \times 10^{13} B_o^2 \quad (\text{keV/cm}^3) \tag{13.30}$$

which when substituted in (13.29) yields

$$P_n = 1.94 \times 10^5 B_o^4 a^2 Q_n \langle\sigma v\rangle/(kT_o)^2 \frac{\text{MW}}{\text{m}} \tag{13.31}$$

where the density does not appear explicitly and "a" is in cm. For $kT_o = 15$ keV, the quantity $Q_n \langle\sigma v\rangle/(kT_o)^2$ has the value 2.2×10^{-14} cm^3/keV-sec.

In order to maximize the reactor power output, equation (13.29) reveals that the density n_o must be made as large as possible which is equivalent to making B_o as large as possible since the power production scales as B_o^4 whereas joule losses (proportional to the square of the current) scale as B_o^2. Magnetic field technology and mechanical limitations will probably limit B_o to 100-200 KG and at $B_o = 150$ KG for kT_o between 10 and 20 keV the plasma density is fixed at 7.8×10^{16} which is in the range encountered in present day θ-pinch experiments.

Being a pulsed system the above reactor is expected to produce power in a repetitive operating mode. The cyclic time history of the main events is shown in Figure 93. The cycle begins with the application of the implosion field B_s. This field

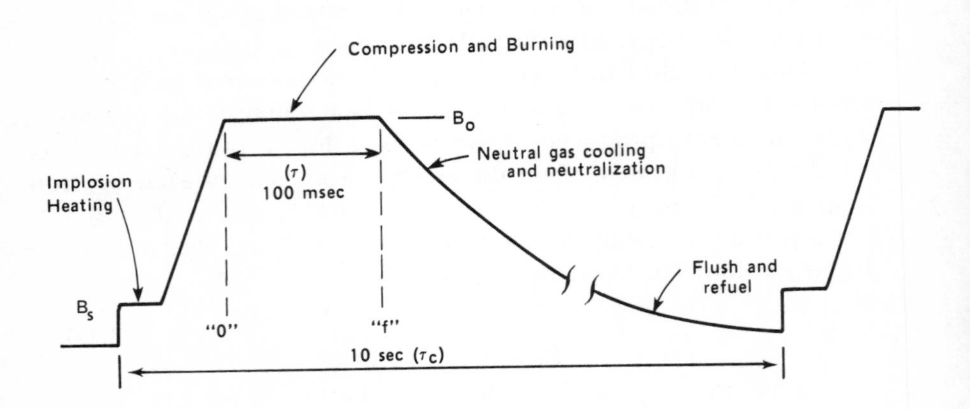

Figure 93. Cycle and burn times in a staged theta pinch reactor.

is applied suddenly and sustained for long enough time for the equilibrated plasma at temperature T_E to be "picked up" by the compression field. The rise time of the compression field B_o is taken as 10 milliseconds and this field is held constant during the burn time τ of 100 sec.

The burning process in the plasma begins at state "0" shown in Figure 93, and ends at "f" at which state the plasma contains approximately 11% helium from the D-T reaction. The magnetic field is subsequently relaxed to some lower value which allows for the plasma column to expand radially to the vicinity of the wall and extinguishes the burn. At that point neutral gas is made to flow between the wall and the plasma boundary, removing heat from the column and neutralizing the plasma. In this "off time" portion of the cycle the plasma and hot gas are flushed out and replaced by fresh plasma with negligible helium content.

A reactor duty factor (ξ^{-1}) is defined such that the quantity ξ represents the ratio of the cycle time τ_c to the burn time τ. This ratio determines the average thermal power loading P_W on the reactor first wall which in this case is the implosion heating coil (see Figure 92). Because of radiation damage limitations such as those delineated in the next chapter, and of heat transfer limitations imposed by lithium as coolant as described in Chapter 15, the cycle time τ_c is chosen to limit P_W to 3.5 MW/m^2 (350 W/cm^2) giving the reactor a duty factor of 1%. During the burning pulse (τ) the wall loading exceeds the above value by about two orders of magnitude and heat transfer may not balance the energy deposition in this short interval of time. Care must therefore be exercized to design for reasonable temperature rises and thermal gradients in the first wall and in the blanket materials. We will note in the next chapter for example that sudden heat deposition could among other things lead to serious wall erosion by evaporation.

One of the most important features of a pulsed reactor with small duty factor is the near absence of electromagnetic pumping losses. As will be observed in Chapter 15, the use of lithium as a heat transfer medium could require a pumping power of the order of 10% of the reactor electrical power output due to the pumping of lithium across magnetic fields. In a pulsed reactor during heat transfer, the lithium need not be pumped across the magnetic field as in a steady-state reactor; and the flow is not along the magnetic lines (where losses are comparatively small) since the field is absent. Hence if the flow is initially turbulent it retains its character which allows the field-free heat transfer coefficient in a pulsed reactor to be substantially higher than is possible in steady-state systems.

The scaling laws presented earlier are accurate for a plasma of unvarying properties, *i.e.,* for negligible D-T burn up. But as burn proceeds energetic (3.5 MeV) alpha particles are produced which constitute a major energy source to the plasma. The large energy deposition and subsequent thermonuclear burn in the θ-pinch reactor is a nonlinear and dynamic process which requires computer solution for realistic answers. Such calculations have been carried out for RTPR by the Los Alamos group and some of the results are shown in Figure 94. This figure shows the electron and ion temperature and densities as a function of the burn time τ. We note that the initial reaction rate produces a rapid buildup of α-particles, and because of their large energy they

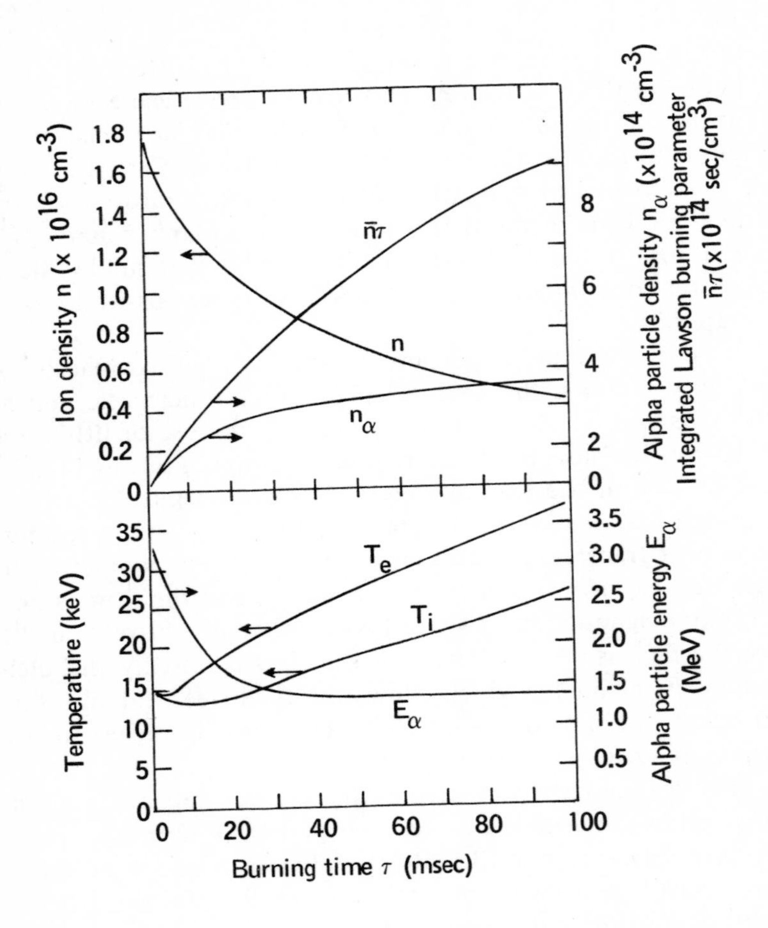

Figure 94. Electron and ion temperatures, alpha particle energy and $n\tau$ as a function of burn time τ.

constitute a significant portion of the total particle pressure. This large partial pressure causes the plasma column to expand and as a result of this essentially adiabatic decompression the electron and ion temperatures drop during the first few msecs of the burn. However, as the burn continues the alpha particles heat the electrons and the ions with an appreciable portion of the energy going preferentially to the electrons as we noted in Chapter 4. Only when the electrons become sufficiently hot does a larger fraction of the energy go to the ions. This is not likely in the short burn time indicated and the electron temperature always leads the ion temperature in spite of the bremsstrahlung radiation cooling by the electrons.

Although we have in Chapter 4 the formula by which we can calculate the α-particle heating, we can estimate this effect by simply calculating the pressure increase in the plasma column which would occur if expansion were prevented (for example by increasing the external field B_o). After the burn the final particle pressure can be written as

$$P_f = P_o + 2/3\, n_{\alpha f} E_{\alpha b} \qquad (13.32)$$

where $P_o = 2n_o kT_o$ is the particle pressure at the start of the burn, $n_{\alpha f}$ is the final number density of the α-particles, and $E_{\alpha b}$ is α-particle energy at birth, *i.e.*, 3.5 MeV. The fractional burn-up of fuel ions is given by

$$f_B = \frac{2n_{\alpha f}}{n_f + 2n_{\alpha f}} \qquad (13.33)$$

where n_f is the final ion density. In the case of zero expansion,

$$n_f = n_o - 2n_{\alpha f} \qquad (13.34)$$

since two ions are required to produce one α-particle. With (13.34), equation (13.33) can be put in the form

$$f_B = \frac{n_{\alpha f}}{n_o} \qquad (13.35)$$

which upon insertion in (13.32) gives

$$\frac{P_f}{P_o} = 1 + \frac{f_B\, E_{\alpha b}}{6\, kT_o} = 1 + 39.1\, f_B \qquad (13.36)$$

for $E_{\alpha b}$ = 3.5 MeV and kT_o = 15 keV.

Equation (13.36) is plotted in Figure 95 where we note that the internal pressure would increase five-fold for a 10% burn-up fraction and about forty-fold for complete burn-up. This pressure

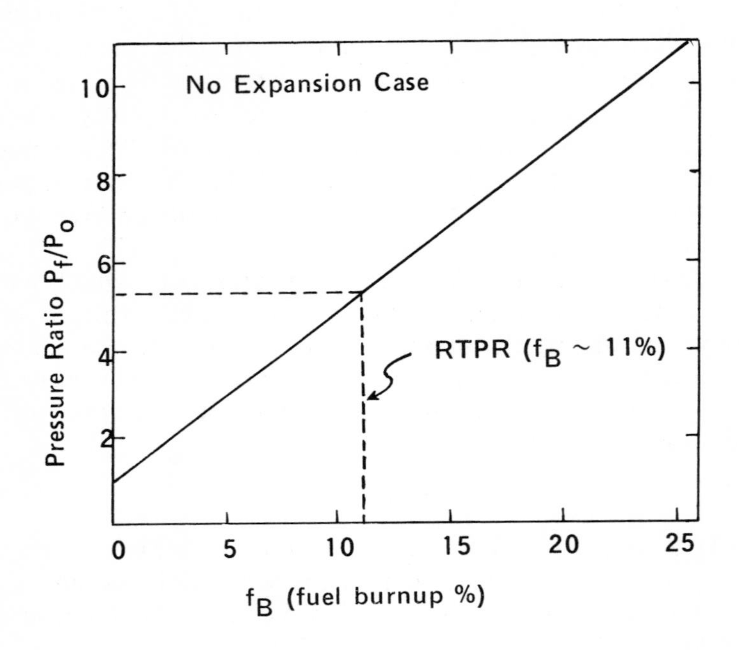

Figure 95. Variation of pressure ratio P_f/P_o with burn-up fraction for no expansion.

is available to do work against the magnetic field. For the case of β = 1 plasma and a constant magnetic field B_o, the α-particle pressure will cause the plasma to expand from an initial radius "a_o" to a final radius "a_f" doing an amount of work given by

$$\Delta W = \pi(B_o^2/8\pi)(a_f^2 - a_o^2) \qquad (13.37)$$

This work is available to the source of the driving magnetic field at essentially 100% efficiency and, as we recall, it amounts to about 7% of the fusion energy thus providing the high β theta-pinch reactor with a significant direct conversion process. The effect of the plasma expansion on the plasma radius, bremsstrahlung power, and thermonuclear power production is shown in Figure 96 where the results have been taken from a burn-up computer code for RTPR. Since the system is pulsed, the power production by fusion and loss by bremsstrahlung are the time integrated values

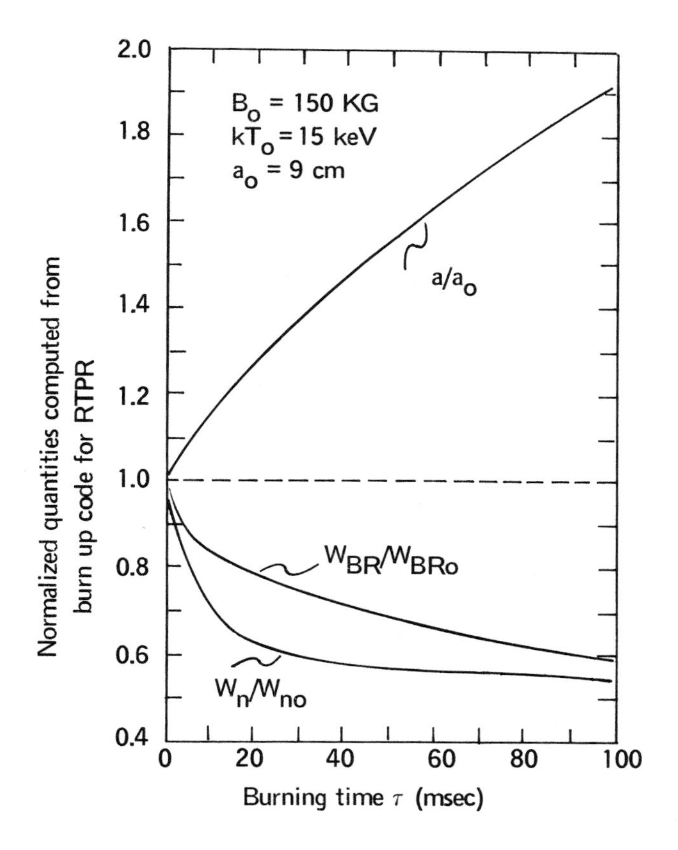

Figure 96. Plasma radius, neutron energy production and bremsstrahlung energy as a function of burn time.

but normalized to values corresponding to the initial conditions. For example the quantity

$$\frac{W_n}{W_{no}} = \frac{\int_0^{\tau} P_n(t)\, dt}{P_{no}\, \tau} \tag{13.38}$$

where the denominator is calculated from equation (13.31) assuming initial conditions a_o and kT_o and multiplying by the burn time (τ). In the case of the radiated power the denominator is obtained from equation (2.16) which for kT_o in keV is given by

$$W_{BRo} = 1.72 \times 10^{-34}\ a_o^2\ n_o^2\ (kT_o)^{1/2}\ \tau \quad \text{MJ/m} \qquad (13.39)$$

A summary of the plasma conditions before and after the 100 msec burn is given in Table 33.

Table 33

RTPR - Plasma Parameters During Burn ($B_o = B_f = 150$ KG)

Symbol	*Definition*	*Value*
a_o	initial plasma radius	9.0 cm
a_f	final plasma radius	17.33 cm
kT_o	initial electron and ion temperature	15 keV
kT_{if}	final ion temperature	26.7 keV
kT_{ef}	final electron temperature	37.5 keV
$E_{\alpha f}$	final α-particle mean energy	1.39 MeV
n_o	initial ion and electron density	1.86×10^{16} cm^{-3}
n_f	final ion density	4.37×10^{15} cm^{-3}
n_{ef}	final electron density	4.91×10^{15} cm^{-3}
$n_{\alpha f}$	final α-particle density	2.72×10^{14} cm^{-3}
$\overline{n\tau}$	integrated Lawson parameter	8×10^{14} sec/cm^3
f_B	fuel burn up fraction	11%
τ_c	cycle time	10 sec
τ	burn time	0.10 sec
ξ^{-1}	duty factor	1%

We turn now to an examination of the power plant based on the theta-pinch reactor concept for which the energy flow diagram is illustrated in Figure 97. The various energies appearing in the diagram are defined in Table 34. In considering the overall energy balance for the reactor, account is taken of the core and compression coil joule losses and an estimate is made of the expenditure of energy W_A used in driving the electrical generating plant and auxiliaries such as coolant pumps. There is also an energy loss W_{sc} in the cryogenic circuit which will depend on the choice of the super-conducting material used.

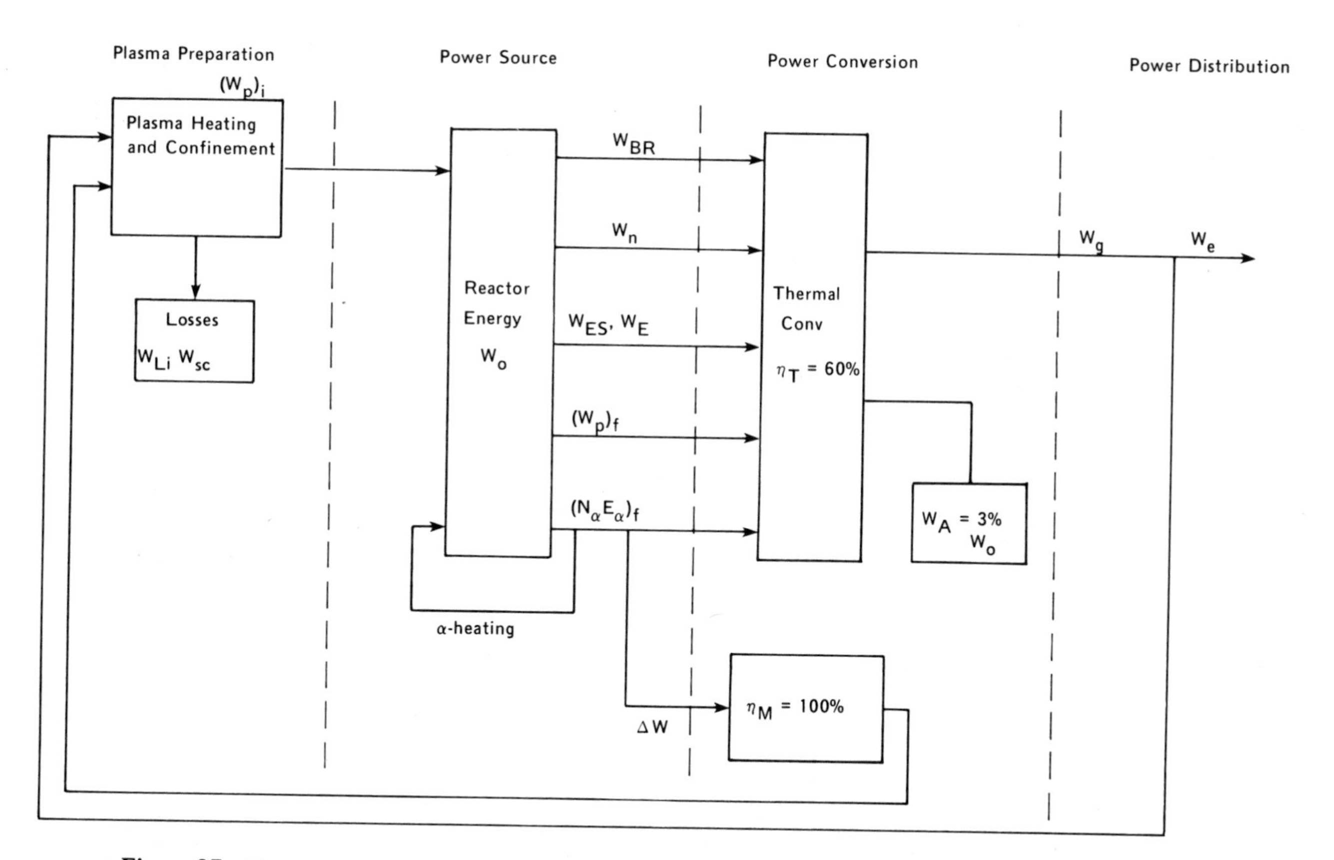

Figure 97. Energy flow diagram for RTPR.

Table 34

Energy Values in the RTPR (MJ/m)

Symbol	*Definition*	*Value*
$(W_p)_i$	Initial plasma energy = $3n_o kT_o (\pi a_o^2)$	3.41
W_{ES}	Joule Eddy current losses in shock heating coil	1.06
$W_E'(Li)$	Joule Eddy current losses in blanket	0.50
$W_E'(C)$	Joule Eddy current losses in C blanket	0.19
W_E	Joule Eddy current losses in compression coil	0.10
W_T	Joule transport current - losses in compression coil	5.24
W_L	Resistive energy losses (Total) = $W_{ES} + W_E'(Li') + W_E'(C) + W_E + W_T$	7.09
W_{SC}	Energy losses in cryogenic magnetic energy storage system	1.57
ΔW	Direct energy conversion (high β plasma expansion)	6.14
W_i	Total plasma preparation energy = $(W_p)_i + W_L + W_{SC} - \Delta W$	5.93
W_{BR}	Bremsstrahlung energy loss	1.08
W_n	Thermonuclear neutron energy (Q_n = 18.9 MeV)	88.40
$(W_p)_f$	Final plasma energy	6.88
$(n_\alpha E_\alpha)_f$	Final α-particle energy	5.70
W_o	Total reactor energy = $W_{BR} + W_n + W_{ES} + W_E'(Li) + W_E'(C) + (W_p)_f + (N_\alpha E_\alpha)_f$	103.81
W_g	Gross electric energy = $\eta_T W_o$	60.21
W_A	Thermal conversion losses = 3% W_g	1.80
W_C	Circulating energy = $W_i + W_A$	7.73
W_e	Net electric energy ($W_g - W_c$)	52.48
W_{Bo}	Peak magnetic energy in cryogenic system	315.00
η_T	Thermal conversion efficiency	58%
η_M	Direct conversion efficiency	100%
η_p	Overall plant efficiency	50.1%
Q	Energy multiplication factor = W_o/W_i	17.5
ϵ	Fractional circulating power = W_c/W_g	12.8%
M	Excess of electrical energy over losses	10.2
τ	Burning time	100
τ_c	Cycle time	10 sec
ξ^{-1}	Duty factor (= τ/τ_c)	0.01

Symbol	*Definition*	*Value*
P_W	First wall nuclear throughput	3.5
P_T	Reactor thermal power = $W_o L/\tau_c$	3.63 GW(Th)
P_e	Reactor net electrical power = $\eta_p P_T$	1.8 GW(e)
L	Reactor length	350 m

If, as shown in Table 34, we use M to indicate the excess of electrical energy over losses then we can write for the energy balance

$$\eta_T W_o = M W_i \tag{13.40}$$

or

$$\eta_T [W_n + W_{BR} + (W_p)_f + (N_\alpha E_\alpha)_f + W_{ES} + W_E'(Li) + W_E'(C)]$$
$$= M[(W_p)_i + (W_L - \Delta W) + W_{SC}] \tag{13.41}$$

The square bracket on the left side of the above equation represents the energy available as high-grade heat (~ 800°C), and the bracket on the right side represents energy losses which must be supplied during each burning pulse. The familiar energy multiplication factor Q can now be used to rewrite equation (13.41) in the form

$$Q = \frac{W_o}{W_i} = \frac{M}{\eta_T} \tag{13.42}$$

where it should be noted that the final plasma energy is given by

$$(W_p)_f = \frac{3}{2} [n_{ef} kT_{ef} + n_f kT_{if}] \pi a_f^2 \tag{13.43}$$

The plant circulating-energy fraction is

$$\epsilon = \frac{W_c}{W_g} = \frac{1}{M}(1 + W_A/W_i) \tag{13.44}$$

where W_A, the energy associated with the auxiliaries, is assumed to be 3% W_g. The losses W_{SC} associated with the super-conducting energy storage of the reactor is a quantity which is not presently known but estimated to be about 0.5% of W_{BO}, the peak energy stored in the cryogenic system. It therefore assumes the value

W_{SC} = 1.57 MJ/m so that ϵ = 12.8%, M = 10.2 and Q = 17.5. If we assume as we did in Chapter 9 that the efficiency of the thermal converter is on the order of 50%, or more specifically 58% then the overall plant efficiency, given by

$$\eta_p = \eta_T (1 - \epsilon) \tag{13.45}$$

is about 50.1% which is clearly quite attractive.

As pointed out earlier, the duty factor ξ^{-1} is related to the burn time (τ = 0.10 sec) by

$$\xi^{-1} = \tau/\tau_c \tag{13.46}$$

for which the cycle time τ_c is determined from the limitations on the wall loading. For a wall with radius "b" we recall that the wall loading can be written as

$$P_W = \frac{W_n}{2\pi b\tau_c} \quad \frac{14.1 \text{ MeV}}{Q_n} \tag{13.47}$$

where as before Q_n = 18.9 MeV, and W_n is the thermonuclear neutron energy. If P_W is limited to 3.5 MW/m^2 and using W_n = 88.4 then equation (13.47) yields τ_c = 10 sec. In this case the average primary neutron current on the first wall (see equation 3.31) neglecting back scattering is 1.4 x 10^{15} neutrons/ cm^2 per pulse or 1.4 x 10^{14} neutrons per cm^2 per second, and during the burn the instantaneous current is 1.4 x 10^{16} per cm^2 per second. For a continuous operation of the reactor over a period of one year the integrated neutron current is 4.4 x 10^{21} neutrons per cm^2 per year. Using L to indicate the length of the reactor, its thermal power can be readily obtained to read

$$P_T = \frac{W_o L}{\tau_c} \text{ MW} \tag{13.48}$$

where W_o is the total reactor energy. A toroidal theta-pinch reactor of major radius of 56 meters has a length of 350 meters and using the value of W_o given in Table 34, equation (13.48) gives P_T = 3.63 gigawatts (GW). The net electrical power output is

$$P_e = \eta_p P_T \tag{13.49}$$

which upon using η_p = 50.1% yields P_e = 1.82 GW.

We note from the above discussion of a conceptual design of a pulsed fusion reactor that regardless of the large power output (and size) it has certain features which make it perhaps more attractive than steady-state systems. The pulsed device operates at a higher beta and hence at higher plasma densities than a steady-state reactor, but since the system is "on" for only a few per cent of the time an acceptable wall loading can be realized with a much smaller vacuum chamber than its counterpart in a steady-state toroidal system such as Tokamak. Moreover the blanket coolant does not flow across magnetic field lines during 98% of the coolant cycle, thus eliminating electromagnetic losses which can be prohibitive in steady-state reactors. The fueling and flushing of the plasma between burning pulses can perhaps be accomplished without "divertors" which appear to be unavoidable in steady-state toroidal systems. In addition, the direct energy conversion by the expansion of the high-beta plasma against the confining magnetic field furnishes the joule losses in the compression coil thus possibly obviating the need of electrical input to this system.

REFERENCES

1. Bell, G. I., W. H. Borkenhage, and F. L. Ribe. *BNES Nuclear Fusion Reactors* (Proc. Int. Conf. Culham), p. 242 (1969).
2. Burnett, S. C., W. R. Ellis, and F. L. Ribe. *Technology of Controlled Fusion Experiments and the Engineering Aspects of Fusion Reactors,* USAEC CONF-721111, pp. 160 (1974).
3. Carruthers, R., P. A. Davenport, and J. T. D. Mitchell. Culham Laboratory Report CLM-R85 (1967).
4. Fowler, T. K. and M. Rankin. *Plasma Physics (J. Nuclear Energy, Part C), 8,* 121 (1966).
5. Golovin, I. N. *Fusion Technology, Energy 70*, Las Vegas, Nevada, p. 1 (1970).
6. Golovin, I. N., Y. N. Dnestrovsky, and D. P. Kostomarov. *BNES Nuclear Fusion Reactors* (Proc. Int. Conf. Culham), p. 194 (1969).
7. Gough, W. C. (Ed.). "CTR Engineering Systems Study Review Meeting," USAEC Report Wash.-1278 (1973).
8. Homeyer, W. G. "Thermal and Chemical Aspects of the Thermonuclear Blanket Problem," M.I.T. Research Laboratory of Electronics TR-435 (1965).
9. Kofoed-Hansen, O. "Introduction to Controlled Thermonuclear Research," Riso Report No. 18, pp. 347-365 (1960).

10. Kuo-Petravic, L. G., M. Petravic and C. J. H. Watson. *BNES Nuclear Fusion Reactors* (Proc. Int. Conf. Culham), p. 144 (1969).
11. Post, R. F. *BNES Nuclear Fusion Reactors* (Proc. Int. Conf. Culham p. 88 (1969).
12. Post, R. F. "Critical Conditions for Self-Sustaining Reactions in the Mirror Machine," Nuclear Fusion Supplement, pt. 1, pp. 99 (1962).
13. Sivukhin, D. V. *Plasma Physics (J. Nuclear Energy Part C), 8,* 607 (1966).
14. Sweetman, D. R. *BNES Nuclear Fusion Reactors* (Proc. Int. Conf. Culham), p. 112 (1969).
15. Wandel, C. F., T. H. Jensen, and O. Kofoed-Hansen. "A Compilation of Some Rates and Cross-Sections of Interest in Controlled Thermonuclear Research," *Nucl. Instrum. Meth., 4,* 249 (1959).
16. Werner, R. W., G. A. Carlson, J. Hovingh, J. D. Lee, and M. A. Peterson. Lawrence Livermore Laboratory Report UCRL-74054-2 (1974).

CHAPTER 14

RADIATION DAMAGE TO MATERIALS IN FUSION REACTORS

It has been said that the problem of radiation damage to materials might well constitute the most severe constraint on the engineering feasibility of fusion reactors. Although some aspects of this problem—namely displacement damage by high energy neutrons—have been examined in Chapter 3, a detailed discussion of radiation-induced damage in materials will be presented in this chapter. We shall focus our attention, however, on the reactor "first wall" because it faces the hot plasma directly and as a result it is subject to simultaneous damage by neutrons, charged particles, and electromagnetic radiation. Although considerable progress has been made in assessing these phenomena true understanding of the underlying physical principles as well as accurate estimates of the extent and nature of the damage are still not completely within our grasp. The results presented here should therefore be viewed as order of magnitude estimates based on present day understanding.

It is generally believed that the "erosion" of the solid wall facing a magnetically-confined hot plasma in a controlled fusion reactor is primarily due to three processes: sputtering, evaporation, and blistering. In the first process the wall temperature remains low and wall material is ejected as result of collision cascades initiated in the solid by incoming energetic particles such as neutrons, plasma ions, and neutrals. When the wall is subjected to very large thermal loading its surface gets heated to such a temperature that a large number of atoms acquire sufficient energy to overcome the surface binding energy and "evaporate."

Blistering is the explosion of gas bubbles formed just beneath the wall surface by injected gas atoms and occurs for gases not soluble in the wall material. This process happens mainly with helium.

Exact determination of the damage in the first wall is essential not only because the life time of that important reactor component must be known but also because of the serious contamination problem it could give rise to in the fusion plasma. In what follows we shall first examine the underlying theories of the damage processes and then apply the results to some conceptual designs to determine the extent of damage one might expect in an actual reactor.

14.1 SPUTTERING THEORY

Because it can be initiated by neutrons, plasma particles and neutrals, sputtering is perhaps the most serious of the damage processes. Most sputtering theories treat the interaction of the incoming particle and the target atoms on the basis of binary collision although recent experimental results indicate that sputtering is a collective phenomena. Since no theory based on the latter mode of interaction is available we shall restrict the present discussion to the former theory and deduce a yield formula which will enable us to estimate the damage by this process; we shall follow closely the analysis of Sigmund (1969).

Before embarking on the mathematical formulation of this problem it would be desirable to present a qualitative picture of the sputtering of a random target by an ion beam. An impinging ion undergoes a series of collisions in the target material, and recoiling atoms with sufficient energy will in turn undergo secondary collisions giving rise to another generation of recoiling atoms thereby leading to the formulation of a cascade. It is then possible for both the ion and the energetic recoil atoms to be scattered back through the surface as result of collisions suffered at a depth that might be a certain fraction of the total ion range. The backscattered ion and energetic recoil atoms account for most of the sputtered energy but not for most of the sputtered atoms. The higher order recoil atoms in the path of the incident ion and energetic recoils have small ranges and can be readily sputtered if they are located originally within two atomic layers from the surface. These are the atoms that account for the major portion of the sputtering yield whose distribution peaks at low energies.

With this picture in mind we may view the sputtering yield calculation as consisting of the following several steps: (1) determining the energy deposited by the energetic particles (ion and recoil atoms) near the material surface; (2) converting this energy into a number of low-energy recoil atoms; (3) determining how many of these particles arrive at the surface; and (4) selecting those atoms that have sufficient energy to overcome the surface-binding forces. This may be accomplished by solving a transport equation for which the essential input quantities include the scattering cross sections for high energy ions and atoms, the cross section for low energy atoms, and the surface-binding forces.

We consider a semi-infinite target as shown in Figure 98 and an atom (or ion) starting its motion at $t=0$ with velocity v at $x=0$.

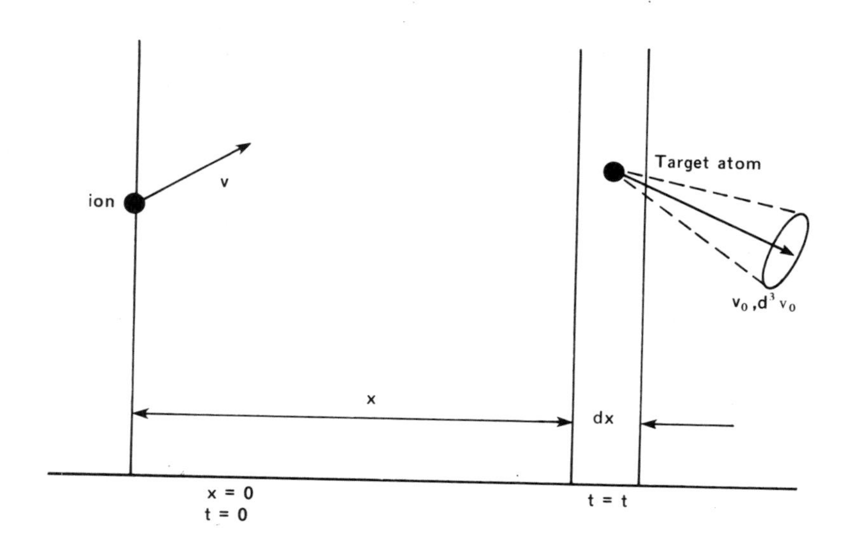

Figure 98. Geometry of sputtering calculation.

The quantity of interest is $G(x, \vec{v}_0, \vec{v}, t)d^3v_0dx$ which gives the expected number of atoms in dx about x, with velocity in d^3v_0 about $\vec{v}_0$ at time t. Noting that $dx = |v_{0x}|dt$, the number of atoms crossing the plane at x in the time interval dt is given by G $d^3 v_0 |v_{0x}| dt$, so that the sputtering yield for backward sputtering at the surface x = 0 can be readily written as

$$S = \int d^3 v_0 |v_{0x}| \int_0^\infty dt\, G(o, \vec{v}_0, \vec{v}, t) \tag{14.1}$$

The first integral extends over all values of v_0 with negative x component large enough to overcome surface-binding forces. By the same token, the yield for transmission sputtering through a surface at x=d is given by

$$S = \int d^3 v_0 v_{0x} \int_0^\infty dt\, G(d, \vec{v}_0, \vec{v}, t) \tag{14.2}$$

The equation which the quantity G must satisfy is the familiar (Boltzmann) transport equation

$$\frac{1}{v}\frac{\partial G}{\partial t} + \vec{\Omega}\cdot\nabla G + N \int d\sigma\,(G - G' - G'') = 0 \tag{14.3}$$

where $\vec{\Omega} = \vec{v}/v$. For the one-dimensional problem at hand the above equation has the form

$$\frac{1}{v}\,\frac{\partial G(x, \vec{v}_0, \vec{v}, t)}{\partial t} + \frac{v_x}{v}\,\frac{\partial G(x, \vec{v}_0, \vec{v}, t)}{\partial x}$$

$$+ N \int d\sigma\, [\, G(x,\vec{v}_0,\vec{v},t) - G(x,\vec{v}_0,\vec{v}',t) - G(x,\vec{v}_0,\vec{v}'',t)] = 0 \tag{14.4}$$

where N is the density of the target atoms,

$$d\sigma = d\sigma(\vec{v}, \vec{v}', \vec{v}'') = K(\vec{v}, \vec{v}', \vec{v}'')\, d^3 v'\, d^3 v''$$

is the differential scattering cross section, v′ is the velocity of the scattered particle and v″ is the velocity of the recoiling atom as illustrated in Figure 99a.

The above equation is merely a balance relationship and can be derived by counting the number of particles expected at the plane B located at x in Figure 99b. That number consists of two parts:

1) $Nv\,\delta t \int d\sigma\, G(c)$ which is the number of particles arriving at B as result of collisions suffered between A and B, and
2) $(1 - Nv\,\delta t \int d\sigma) G(A)$ which is the number of particles that has streamed from A to B, *i.e.*, those which did not suffer a collision between those two points.

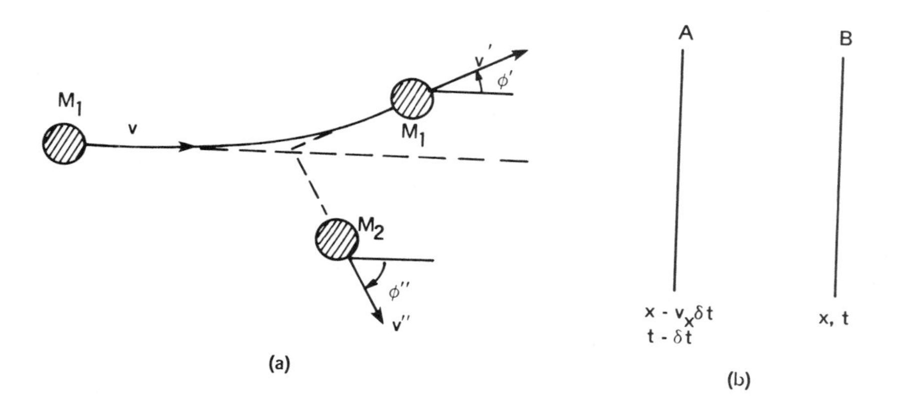

Figure 99. A scattering event in the laboratory system.

If we now note that the expected number of particles at B, *i.e.*, G(B) is simply $G(x, \vec{v}_0, \vec{v}, t)$ the quantity of interest, and that

$$G(A) = G(x-v_x\delta t, \vec{v}_0, \vec{v}, t-\delta t)$$
$$G(c) = G(x, \vec{v}_0, \vec{v}', t) + G(x, \vec{v}_0, \vec{v}'', t)$$

where $\vec{v}'$ and $\vec{v}''$ are the velocities shown in the collision geometry (Figure 99a) then the balance relation states

$$G(x,\vec{v}_0,\vec{v},t) = Nv\delta t \int d\sigma[G(x,\vec{v}_0,\vec{v}',t) + G(x,\vec{v}_0,\vec{v}'',t)] + [1 - Nv\delta t \int d\sigma]\, G(x-\eta v\delta t,\vec{v}_0,\vec{v},t-\delta t) \qquad (14.5)$$

having used $\eta = v_x/v$, and $\eta v\delta t = v_x\delta t$. Expanding the last term about x and t, and dividing the result by $v\delta t$ we obtain

$$-\frac{1}{v}\frac{\partial}{\partial t} G(x,\vec{v}_0,\vec{v},t) - \eta G(x,\vec{v}_0,\vec{v},t) = N\int d\sigma\, [G(x,\vec{v}_0,\vec{v},t) - G(x,\vec{v}_0,\vec{v}',t) - G(x,\vec{v}_0,\vec{v}'',t)] \qquad (14.6)$$

This is the transport equation in an isotropic homogeneous medium which is the equation of interest in this discussion. It differs from the neutron transport equation in two aspects:

(1) here, the variable is v and v_0 is a parameter, whereas the opposite is true in the neutron case; and (2) in this case we deal with two scattering terms rather than one since we include the recoil atom. Equation (14.6) applies when the incident particle is of the same species as the target atom. For the more general case where the impinging particle is different than the target atom we define a similar function $G_{(1)}(\vec{x},\vec{v}_0,\vec{v},t)$ such that when multiplied by $d3v_0 dx$ it gives the expected number of target atoms in dx about x with velocities in $d3v_0$ about v_0 moving as a result of an ion starting with $\vec{v}$ in x = 0 at t = 0 (see Figure 98). If correspondingly we call the ion-target cross section $d\sigma_{(1)}$ then $G_{(1)}$ satisfies the following equation

$$-\frac{1}{v}\frac{\partial}{\partial t}G_{(1)} - \eta\frac{\partial}{\partial x}G_{(1)} = N\int d\sigma_{(1)}\,[G_{(1)}(x,\vec{v}_0,\vec{v},t)$$

$$- G_{(1)}(x,\vec{v}_0,\vec{v}\,',v) - G(x,\vec{v}_0,\vec{v}\,'',t)] \qquad (14.7)$$

which is analogous to equation (14.6) except that G and not $G_{(1)}$ represents the recoil term. Of more interest is the quantity F defined such that F $d3v_0 dx$ gives the total number of atoms that penetrate the plane x with velocities in $d3v_0$ about v_0. It is given by

$$F(x,\vec{v}_0,\vec{v}) = \int_0^{\infty} G(x,\vec{v}_0,\vec{v},t)\,dt \qquad (14.8)$$

and the equation it obeys is simply (14.6) integrated over time. If we assume that the (cascading) event starts with one particle at the surface x = 0 then we can write

$$G(x,\vec{v}_0,\vec{v},t=0) = \delta(x)\,\delta(\vec{v}-\vec{v}_0)$$

and if we further assume that all atoms have slowed down beyond any finite velocity, we can also write

$$G(x,\vec{v}_0,\vec{v},t=\infty) = 0$$

We note therefore that

$$\int_0^{\infty}\frac{\partial}{\partial t}G(x,\vec{v}_0,\vec{v},t)dt = G(x,\vec{v}_0,\vec{v},\infty) - G(x,\vec{v}_0,\vec{v},0) = -\delta(x)\,\delta(\vec{v},\vec{v}_0)$$

so that the equation for F becomes

$$\frac{1}{v}\delta(x)\,\delta(\vec{v}-\vec{v}_0) - \eta\,\frac{\partial}{\partial x}\,F(x,\vec{v}_0,\vec{v}) = N\int d\sigma\,[F(x,\vec{v}_0,\vec{v}) - F(x,\vec{v}_0,\vec{v}\,')$$

$$- F(x,\vec{v}_0,\vec{v}\,'')] \tag{14.9}$$

If equation (13.7) is used then $F_{(1)}$ defined similarly, satisfies

$$-\eta\,\frac{\partial}{\partial x}\,F_{(1)}(x,\vec{v}_0,\vec{v}) = N\int d\sigma_{(1)}\,[F_{(1)}(x,\vec{v}_0,\vec{v}) - F_{(1)}(x,\vec{v}_0,\vec{v}\,') -$$

$$- F(x,\vec{v}_0,\vec{v}\,'')] \tag{14.10}$$

In this case, however, there are no moving target atoms at $t=0$ and hence there is no source term equivalent to that in equation (14.9), *i.e.*, $G_{(1)}(t{=}0) = G_{(1)}(t{=}\infty) = 0$. As in the case of many other applications of transport phenomena, equations (14.9) and (14.10) are difficult to solve and are much too comprehensive for all practical purposes. For the calculation of sputtering yields it is sufficiently adequate to deal with yet another quantity which we obtain by integrating these equations over $\vec{v}_0$.

We consider only backward sputtering for which we introduce the new functions

$$H(x,\vec{v}) = \int d^3v_0\,|v_{0x}|\,F(x,\vec{v}_0,\vec{v}) \tag{14.11a}$$

$$H_{(1)}(x,\vec{v}) = \int d^3v_0\,|v_{0x}|\,F_{(1)}(x,\vec{v}_0,\vec{v}) \tag{14.11b}$$

where the integrations obey the conditions

$$\eta_0 = \frac{v_{0x}}{v} \leqslant 0$$

$$E_0 = \tfrac{1}{2} M_2 v_0^2 \geqslant U(\eta_0) \tag{14.12}$$

and $U(\eta_0)$ is the surface-binding energy characterized by the direction of ejection, $\cos^{-1}\eta_0$. The quantities H and $H_{(1)}$ represent the sputtering yields of a target atom or an arbitrary ion when the source is at $x=0$, and the sputtering surface is in the plane x. Multiplying equation (14.9) by $|v_{0x}|$ and integrating we get

$$\eta\delta(x) - \eta \frac{\partial}{\partial x} H(x,\vec{v}) = N \int d\sigma \, [H - H' - H'']$$

where again we require that $\eta < 0$ and the energy of the impinging particle, $E = \frac{1}{2}M_2 v^2$ be larger than $U(\eta)$. To insure that these conditions are satisfied we employ the step functions

$$\begin{aligned} \theta(-\eta) &= 1 , \quad \eta < 0 \\ &= 0 , \quad \eta > 0 \\ \theta\,[E - U(\eta)] &= 1 , \quad E > U(\eta) \\ &= 0 , \quad E < U(\eta) \end{aligned} \tag{14.13}$$

with which the equation of interest becomes

$$- \delta(x)\,\eta\theta(-\eta)\,\theta\,[E - U(\eta)] - \eta \frac{\partial}{\partial x} H(x,\vec{v}) =$$

$$N \int d\sigma [H(x,\vec{v}) - H(x,\vec{v}') - H(x,\vec{v}'')]$$

We note that the function H already contains the conditions given by (14.13), and therefore there is no need to include them explicitly in these terms. Moreover, in a random medium the function H depends only on x, v and η where $\eta = \cos\alpha = v_x/v$, so that it is convenient to transform from v to E, variables and rewrite the transport equation as

$$- \delta(x)\eta\theta(-\eta)\,\theta[E-U(\eta)] - \eta \frac{\partial}{\partial x} H(x,E,\eta) = N \int d\sigma \, [H(x,E,\eta)$$

$$- H(x,E',\eta) - H(x,E'',\eta)] \tag{14.14}$$

In this equation E′ and η' denote the energy and direction of the scattered particle, and E″ and η'' represent the corresponding quantities for the recoil particle. In view of equation (14.2) we see that the sputtering yield can now be written as

$$S(E,\eta) = H(x=0,E,\eta) \tag{14.15}$$

For forward or transmission sputtering $\eta>0$, and equation (14.14) will hold provided $(-\eta)$ is replaced by η and $\theta(-\eta)$ by $\theta(\eta)$ so that the first term will read $\delta(x)\eta\theta(\eta)\theta[E-U(\eta)]$. The corresponding yield equation for forward sputtering then becomes

$$S(E,\eta) = H(x{=}d,E,\eta) \tag{14.16}$$

The determination of the yield requires solving equation (14.14) for H and here we will use the standard procedure of solving this type of transport equation, namely the use of expansion in Legendre polynomials. The solutions will converge rapidly if the dependence on η is smooth and experiments (Kaminsky, 1965) have shown no dramatic fluctuations as function of η, at least for not too oblique incidence. We therefore write

$$H(x,E,\eta) = \sum_{\ell=0}^{\infty} (2\ell+1)\, H_\ell(x,E) P_\ell(\eta) \tag{14.17}$$

where $P_\ell(\eta)$ are Legendre polynomials, and substitute this expansion in equation (14.14). If we multiply the result by $(2\ell + ½)$ $P_\ell(\eta)$ and integrate over η utilizing the well known relations

$$(\ell+1)P_{\ell+1}(x) - (2\ell+1)\, x\, P_\ell(x) + \ell P_{\ell-1}(x) = 0$$

$$\frac{2\ell+1}{2} \int_{-1}^{+1} P_\ell(\eta)\, P_k(\eta) d\eta = \delta_{\ell k} \tag{14.18}$$

we obtain for the first term in the equation

$$\frac{2\ell+1}{2} \int_{-1}^{+1} -\eta\theta(-\eta)\theta\,[E-U(\eta)]\; P_\ell(\eta) d\eta = \frac{2\ell+1}{2} \left[\int_{-1}^{0} \cdots + \int_{0}^{1} \cdots\right]$$

where the last term is zero by virtue of (14.13). Denoting by Q_ℓ the first integral in the right-hand side, *i.e.*,

$$Q_\ell(E) = \frac{2\ell+1}{2} \int_{-1}^{+1} (-\eta) d\eta\; [E - U(\eta)]\; P_\ell(\eta)$$

we see that the first term in (14.14) simply becomes $\delta(x) Q_\ell(E)$. The second term becomes by virtue of (14.18)

$$\text{Second term} = \ell H_{\ell-1}(x,E) + (\ell+1)\, H_{\ell+1}(x,E)$$

In order to obtain the remaining terms we first note from the collision geometry shown in Figure 100 that φ' and φ'' are respectively the laboratory angles of scattered and recoil atoms.

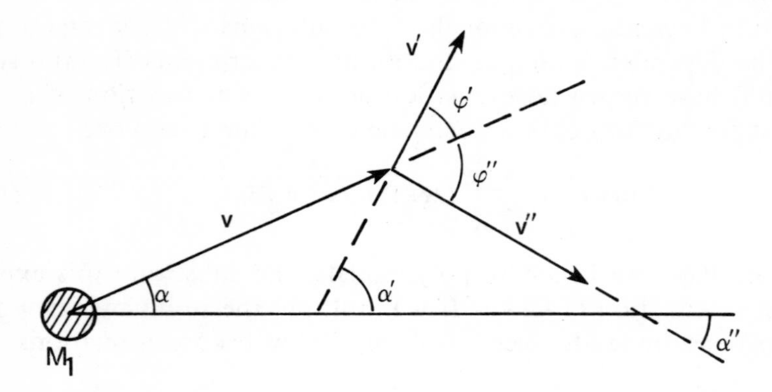

Figure 100. Collision geometry.

We note that $\eta = \cos\alpha$, $\eta' = \cos\alpha'$, and $\eta'' = \cos\alpha''$, and if we call

$\vec{n}$ = unit vector along $\vec{v}$
$\vec{n}'$ = unit vector along $\vec{v}'$
$\vec{n}''$ = unit vector along $\vec{v}''$

then we can write

$$\eta' = \eta\, \vec{n} \cdot \vec{n}' - \sqrt{1-\eta^2}\ \sqrt{1-(\vec{n} \cdot \vec{n}')^2}$$

or more generally, if $\vec{v}$ and $\vec{v}'$ have azimuthal angles ϕ and ϕ', then

$$\eta' = \eta \cos\varphi' + \sqrt{1-\eta^2}\ \sqrt{1-(\vec{n} \cdot \vec{n}')^2} \cos(\phi' - \phi)$$

By the same token we have

$$\eta'' = \eta \cos\varphi'' + \sqrt{1-\eta^2}\ \sqrt{1-(\vec{n} \cdot \vec{n}'')^2} \cos(\phi'' - \phi)$$

so that we can write

$$P_\ell(\eta') = P_\ell(\eta)\, P_\ell(\cos\varphi') + \text{cross terms}$$

and upon integration over η the cross terms vanish. In view of this the transport equation becomes

$$\delta(x)\,Q_\ell(E) - \frac{\partial}{\partial x}\,[\ell H_{\ell-1}(x,E) + (\ell+1)\,H_{\ell+1}(x,E)]$$

$$= (2\ell+1)\,N \int d\sigma\,[H_\ell(x,E) - P_\ell(\cos\varphi')\,H_\ell(x,E')$$

$$- P_\ell(\cos\varphi'')\,H_\ell(x,E'')] \qquad (14.19)$$

For the case of transmission sputtering we first note that

$$P_\ell(-\eta) = (-1)^\ell\,P_\ell(\eta)$$

and if, as before, we replace η by $(-\eta)$ we note that equation (14.19) remains applicable provided we change $Q_\ell(E)$ to $\bar{Q}_\ell(E) = (-1)^\ell\,Q_\ell(E)$.

Rather than solving equations (14.19) for H_ℓ it is customary and convenient to calculate the moments of this function by multiplying the above equation by $x^n (n=0,1,2,...)$ and integrating over x. If we now define the moments

$$H_\ell^n(E) = \int_{-\infty}^{+\infty} x^n\,H_\ell(x,E)\,dx \qquad (14.20a)$$

$$H_{(1)\ell}^n(E) = \int_{-\infty}^{+\infty} x^n\,H_{(1)\ell}(x,E)dx \qquad (14.20b)$$

and note that

$$\delta_{no}(x) = \int_{-\infty}^{+\infty} x^n\,dx\,\delta(x) = \begin{cases} 1, & n=0 \\ 0, & n\neq 0 \end{cases}$$

then equation (14.19) becomes

$$\delta_{no}\,Q_\ell(E) - n\,[\ell\,H_\ell^{n-1}(E) + (\ell+1)\,H_{\ell+1}^{n-1}(E)] =$$

$$(2\ell+1)N\int d\sigma[H_\ell^n(E) - P_\ell(\cos\varphi')H_\ell^n(E') - P_\ell(\cos\varphi'')\,H_\ell^n(E'')]$$

If we further let

E = energy of incoming particle = E
E-T = energy of scattered particle = E′
T = energy of recoil particle = E″

and also separate the elastic from inelastic collisions, *i.e.*, write the right-hand side of the above equation as

$$\text{R.H.S.} = (2\ell+1)N \left\{ \int d\sigma_{inel} [\ ] + \int_{T=0}^{E} d\sigma(E,T) \, [H_\ell^n(E) - P_\ell(\cos\varphi')H_\ell^n(E-T) - P_\ell(\cos\varphi'')H_\ell^n(T)] \right\}$$

then according to Lindhard *et al.* (1963), the inelastic term may be expressed as

$$\int d\sigma_{inel} [\ ] = S_e(E) \frac{d}{dE} H_\ell^n(E)$$

where S_e is the electronic stopping cross section, and $d\sigma(E,T)$ is the differential cross section for elastic scattering. With these, the moment equation assumes the form

$$\delta_{no} Q_\ell(E) - n [\ell H_{\ell-1}^{n-1}(E) + (\ell+1) H_{\ell+1}^{n-1}(E)] =$$

$$(2\ell+1) N S_e(E) \frac{d}{dE} H_\ell''(E) + (2\ell+1) N \int d\sigma(E,T) \, [H_\ell^n(E)$$

$$- P_\ell (\cos\varphi') H_\ell^n(E-T) - P_\ell (\cos\varphi'') H_\ell^n(T)] \tag{14.21a}$$

and for the case with different impinging and medium particles it takes on the form

$$- n [\ell H_{(1)\ell-1}^{n-1} (E) + (\ell+1) H_{(1)\ell+1}^{n-1}(E)] = (2\ell+1) N S_{e(1)} \frac{d}{dE} H_{(1)\ell}^n(E)$$

$$+ (2\ell+1) N \int_{T=0}^{T_M} d\sigma_{(1)}(E,T) \, [H_{(1)\ell}^n (E) - P_\ell (\cos\varphi') H_{(1)\ell}^n(E-T)$$

$$- P_\ell (\cos\varphi'') H_\ell^n(T)] \tag{14.21b}$$

To evaluate the limits in the above equations let's observe from Figure 99 that the conservation of momentum in component form gives

$$M_1 v = M_1 v' \cos\varphi' + M_2 v'' \cos\varphi''$$

$$o = M_1 v' \sin\varphi' - M_2 v'' \sin\varphi''$$

where $v = \sqrt{2E/M_1}$, $v' = \sqrt{2(E-T)/M_1}$, and $v'' = \sqrt{2T/M_2}$. Eliminating φ'' in favor of φ' and expressing the velocities in terms of energies we find that

$$\cos\varphi' = (1-\frac{T}{E})^{1/2} + \tfrac{1}{2}(1-\frac{M_2}{M_1})(\frac{T}{E})(1-\frac{T}{E})^{-1/2}$$

and similarly

$$\cos\varphi'' = \left[\frac{T}{E}\;\frac{(M_1+M_2)^2}{4M_1M_2}\right]^{1/2} = (T/T_m)^{1/2}$$

where

$$T_m = \gamma E$$

$$\gamma = \frac{4M_1M_2}{(M_1+M_2)^2} \qquad (14.22)$$

Equation (14.21a) applies when the incident and target atoms are of the same species, *i.e.*, $M_1 = M_2$ and for that case

$$\cos\varphi' = (1 - T/E)^{1/2}$$

$$\cos\varphi'' = (T/E)^{1/2}$$

It must be noted before proceeding to the question of analytic solutions that the use of the moment equations is somewhat limited and useful only when there is a procedure to reconstruct a function from its moments. This is quite feasible when the distribution functions are Gaussian or close to it. Moreover, it must also be observed that neither a bulk displacement threshold, E_d, nor a bulk binding energy, V, lost to the lattice by a recoiling atom have to be included in the derivation of equation (14.21). Although the first quantity might be neglected without introducing serious error, the latter may not always be neglected especially in covalent crystals. In those cases the recoil term $H_\ell^n(T)$ has to be replaced by $H_\ell^n(T-V)$. It must also be kept in mind that bulk energies have to be introduced or else the yield would be infinite; Energy conservation requires that even for $E_d = V = 0$ only a finite number of atoms can penetrate the surface and overcome the surface barrier. The competition between surface and bulk-binding forces shows up in the energy spectrum of sputtered atoms.

If the source is isotropic, such as we would expect for the recoils from elastic collisions caused by an isotropic flux of fast neutrons, then the sputtering yield of a source at x = 0 with the surface at x is given by

$$S(x,E) = \frac{1}{2} \int_{-1}^{+1} d\eta \; H(x,E,\eta)$$

where ½ accounts for the half region. If we substitute for H from equation (14.17) and utilize the fact that

$$\int_{-1}^{+1} d\eta \; P_\ell(\eta) = \frac{2}{2\ell+1} \; \delta_{\ell o}$$

then the yield equation becomes

$$S(x,E) = H_o(x,E) \tag{14.23a}$$

and for different species it acquires the form

$$S_{(1)}(E,x) = H_{(1)o}(x,E) \tag{14.23b}$$

We proceed now to explore the analytic solutions to equations (14.21a and b), and before we can do that we must specify the input quantities S_e, $d\sigma$, and the surface-binding energy U. Only for simple expressions of these quantities can we hope to obtain analytic solutions and we shall use these forms. For the stopping cross section S_e we shall use Lindhard's (1963) expression

$$S_e(E) = K\,E^{1/2}, \quad S_{(1)e}(E) = K\,E^{1/2} \tag{14.24a}$$

except when the incident ion velocity is so high that the Bethe-Block (1953) formula applies, *i.e.*,

$$S_{(1)e}(E) = \text{const}/E \tag{14.24b}$$

The constants in these expressions depend on the incident particle and the properties of the target medium. For the elastic scattering cross sections we shall use the power approximation of the Thomas-Fermi cross sections as given by Lindhard, namely

$$d\sigma = C\,E^{-m}\,T^{-1-m}\,dT \tag{14.25a}$$

$$d\sigma_{(1)} = C_{(1)}\,E^{-m}\,T^{-1-m}\,dT \tag{14.25b}$$

where m is between zero and unity. For m = 1 we note that

$$d\sigma = \frac{C}{E}\,\frac{dT}{T^2}$$

which is the Rutherford cross section, and if we compare it with equation (3.58) of Chapter 3 we see that the constant C = $(M_1/M_2)\pi Z_2^{\,2} e^4$. In general, however, these constants depend on the range of energy of the incident ion as well as on the masses of the incident and target particles or, more explicitly, on the appropriate interaction potential. They are commonly given by

$$C = \tfrac{1}{2}\pi\,\lambda_m\,(a_{22})^2\left(\frac{2Z_2^{\,2}e^2}{a_{22}}\right)^{2m} \tag{14.26a}$$

$$C_{(1)} = \tfrac{1}{2}\pi\,\lambda_m\,(a_{12})^2\left(\frac{M_1}{M_2}\right)^{m}\left(\frac{2Z_1Z_2e^2}{a_{12}}\right)^{2m} \tag{14.26b}$$

where Z_1 and Z_2 are the atomic numbers of the impinging and target particles, a_{12} and a_{22} are the Thomas-Fermi screening radii (similar to the screening radius, a, appearing in the Born-Mayer potential given below), and λ_m are dimensionless constants given by

$$\lambda_1 = 0.5;\quad \lambda_{1/2} = 0.327;\quad \lambda_{1/3} = 1.309 \tag{14.27}$$

In the eV range we shall characterize collisions by a power cross section which for m = 0, from (14.25) leads to

$$d\sigma = C\,dT/T;\qquad d\sigma_{(1)} = C_{(1)}\,dT/T \tag{14.28}$$

and assuming that (14.26) remains valid in this region they become

$$C = \tfrac{1}{2}\pi\,\lambda_o\,a_{22}^{\,2};\qquad C_{(1)} = \tfrac{1}{2}\pi\,\lambda_o\,a_{12}^{\,2} \tag{14.29}$$

At very low energies corresponding to $\epsilon \leqslant 0.005$ where

$$\epsilon = \frac{M_2}{M_1 + M_2}\,\frac{E}{A}$$

with A being the constant in the Born-Mayer potential

$$V(r) = A\,e^{-r/a}$$

$$A = 52\,(Z_1Z_2)^{3/4}\ \text{eV}\qquad a = 0.219\ \text{A}^\circ \tag{14.30}$$

the validity of most existing potentials becomes questionable and so does the concept of two-particle collision. At these energies we shall assume that the constants given by (14.29) will be replaced by

$$C_o = \tfrac{1}{2}\pi\, \lambda_o\, a^2; \quad \lambda_o = 24; \quad a = 0.219 \text{ A}^\circ \tag{14.31}$$

We turn now to the elastic stopping powers which can be written in terms of the elastic scattering cross sections discussed above, as

$$S_n(E) = \int_o^E T\, d\sigma = C\, E^{-m} \int_o^E T^{-m}\, dT = \frac{C}{1\text{-}m} E^{1-2m} \tag{14.32a}$$

and in a similar manner,

$$S_{(1)n}(E) = \int_o^{T_m} T\, d\sigma_{(1)} = C_{(1)}\, E^{-m} \left[\frac{T^{1-m}}{1\text{-}m}\right]^{T_m=\gamma E}$$

$$= \frac{C_{(1)}}{1\text{-}m}\, \gamma^{1-m}\, E^{1-2m} \tag{14.32b}$$

These expressions can be used to estimate the energy limits within which cross sections for various values of m apply especially the limiting energy E* up to which the cross section with m = 0 is feasible. We assume that m = 1/3 is valid for E > E* so that

$$E^* = \left(\frac{3\,\lambda_{1/3}}{2\,\lambda_0}\right)^{3/2} \frac{M_1 + M_2}{M_2} \left(\frac{a_{12}}{a}\right)^3 \frac{Z_1 Z_2 e^2}{a_{12}} \tag{14.33}$$

which has values of the order of some hundred eV. The stopping powers as given by equation (14.32) become proportional to the energy for m = 0, and this leads to divergence in the total range of an ion or atom. This will not, however, introduce any divergences in the sputtering calculations, but in order to avoid drastic over- or underestimates in the sputtering yield one might prefer to use a low-energy cross section with m = 0.05. This has been done and the resulting change in the sputtering yield, though measurable, was small compared with the uncertainty introduced by lack of knowledge of surface-binding conditions.

The two basically simple surface conditions that are commonly used are the planar potential barrier or work-function leading to

$$U_{(\eta o)} = U_o/\eta_o^{\,2} \tag{14.34}$$

and the spherically symmetric potential barrier given by

$$U_{(\eta o)} = U_o \tag{14.35}$$

Much discussion has gone into which of the above forms is the more appropriate; but it can be said that while equation (14.34) is not as accurate, equation (14.35) is not as unphysical as one might expect. In general there is a distribution of binding energies, U_o, and the average surface-binding energy may depend somewhat on the crystal surface considered and on the state of the damage of the surface. Nevertheless we will assume that equation (14.34) is the most realistic expression, at least for metals, wtih U_o equal to the cohesive energy per atom or the measured energy of sublimation. We shall also calculate sputtering yields using equation (14.35) to assess the sensitivity of this quantity to the form of $U(\eta_o)$.

As pointed out earlier the complete solutions of equation (14.21), while difficult, are not truly needed since only very few of the $H_\ell^n(E)$ are significant at high ion energies, namely at $E \gg U_o$. Since U_o is typically of the order of few eV we will assume that $E > 100\text{-}200$ eV and therefore sputtering near threshold will be excluded. For $n = 0$ equation (14.21a) becomes

$$Q_\ell(E) = (2\ell+1)N\,S_e(E)\,\frac{d}{dE}\,\overset{\circ}{H}_\ell(E) + (2\ell+1)N \int_{T=0}^{E} d\sigma(E,T)$$

$$[H_\ell^\circ(E) - P_\ell\,(\cos\varphi')\,H_\ell^\circ(E\text{-}T) - P_\ell(\cos\varphi'')\,H_\ell^\circ(T)]$$

Recalling that for $M_1 = M_2$

$$\cos\varphi' = (1 - T/E)^{1/2}, \quad \cos\varphi'' = (T/E)^{1/2}$$

and assuming $S_e(E) = 0$, then with

$$d\sigma = C\,E^{-m}\,T^{-1-m}\,dT$$

we have

$$Q_\ell(E) = (2\ell+1)N\,C\,E^{-m} \int_o^E \frac{dT}{T^{1+m}} [H_\ell^\circ(E) - P_\ell(1 - T/E)^{1/2}$$

$$H_\ell^\circ(E-T) - P_\ell(T/E)^{1/2} H_\ell^\circ(T)] \qquad \textbf{(14.36a)}$$

Moreover, we also recall that Q_ℓ can be written as

$$Q_\ell(E) = \frac{2\ell+1}{2} \int_{-1}^{0} (-\eta)d\eta\, \theta[E - U_{(\eta)}]\; P_\ell(\eta)$$

and if we use equation (14.34) for U, *i.e.*, $U_{(\eta_o)} = U_o/\eta_o{}^2$, and note that

$$P_{o(\eta)} = 1; \quad P_{1(\eta)} = \eta$$

$$P_{2(\eta)} = \tfrac{1}{2}(3\eta^2 - 1); \quad P_{3(\eta)} = \tfrac{1}{2}(5\eta^3 - 3\eta)$$

then for $\ell = 0$ the above expression becomes

$$Q_o(E) = \tfrac{1}{2} \int_{-1}^{0} (-\eta)d\eta\; \theta\, [E - U_o/\eta_o{}^2]$$

But $\theta\, [E - U_o/\eta_o{}^2] = 1$ if $\eta^2 > U_o/E$ or $\eta = -\sqrt{U_o/E}$ which constitutes the upper bound, thus

$$Q_o(E) = -\tfrac{1}{4}\,\eta^2 \Big|_{-1}^{-\sqrt{U_o/E}}$$

or

$$Q_o(E) = -\tfrac{1}{4}\,[U_o/E - 1] = \tfrac{1}{4}\,[1 - U_o/E] \qquad \textbf{(14.36b)}$$

By the same token we note that $Q_1(E)$ is given by

$$Q_1(E) = 3/2 \int_{-1}^{-\sqrt{U_o/E}} (-\eta)\,\eta\; d\eta = 1/2[-1 + (U_o/E)^{3/2}] \qquad (14.36c)$$

and Q_2 and Q_3 are given by the expressions:

$$Q_2(E) = 5/2 \int_{-1}^{-\sqrt{U_o/E}} (-\eta)\; 1/2(3\eta^2 - 1)\; d\eta$$

$$= 5/16\,[1 + 2(U_o/E) - 3(U_o/E)^2] \qquad (14.36d)$$

$$Q_3(E) = 7/4\,[-(U_o/E)^{3/2} + (U_o/E)^{5/2}] \qquad (14.36e)$$

We observe that equations (14.36b-e) give zero if $E < U_o$ or

$$Q_\ell(E) = 0; \qquad E < U_o$$

which leads to

$$H_\ell^\circ(E) = 0, \qquad E < U_o \qquad (14.37)$$

If in place of equation (14.34) we use equation (14.35) for $U_{(\eta o)}$ we find that

$$Q_o \approx 1/4; \qquad Q_1 \approx -1/2; \qquad Q_2 \approx 5/16; \qquad Q_3 \approx 0$$

i.e., independent of U_o/E. Following Robinson (1965) one can calculate the asymptotic expansion in powers of E, and writing only terms with positive powers of E we get

$$H_0^\circ(E) \sim \frac{1}{\psi(1) - \psi(1-m)} \; \frac{m}{1-2m} \; \frac{E}{8NC\,U_o^{\,1-2m}} + \frac{E^{2m}/8NC}{-1/m - B(-m,2m)} \qquad (14.38a)$$

$$H_1^\circ(E) \sim \frac{-1}{\psi(1) - \psi(1-m)} \; \frac{m}{1\text{-}4m} \; \frac{E^{1\,2}}{4NC\,U_o^{\,1/2-2m}} + \frac{E^{2m}/8NC}{1/m + 2/(1+2m) + B(-m, 3/2 + 2m)} \qquad (14.38b)$$

$$H_2^\circ(E) \sim \frac{E^{2m}/8NC}{1/m + 3/(1+m) + 3B(-m,2+2m) + 2B(-m,2m)} \qquad (14.38c)$$

$$H_3^\circ(E) \sim 0 \qquad (14.38d)$$

where the functions B(m,n) and $\psi(m)$ are the beta and digamma functions, respectively. The familiar form of the gamma function is

$$\Gamma(m) = \int_0^\infty t^{m-1}\, e^{-t}\, dt, \quad m > 0 \qquad (14.39)$$

and

$$\Gamma(m) = m\ \Gamma\ (m\text{-}1)$$

while the beta function is defined by

$$B(m,n) = \int_0^1 t^{m-1}\ (1\text{-}t)^{n-1}\ dt,\ \ m > 0;\ \ n > 0$$

$$B(m,n) = \Gamma(m)\ \Gamma(n)\ /\ \Gamma(m+n) \tag{14.40}$$

The function $\psi(m)$ is given by

$$\psi(m) = \frac{d}{dm}\ \ell n\ \Gamma(m) = \frac{1}{\Gamma(m)}\ \frac{d}{dm}\ \ \Gamma(m) = \frac{\Gamma'(m)}{\Gamma(m)} \tag{14.41}$$

where prime denotes differentiation with respect to the argument. We note that

$$\frac{\Gamma'(m)}{\Gamma(m)} = -\ \gamma + \left(1 - \frac{1}{m}\right) + \left(\frac{1}{2} - \frac{1}{m+1}\right) + \ \ldots . + \left(\frac{1}{n} - \frac{1}{m+n-1}\right) + \ldots .$$

therefore the quantity $\psi'(m)$ can be written as

$$\psi'(m) = \frac{d}{dm}\left[\frac{\Gamma'(m)}{\Gamma(m)}\right] = \left(\frac{1}{m}\right)^2 + \left(\frac{1}{m+1}\right)^2 + \ldots . + \left(\frac{1}{m+n-1}\right)^2 + \ldots .$$

which for m = 1 becomes (with γ = Eulers Constant = 0.5772)

$$\psi'(1) = 1 + (1/2)^2 + (1/3)^2 + \ldots . + (1/n)^2 + \ldots .$$

but the sum

$$\frac{1}{1^{2p}} + \frac{1}{2^{2p}} + \frac{1}{3^{2p}} + \ldots . + \frac{1}{n^{2p}} + \ldots . = \frac{2^{2p-1}\ \pi^{2p}\ B_p}{(1\text{-}p)!}$$

where B_p is Bernoulli's number which has a value for p = 1 of B_1 = 1/6. In view of this we see that

$$\psi'(1) = \frac{2\pi^2\ 1/6}{2!} = \frac{\pi^2}{6} \tag{14.42}$$

Bernoulli's numbers are defined in accordance with the series expansion

$$\frac{x}{e^x - 1} = 1 - \frac{x}{2} + B_1 \frac{x^2}{2!} - \frac{B_2 x^4}{4!} + \frac{B_3 x^6}{6!} - \ldots$$

from which we readily observe that

$$B_1 = 1/6; \quad B_2 = 1/30; \quad B_3 = 1/42 \tag{14.43}$$

Returning to equations (14.38) and substituting for the functions ψ and B, we find that for m = 0, *i.e.*, $E < E^*$

$$H_0^\circ(E) \sim \frac{1}{8\psi'(1)} \frac{E}{NC_o U_o} \tag{14.44a}$$

$$H_1^\circ(E) \sim - \frac{1}{4\psi'(1)} \frac{E^{1/2}}{NC_o U_o^{1/2}} \tag{14.44b}$$

$$H^\circ_2(E) \sim 0 \tag{14.44c}$$

$$H_3^\circ(E) \sim 0 \tag{14.44d}$$

with C_o given by equation (14.31). The constant terms have been dropped from the above equations and because $H_0{}^0(E)$ is the leading and dominant term it will be the only one to be kept for m=0.

It is useful to pause briefly and examine the meaning of the moment $H_0{}^0(E)$. According to equation (14.23a) it determines the sputtering yield from a surface in x of an isotropic source in x = 0. According to (14.20a) the moment $H_0{}^0(E)$ is the integral of $H_0(x,E)$ over all x, *i.e.*, it gives the number of atoms penetrating a plane at an arbitrary position x with a certain minimum energy, when there is a homogeneous isotropic source of recoiling atoms throughout an infinite medium. The first term in $H_0{}^0(E)$ according to equation (14.38a) is proportional to $(E/U_o)(U_o^{2m}/NC)$ where (E/U_o) is roughly the number of atoms per ion that is set in motion with an energy greater than U_o, and (U_o^{2m}/NC) is effectively the range of an atom with energy U_o. The latter quantity is determined by the scattering law at $E = U_o$, *i.e.*, m = 0 while the former is insensitive to the scattering law of the high energy ions. Because of its dependence on (E^{2m}/NC) the second term in (14.38a) is proportional to the range of a particle of energy E and thus gives the probability that the (impinging) ion itself penetrates the plane under consideration. In order to assess the importance of this term one has to insert the values of m and C that are appropriate at the energy E, and for $m \geqslant 1/2$ the second term of equation (14.38a) could compete with the first one. Generally, however, the second term is (numerically) negligibly small.

To appreciate the importance of both these terms let's consider a medium mass target such that the range (E^{2m}/NC) is about 100 Å at 100 keV with $m = 1/2$. The range (U_o^{2m}/NC) at $m = 0$ is about 1 Å while (E/U_o) is about 10^4-10^5. Thus the ratio between the two competing terms is about 100-1000 and does not depend on E for $m = 1/2$. At high energies such effects as inelasticity reduce both terms while at lower energies the situation as given by equation (14.44a) is approached. The second term in equation (14.38b) can be dropped on the basis of a similar argument and once this has been done the first term can be ignored relative to the leading term since it is smaller by $(U_o/E)^{1/2}$. Similar considerations also apply to the moments for $\ell \geqslant 2$.

It is also of interest to observe that quantities characterizing low-energy recoils (C_o, U_o) enter only $H_0{}^0(E)$ and $H_1{}^0(E)$ while those characterizing the ion enter $H_2{}^0(E)$ and some of the higher moments, $H_\ell{}^0(E)$, in addition to $H_0{}^0$ and $H_1{}^0$. This is related to the fact that the velocity distribution of the low-energy recoils is essentially isotropic (apart from a weak anisotropy required by the conservation of momentum as expressed by the $\ell = 1$ term) while the velocity distribution of the ion is anisotropic during the slowing down.

Since the above results have been obtained ignoring the effect of the electronic stopping power, we can include the effect of $S_e(E)$ on $H_0{}^0(E)$ by replacing E by $\nu(E)$, *i.e.*,

$$H_0^\circ(E) = \frac{1}{8\psi'(1)} \frac{\nu(E)}{NC_oU_o} = \frac{3}{4\pi^2} \frac{\nu(E)}{NC_oU_o} \tag{14.45}$$

where $\nu(E)/E$ represents the fraction of the energy that is not lost to ionization during the slowing down process.

We turn now to the solution of equation (14.21b) for $M_1 \neq M_2$, and for $n = \ell = 0$, neglecting once again $S_{(1)e}$, we obtain

$$0 = N \int_{T=0}^{T_m} d\sigma_{(1)} [H^{\circ}_{(1)0}(E) - P_o(\cos\varphi')\, H^{\circ}_{(1)o}(E - T)$$

$$- P_o(\cos\varphi'')\, H^{\circ}_o(T)]$$

Noting that the zero-order Legendre polynomials are unity, the above equation reduces to

$$0 = N \int_0^{T_m} d\sigma_{(1)}\, [H^{\circ}_{(1)o}(E) - H^{\circ}_{(1)o}(E - T) - H^{\circ}_o(T)] \qquad (14.46)$$

Because of linear dependence on E, as shown by equation (14.44a), $H_0{}^0(0) = 0$ and the above equation is satisfied for T = E, *i.e.*,

$$H^{\circ}_{(1)o}(E) = H^{\circ}_0(E) \qquad (14.47a)$$

Thus, (14.44a) is also a solution of (14.46). If we include the stopping power then we can write by analogy to equation (14.45)

$$H^{\circ}_{(1)o}(E) = \frac{1}{8\psi'(1)} \frac{\nu_{(1)}(E)}{NC_o U_o} \qquad (14.47b)$$

Before concluding this section we wish to estimate the effect of the bulk-binding energy V on the solution of equation (14.36). We recall that in this case the recoil term $M_\ell{}^0(T)$ has to be replaced by $H_\ell{}^0(T\text{-}V)$ as result of which the solution of the equation of interest becomes somewhat inconvenient. An easier way will be to go back to $F(x, \vec{v}_o, \vec{v})$ in equation (14.8) but since we will not make much use of $H_0{}^0(E)$ for $V \neq 0$ we will only mention the results. Using the surface-binding energy (14.34) we simply make the following substitutions in equations (14.44a), 45, and 47a):

$$\frac{1}{U_o} \rightarrow \frac{1}{V} \left[\left(1 + \frac{U_o}{2V}\right) \ln \left(1 + \frac{2V}{U_o}\right) - 1 \right] \qquad (14.48a)$$

and if equation (14.35) is used the proper substitution becomes

$$\frac{1}{U_o} \rightarrow \frac{1}{2V} \ln \left(1 + \frac{2V}{U_o}\right) \qquad (14.48b)$$

Both of these results with the aid of L'Hospital's rule reduce to $1/U_o$ in the limit of $V \to 0$. Moreover, both of the above expressions are exact for $V \ll U_o$ and very accurate for $V \leqslant 2\ U_o$ but they are inaccurate for $V \gg 2\ U_o$.

Finally we are in a position to write down a general sputtering yield formula, and to do so we first make the substitution

$$H_\ell^n(E) = \frac{1}{8\psi'(1)} \cdot \frac{F_\ell^n(E)}{NC_oU_o} = \frac{3}{4\pi^2}\frac{F_\ell^n(E)}{NC_oU_o} \tag{14.49}$$

With this, equation (14.21a) reads for $n \geqslant 1$

$$n\ell\, F_{\ell-1}^{n-1}(E) + n(\ell+1)\, F_{\ell+1}^{n+1}(E) = (2\ell+1)N\, S_e(E)\frac{d}{dE}F_\ell^n(E)$$

$$+ (2\ell+1)N \int d\sigma\, [F_\ell^n(E) - P_\ell(\cos\varphi')\, F_\ell^n(E-T)$$

$$- P_\ell(\cos\varphi'')\, F_\ell^n(T)] \tag{14.50a}$$

and from equation (14.45), when compared to (14.49), we see that

$$F_\ell^\circ(E) = \delta_{\ell o}\, \nu(E)$$

$$F_o^\circ(E) = \nu(E) \tag{14.50b}$$

Equations (14.50a) and (14.50b) determine moments of the depth distribution of deposited energy $F(x,E,\eta)$ which is defined such that $F(x,E,\eta)dx$ gives the amount of energy deposited in a layer dx at x, by an ion of energy E starting at $x = 0$ and all generations of recoil atoms. Equation (14.50b) is the normalization associated with

$$\int_{-\infty}^{+\infty} dx\, F(x,E,\eta) = \nu(E) \tag{14.50c}$$

The most striking feature of F is that it depends only on the slowing-down characteristics of the primary particle and the high-energy recoil atoms. The depth distribution of deposted energy extends over a distance of the order of the ion range and thus is not influenced by the ranges of the low-energy recoils. For this reason one might be justified in neglecting the cutoff energy U_o that arises from equation (14.37) in the integral (14.50a) and also in neglecting the bulk-binding energy V in the recoil term $F_\ell^n(T)$. Asymptotic expansion for large E of the correct equations shows that neither U_o nor V enters the highest terms. We recall from equation (14.20a) that

$$H_\ell^n(E) = \int dx\, x^n\, H_\ell(x,E)$$

but from the orthogonality condition and from the fact that

$$\int_{-1}^{+1} P_\ell(\eta)d\eta = 0, \qquad \ell \neq 0$$

we see that

$$H_\ell(x,E) = \tfrac{1}{2} \int P_\ell(\eta)d\eta\, H(x,E,\eta)$$

Therefore we observe that

$$H_\ell^n(E) = \int x^n\, dx\, \tfrac{1}{2} \int P_\ell(\eta)d\eta\, H(x,E,\eta)$$

and going back to equation (14.49) we also note that

$$H(x,E,\eta) = \frac{3}{4\pi^2}\,\frac{F(x,E,\eta)}{NC_oU_o}, \qquad M_1 = M_2 \tag{14.51a}$$

and

$$H_{(1)}(x,E,\eta) = \frac{3}{4\pi^2}\,\frac{F_{(1)}(x,E,\eta)}{NC_oU_o}, \quad M_1 \neq M_2 \tag{14.51b}$$

These are the sputtering-yield formulae which will be used in estimating this kind of damage in fusion reactors.

14.2 SPUTTERING DAMAGE IN FUSION FIRST WALL

Before applying the above analysis to the calculation of sputtering yield from the first wall of conceptually designed fusion reactors, it would be desirable to briefly review the underlying physical phenomena of this type of damage and to put the yield formulae in a more readily usable form. We have seen that sputtering occurs when particles impinging on the surface of a solid with energies larger than threshold (20-100 eV) penetrate into the surface and are stopped initiating a collision cascade. This cascade reaching the surface with an energy larger than the surface binding energy U_o [see equation (14.34) or (14.35)] leads to ejection of surface atoms. The most general theory of sputtering, presented in the previous section, assumes randomly distributed atoms in a semi-infinite solid and uses the transport equation to calculate the yield. The result, *i.e.*, equations (14.51a, b) show that the yield is proportional to the distribution function $F(x,E,\eta)$ of the energy deposited in nuclear motion from the incoming particles in the surface layers of the solid, and inversely proportional to the surface-binding energy U_o. With equation (14.15), equation (14.51) can be put in the form

$$S(E,\eta) = \lambda \frac{F(x=0,E,\eta)}{U_o} \tag{14.52}$$

giving the number of (backward) sputtered atoms per incoming particle. We recall that E is the energy of the impinging particle, η is the (cos of) angle between the direction of incidence and the surface normal, and λ is the constant that depends on the target material, *i.e.*,

$$\lambda = \frac{3}{4\pi^2} \frac{1}{NC_o} \tag{14.53}$$

In this expression N is the number of atoms per cm^3 of the solid and C_o, which is related to the interaction potential, has an approximate numerical value of 1.81 [$Å^2$] according to equation (14.31); thus $\lambda = 0.042/N[Å]$. Although the function $F(x,E,\eta)$ has been calculated in detail only the approximate values have been used in sputtering calculations. For example, if scattering of the ion back and forth in the solid is negligible and the mean energy transfer to the target atom is small then one may put

$$F(x,E,\eta) \approx \alpha N S_n(E) \tag{14.54}$$

where $S_n(E)$ is the nuclear stopping power of the material with respect to the incident particles, as given by equation (14.32b) or

$$S_n(E) = \frac{C_{(1)}}{1=m} \gamma^{1-m} E^{1-2m} \tag{14.55}$$

with

$$\gamma = 4 M_1 M_2/(M_1 + M_2)^2$$

and α contains mainly the dependence of the sputtering yield on the angle of incidence and on the ratio M_1/M_2 as well as on the form of the interaction potential through m. Numerical values of this factor are shown in Figure 101.

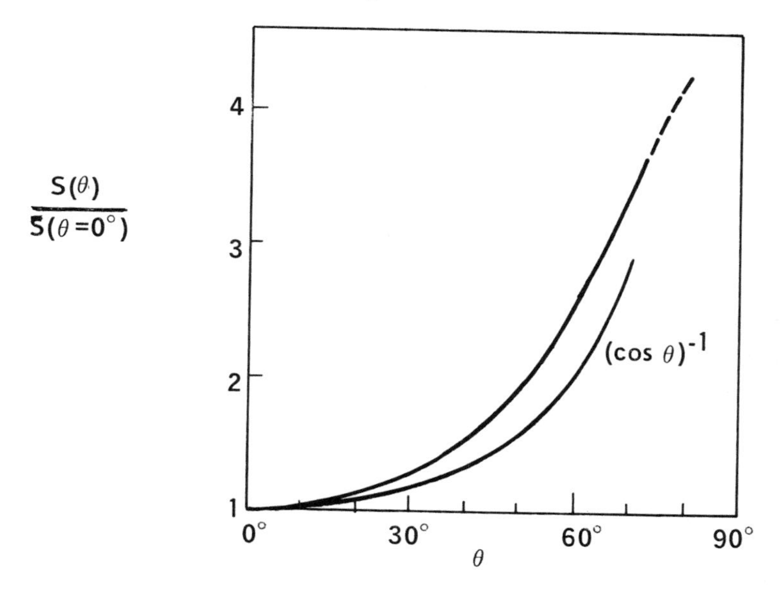

Figure 101. Variation of sputtering yield with angle of incidence for A_r^+ ions on copper (Sigmund 1969).

Another approximate form of $F(x,E,\eta)$ which is useful for neutron sputtering is

$$F(x,E,\eta) \approx \alpha N \langle \sigma E_D \rangle \tag{14.56}$$

where σ is the cross section between the incident particle and target atoms and E_D is the displacement energy of the target atom as explained in Chapter 3 (see also M. T. Robinson, 1969). The energy E_D may be transferred to a lattice atom in a binary collision with a fast neutron or it may be the recoil energy of a (n,γ) reaction with a thermal neutron.

The energy dependence of the sputtering yields appears through the energy dependence of $\langle \sigma E_D \rangle$ or of the stopping power $S_n(E)$. We recall that the latter increases with energy up to a maximum given by equation (14.33) which can be put in the form

$$E_{max} = 0.35\ Z_1 Z_2 e^2 \ \frac{M_1 + M_2}{aM_2} \tag{14.57}$$

with

$$a = 0.8853\ a_o/(Z_1^{2/3} + Z_2^{2/3})^{1/2}$$

being the (Fermi-Thomas) screening radius and a_o = 0.529 A° the first Bohr radius. For $E > E_m$ the nuclear stopping power and correspondingly the sputtering yield decrease with E as 1/E. The dependence of the yield on the angle of incidence is contained in the factor α and for ions with small angles of incidence, *i.e.*, $0 < \eta \leqslant 1$ the dependence is given by

$$\alpha(\eta) = \alpha(1)\eta^{-f} \tag{14.58}$$

where f = 1 for very light ions, and f = 3/2 → 5/3 for heavy ions. It might be noted from Figure 101 that the sputtering yield reaches a maximum at about 70-80° (η = 0.25) and decreases after that. Moreover, if the mean range of the incident particles is much larger than the target thickness–as is the case with fast neutrons–then sputtering occurs in the forward direction, *i.e.*, one obtains forward sputtering for which it has been shown that

$$1 \approx \alpha_{forward} = 17\ \alpha_{backward} \tag{14.59}$$

The dependence of sputtering on the material temperature is contained in U_o which is generally put equal to the heat of sublimation from the surface. This quantity generally changes only 10-20% from room temperature up to the melting point. However the higher the temperature the more target atoms there are with energies larger than the heat of sublimation, thus lowering the effective surface-binding energy. Moreover, as we shall see later, at higher temperatures the residual local temperature increase leads to ejection of atoms through evaporation.

Of special interest to controlled fusion reactors is the sputtering of niobium by different ions and neutrons. These are shown in Figure 102 for incident particles of Nb, He, T, D and H in addition to neutrons. In the case of fast neutrons the yield was computed using equation (14.56) with $\langle \sigma E_D \rangle$ as calculated by Robinson.

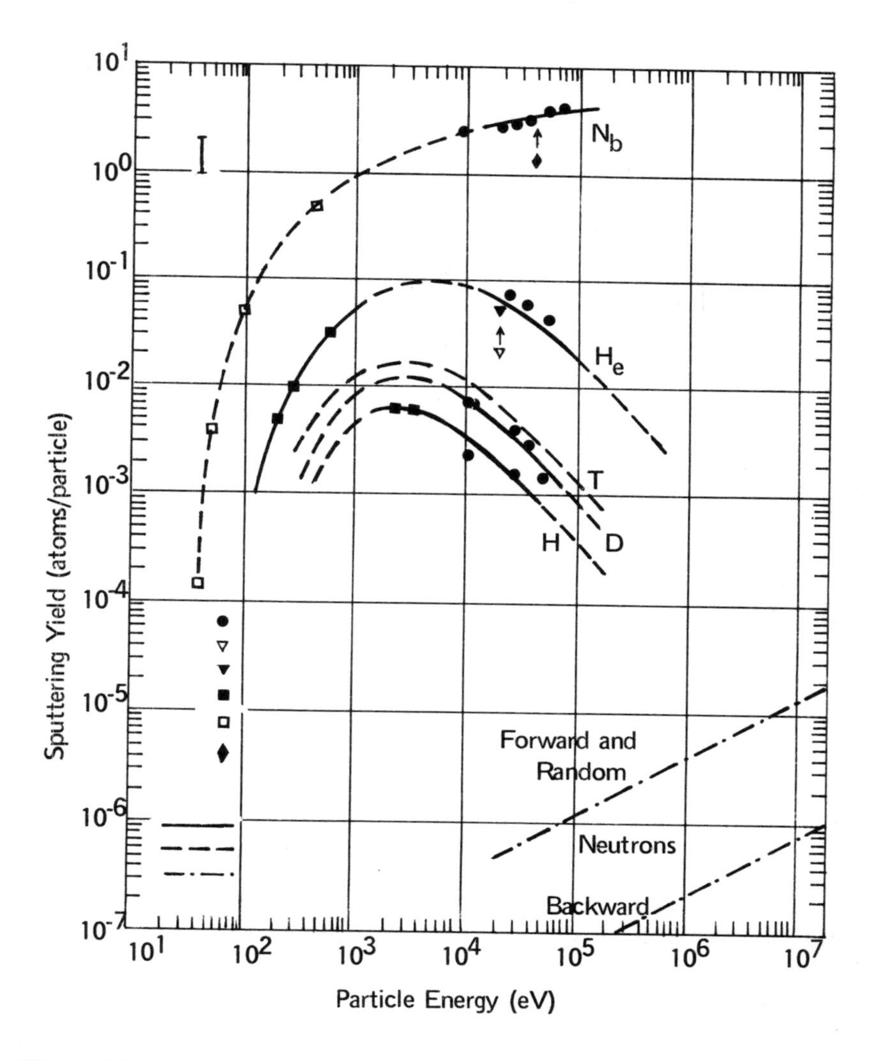

Figure 102. Sputtering yields at normal incidence on polycrystalline Nb (Behrisch 1972).

Experimental verification of neutron sputtering is poor because it is difficult to obtain well defined experimental conditions. Nevertheless experiments have shown in almost all cases that sputtering yield is independent of target temperature as long as the temperature is less than 0.7 of the melting temperature. The dependence of the yield on the angle of incidence for Nb is shown in Figure 103, and the energy distribution of sputtered atoms is shown in Figure 104.

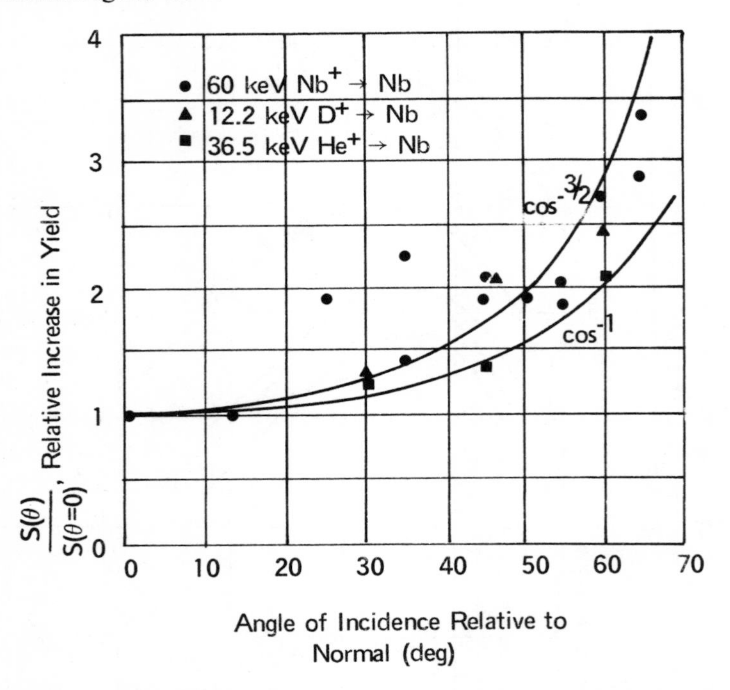

Figure 103. Dependence of sputtering yields on angle of incidence for different ions on Nb (Behrisch 1972).

We turn now to the calculation of sputtering from the first wall of a full-scale fusion reactor. A toroidal reactor of the Tokamak or Stellarotor type will be considered and a D-T fuel cycle will be assumed. Two cases will be examined: the first will be a reactor with a short particle confinement time, *i.e.*, $\tau_c \sim 1$ sec, and thus an ion density of about 10^{14} particles per cm^3 with a burn-up fraction of 2-5% per second. In this system

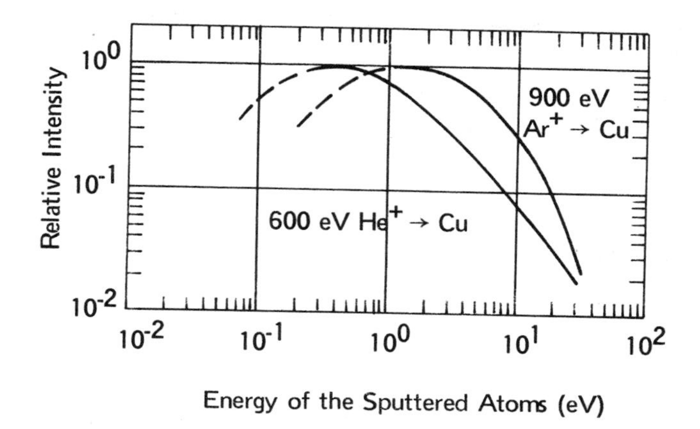

Figure 104. Energy distribution of sputtered material emitted normal to the surface (Behrisch 1972).

the particles leaking from the plasma surface will either strike the wall or will be collected in a "divertor" and ultimately be reinjected into the plasma. The presence of the divertor leads to a reduction of the number of particles striking the wall and hence to lower sputtering damage. For the second case we will consider a reactor with a reasonably long confinement time, *i.e.*, $\tau_c \gg 1$, and an ion density of about 10^{13} per cm^3 with a burn-up fraction of less than 1% per second. For this reactor there will be no refueling during one burning cycle at the end of which the plasma becomes unstable and goes to the wall. The reactor will have to be started again. The two reactors will also be characterized by the wall loading which we take for type 1 ($\tau_c \sim 1$) to be 1000 W/cm^2 while for type II it will be taken as 100 W/cm^2. The blanket will be of Li for tritium breeding, and the first wall will be of Nb with 0.5 cm thickness. The parameters for both types of reactors are shown in Table 35. For the purpose of damage calculation we shall assume that the wall loading is uniform except at such locations as the mouth of the divertor where a ten-fold increased load will be assumed. Although some of the particles ejected from the wall may get redeposited, we will assume in these calculations that no redeposition takes place.

Table 35

Parameters Relevant to First-Wall Load for Two Tokamak Type Reactors (Behrisch 1972)

	Type I P_w=1000 W/cm²	*Type II* P_w=100 W/cm²
Primary neutron current incident on first wall neuts/cm² sec	3×10^{14}	3×10^{13}
J /cm² sec	677	67.7
Total neutron flux in first wall	see Fig. 105	see Fig. 105
- heat transferred to plasma if distributed uniformly onto first wall (b6 radiation and particle flux) J/cm² sec	170	17
Additional heat produced by neutron reactions in blanket J/cm² sec	160	16
Total thermal-energy output J/cm² sec	1000	100
Ratio of plasma volume to first wall area cm³/cm²	50-100	50-100
Plasma parameters		
density cm^{-3}	$1\text{-}2 \times 10^{14}$	$0.1\text{-}1 \times 10^{14}$
burn-up fraction sec^{-1}	2-5%	0.5-1%
temp. at end of discharge keV	20-30	10-20
First-wall material	Nb(+Zr)	Nb(+Zr)
First-wall thickness cm	0.5	0.5
Energy dumped in first wall at end of discharge (uniform distribution) J/cm² sec	20-40	15-30

We examine first sputtering by neutrons and for an incident neutron current of 3×10^{14} (type II) the energy distribution of the neutron flux in the first wall is shown in Figure 105. If we normalize to neutron flux power eV then the value of 2.12×10^8 neutrons per eV per cm² per sec is given to the incident neutron current as indicated in the figure. The sputtering rates are obtained by multiplying the distribution of Figure 105 by the calculated yields of Figure 102 and the results are shown in Figure

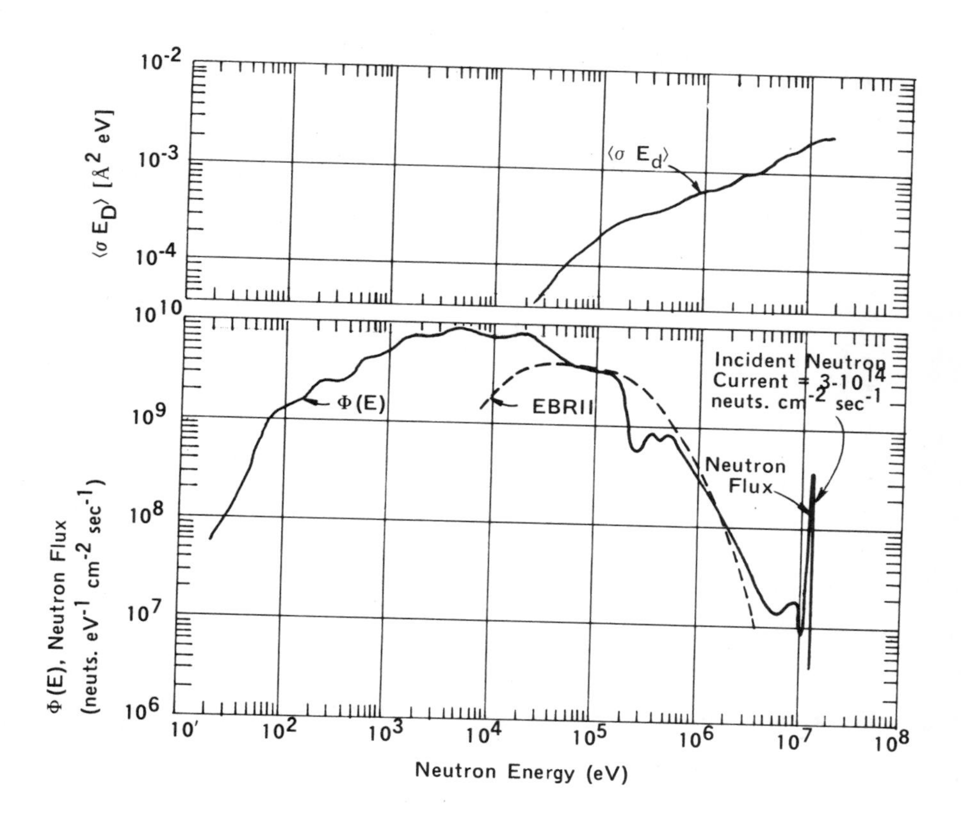

Figure 105. Mean value of σ E_D and neutron flux as a function of neutron energy assuming an incident neutron current of 3 x 10^{14} cm^{-2} sec^{-1}.

106. We note that the upper dashed curve of this figure represents the integration of the histogram starting from the highest energy, *i.e.*, the sputtering by neutrons with energies larger than E. It shows that the contribution to fast neutron sputtering is about 2.6% due to the incident neutron current, 41% due to the random neutron flux around 14.1 MeV, and 56% due to the lower energy neutron flux. The total erosion is about 10^{10} atoms per cm^2 per sec which corresponds to 320 Å per year if we assume 60% operating time (see Table 36). Even if this theory has underestimated the yields by two orders of magnitude it might be concluded that sputtering by fast neutrons is not the major cause of erosion

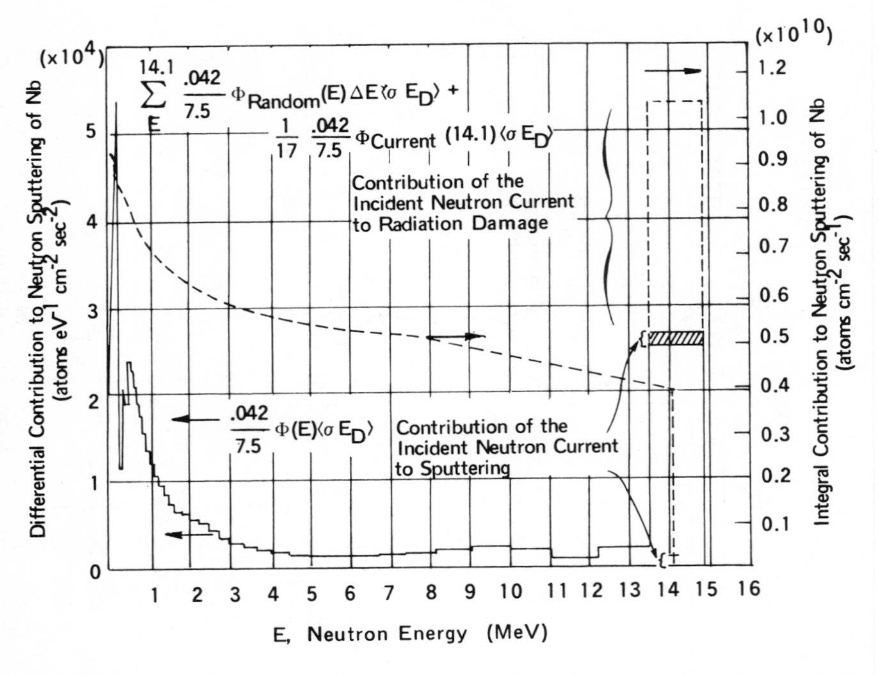

Figure 106. Differential and integral contribution to fast neutron sputtering of Nb for incident neutron current of 3 x 10^{14} cm^{-2} sec^{-1}.

in the first wall of a fusion reactor. This is consistent with the fact that sputtering has not been observed to any appreciable degree in fission reactors.

We turn now to the case of sputtering by plasma particles and recall that 20% of the energy released in the D-T reaction is carried by the α-particles. If the plasma temperature is to remain constant the α-energy must ultimately be released to the first wall although in the case of a steadily refueled reactor some of this energy goes into heating the cold fuel. The latter fact does not, however, reduce the wall loading, and particle bombardment of the first wall can be estimated from the α-energy if the energy losses by radiation are known. This is based on the assumption that the neutrons produced by the D-T reaction leave the plasma and that the α-energy is completely transferred to the fusion plasma.

Table 36

First Wall Erosion per Second and per Year for 60% Operating Time (Behrisch 1972)

	Radiation Damage		*Type I*		*Type II*	
			$\frac{Atoms}{cm^2\ sec}$	$\frac{A}{year}$	$\frac{Atoms}{cm^2\ sec}$	$\frac{A}{year}$
Sputtering	Neutron Sputtering		10^{10}	320	10^{9}	32
	Particle Sputtering, assuming a total particle energy of 100J/cm^2 sec, Type I and 10J/cm^2 sec, Type II	E=50eV	$1.2x10^{14}$	$4x10^{6}$	$1.2x10^{13}$	$4x10^{5}$
		E=100eV	10^{15}	$3.5x10^{7}$	10^{14}	$3.5x10^{6}$
		E=500eV	$5.4x10^{15}$	$2x10^{8}$	$5.4x10^{14}$	$2x10^{7}$
		E=1000eV	$4x10^{15}$	$1.5x10^{8}$	$4x10^{14}$	$1.5x10^{7}$
		E=5000eV	$1.5x10^{15}$	$5x10^{7}$	$1.5x10^{14}$	$5x10^{6}$
Evaporation	Uniform load evaporation at ~1400°K (~1258°K respectively)		10^{5}	1	1	1
	10-fold increased load llocally ~2580°K (~1400°K respectively)		$2x10^{16}$	$7x10^{8}$	10^{5}	1
	Erosion due to dumping of the plasma (40J/cm^2 sec 30J/cm^2 sec, respectively) in time τ	=1 sec		1		1
	Uniform dist 10^5 disch/yr	$=10^{-1}$ sec	10^{4}	1	10^{3}	1
		$=10^{-2}$ sec	$3x10^{6}$	1	10^{5}	1
	10-fold increased load locally 10^5 disch/yr	=1 sec	$3x10^{8}$	1	$3x10^{7}$	1
		$=10^{-1}$ sec	10^{14}	$2x10^{4}$	$3x10^{12}$	$5x10^{2}$
		$=10^{-2}$ sec	$3x10^{19}$	$5x10^{9}$	10^{18}	$2x10^{8}$
	Blistering after total dose of 10^{18}-10^{19} He-particles cm^{-2}		?	?	?	?

Note: 1 Nb atom cm^{-2} sec^{-1} $\cong$ $1.8x10^{-15}$ Å cm^{-2} sec^{-1}.

So long as the impurity concentration in the plasma remains small, the bremsstrahlung radiation accounts for about 10% of the α-energy and the synchrotron radiation accounts for a similar fraction. This means that about 60-80% of the α-energy, *i.e.*, about 100 W/cm^2 in one reactor type (or 10 W/cm^2 in the other), must leave the plasma to the wall through particles. This transport of energy to the wall can be divided into that carried by the electrons and that transported by the ions and neutrals, although there is a great deal of uncertainty as to the extent of the electron contribution. Since the plasma must remain quasi-neutral the number of electrons going to the wall must equal the number of ions but the electrons may lose part of their energy ionizing the neutrals released from the wall; their contribution may therefore be negligible.

For a given total energy the number of ions and neutrals going to the wall depends on their mean energy and the energy distribution. Since very little is known about particle flow in the region between the plasma and the first wall, assumptions must be made and here we will consider the following model. We shall assume that the plasma has a sharp boundary and that a neutral residual gas exists between this boundary and the first wall. The atoms of this gas undergo charge exchange collisions at the plasma boundary producing fast neutrals which are either backscattered at the first wall or trapped and later released, thereby returning to the plasma boundary. In this model, therefore, part of the neutrals is ionized at the plasma boundary and part is trapped at the first wall and, unless replenished, the gas is pumped away in times of the order of 10^{-4} seconds (Behrisch and Heiland, 1970).

Because of the above uncertainties we shall assume that the total energy, E_{tot}, carried by the neutrals and the ions will arrive at the first wall with a mean particle energy $\bar{E}$. We take E_{tot} to be 100 J/cm^2 sec and assume that the energy distribution of the ions or neutrals is given by

$$f(E)dE = \frac{E_{tot}}{\bar{E}^2} \; \frac{K+1}{\Gamma(K+1)} \left(\frac{(K+1)E}{\bar{E}}\right)^K \exp\left[-\frac{(K+1)E}{\bar{E}}\right] dE \tag{14.60}$$

where K is a dimensionless parameter, and the normalization is such that

$$E_{tot} = \int E\, f(E)dE = \overline{E} \int f(E)dE \tag{14.61}$$

We observe that for K = ½ this distribution is Maxwellian, while for larger K the low energy part falls off faster. For wall damage calculations K > ½ appears to be more realistic since very low-energy neutrals are likely to be ionized and low-energy ions are more likely to be confined by the magnetic field than high-energy ions. Assuming K = 1 the distribution given by (14.60) for different mean energies $\overline{E}$ is shown in Figure 107 where the curve for $\overline{E}$ = 180 eV can be fitted to fast neutrals from plasmas in a number of Tokamak devices. The figure also

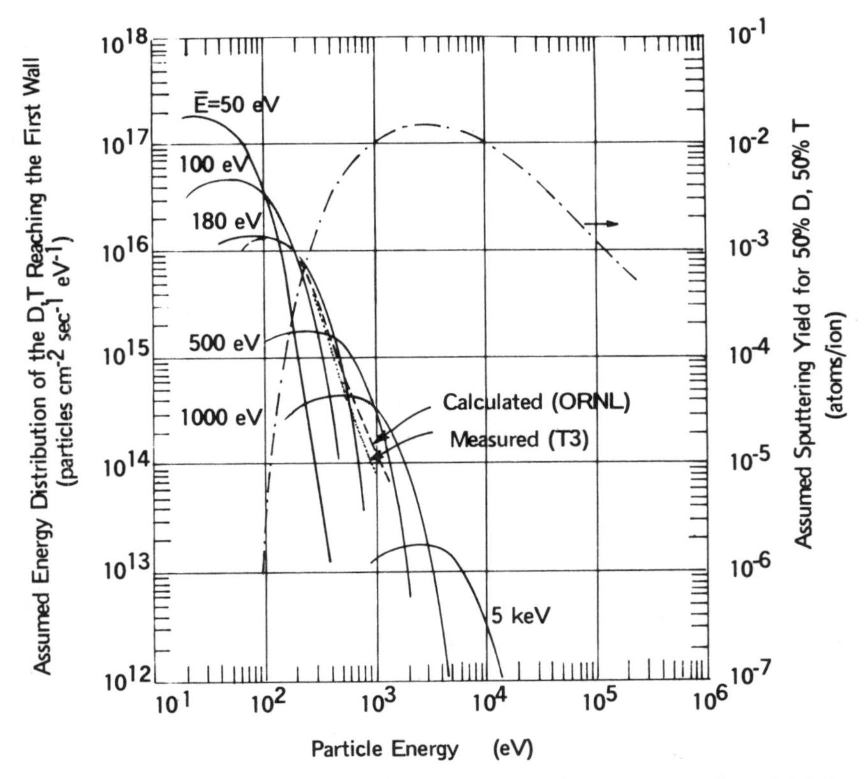

Figure 107. Assumed ion energy distribution and mean sputtering yield for calculating the results of Figure 108.

shows the mean sputtering yields for a 50-50 mixture of D^+ and T^+ on Nb using data from Figure 102. For the same mixture the sputtering yield for a total particle load of 100 J/cm^2 sec assuming the same mean energy $\bar{E}$ for all particles is shown in Figure 108. From the values in this figure first wall erosion due to particle sputtering for type I and type II reactors without diverters is given in Table 36. We might conclude from this data that

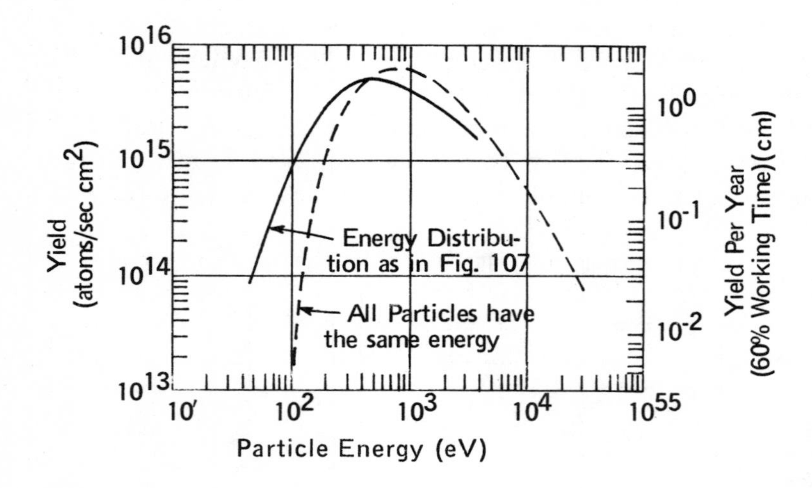

Figure 108. First wall erosion of Nb by sputtering for a total load of 100 J cm^{-2} sec^{-1}.

sputtering by plasma particles is a serious first wall erosion problem if the mean energy of the particles striking the wall is smaller than that corresponding to the plasma temperature (see Vernickel, 1972). The reason for this is (1) that the number of particles bombarding the wall has to increase in order to carry the same total energy E_{tot} at lower mean energy $\bar{E}$, and (2) that the sputtering yields increase for lower $\bar{E}$ with a maximum at 2-3 keV as shown in Figure 108. For further decrease in $\bar{E}$ the sputtering yield decreases and this decrease is much faster if all particles are assumed to have the same energy. As for the effect on the sputtering yield by Nb atoms released from the wall, heated in the plasma and then returned to the wall, it has been shown (Carruthers, 1969) that this effect becomes important only if the impurity concentration exceeds 5×10^{-3}. This, of course, will also result in enhanced energy loss due to radiation.

14.3 EROSION DUE TO EVAPORATION

When a solid is heated to a temperature well above 0°K some of the atoms that are in the high-energy tail of the thermal distribution will have sufficient energy to overcome the surface-binding energy; if the momentum of these atoms at the surface is directed away from the surface they will evaporate. The rate of evaporation can be estimated from the vapor pressure P of the solid material above the surface and is given by

$$\dot{n}(T) = \frac{dn}{dt} = \tilde{\alpha} \frac{N\bar{v}}{4} \qquad (14.62)$$

where $\tilde{\alpha}$ is the probability that an atom from the gas phase sticks at the surface (generally $\tilde{\alpha} = 1$). Noting that $P = Nk_BT$ with k_B being the Boltzman constant and T being the equilibrium temperature, in this case the surface temperature, then in view of the fact that the mean velocity $\bar{v}$ is proportional to $\sqrt{T/M}$ equation (14.62) can be put in the form

$$\dot{n}(T) = \tilde{\alpha} \frac{3.5 \times 10^{22}}{\sqrt{M}} \frac{P(T)[\text{Torr}]}{\sqrt{T\,[^\circ K]}} \frac{\text{atoms}}{\text{cm}^2\ \text{sec}} \qquad (14.63)$$

where M is the mass number. Vapor pressure curves for metals are well known (see Honig and Kramer, 1969) and may be described by

$$P = P_o \exp\left[-\frac{\Delta H}{k_B T}\right] \qquad (14.64)$$

where ΔH is the heat of sublimation which decreases slightly with increasing temperature and $k_B = 8.4 \times 10^{-5}$ eV/deg. For materials of interest as first wall material in a controlled fusion reactor the vapor pressure as well as the evaporation rates are shown in Figure 109. It turns out that for these materials the evaporation rate to within a factor of about 2 can be expressed as

$$\dot{n}(T)\left[\frac{\text{atoms}}{\text{cm}^2\ \text{sec}}\right] \approx 10^{20}\ P(T)\ [\text{Torr}] \qquad (14.65)$$

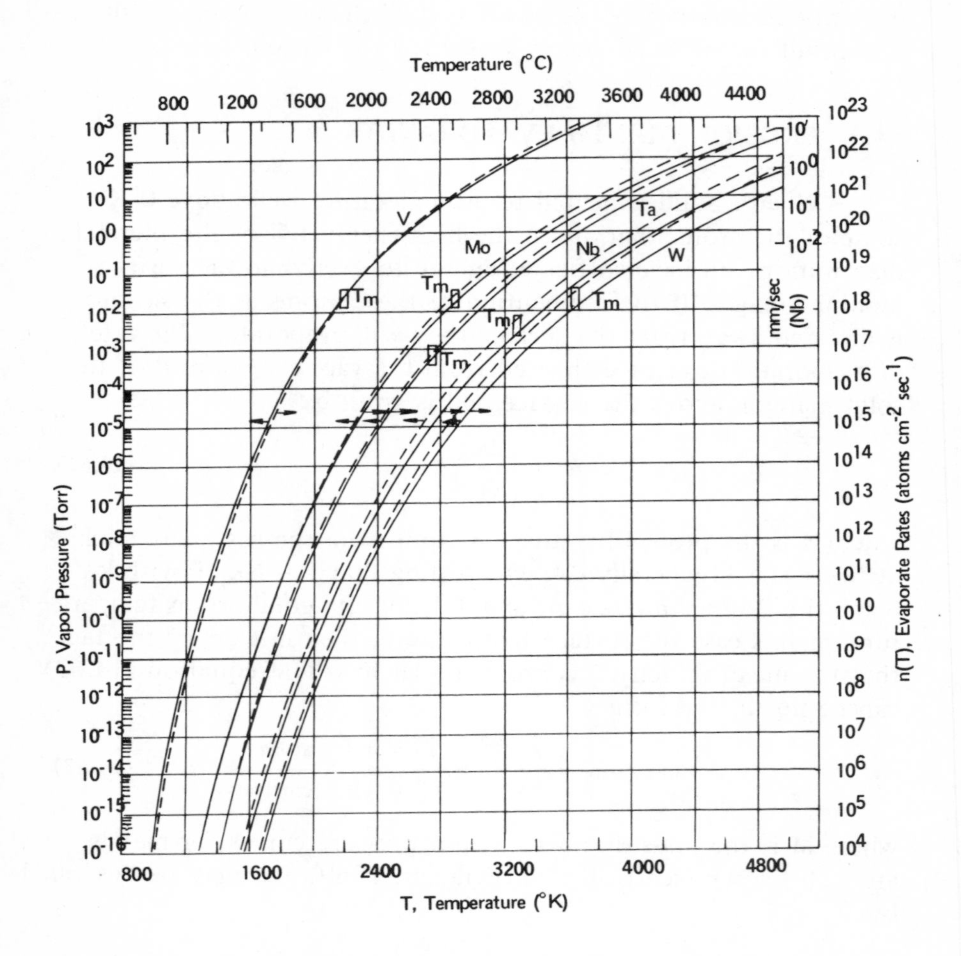

Figure 109. Evaporation rates and vapor pressure for first-wall materials.

The most severe thermal loading on the first wall occurs when the plasma becomes unstable or at the end of a discharge when the plasma is dumped on the wall in a very short period of time. In order to calculate the number of atoms evaporated during one heat pulse we must first calculate the temperature increase in the solid as a result of sudden heating.

The equation of interest in this case is the heat conduction equation which we apply to a semi-infinite solid so that the temperature is a function of one spatial dimension only, *i.e.*,

$$\frac{\partial^2 T}{\partial x^2} = \frac{1}{\alpha}\frac{\partial T}{\partial t} \tag{14.66}$$

where

$$\alpha = \frac{k}{\rho C_p} \tag{14.67}$$

is the thermal diffusivity, k is the thermal conductivity, C_p is the specific heat at constant pressure, and ρ is the density of the solid. If the temperature of the solid initially is $T(x,0) = T_0$ then the temperature in equation (14.66) can be viewed as the difference between the instantaneous temperature $T(x,t)$ and T_0, *i.e.*, $T - T_0$. The boundary conditions for the problem are

(a) $T = 0$ at $t = 0$ (14.68a)

(b) $\partial T/\partial x = -C = -E/\tau\, kA$ at $x = 0$ (14.68b)

(c) $T = 0$ at $x = \infty$ (14.68c)

The second condition represents the heat flow into the solid whereby a total energy E is dumped in the area A in the time τ. We solve equation (14.66) by the Laplace transform method which puts it in the form

$$\frac{d^2 T(x,s)}{dx^2} = \frac{S}{\alpha} T(x,S)$$

whose solution can readily be written as

$$T(x,S) = A(S)\, e^{-x\sqrt{S/\alpha}} \tag{14.69}$$

where $A(S)$ is a constant of integration. Differentiating the above equation with respect to x and evaluating the result at $x = 0$ we get

$$\left.\frac{\partial T}{\partial x}\right|_{x=0} = -\sqrt{S/\alpha}\, A(S) \tag{14.70}$$

If we now write the Laplace transform of equation (14.68b) we get

$$\frac{\partial T(x,S)}{\partial x} = C(S)$$

which upon comparing with equation (14.70) we get the value of the constant A or

$$A(S) = \sqrt{\alpha}\,\frac{C(S)}{\sqrt{S}}$$

so that the solution, equation (14.69) finally becomes

$$T(x,S) = \sqrt{\alpha}\,\frac{C(S)}{\sqrt{S}}\,\exp\left[-x\sqrt{S/\alpha}\right] \tag{14.71}$$

Before inverting the above solution we must first obtain C(S) and to do that we recall that sudden heating simply means that the function C(t) is a step function given by

$$C(t) = C\,[u(t) - u(t-\tau)] \tag{14.72}$$

and illustrated in Figure 110. For $t < \tau$

$$C(t) = C\,u(t)$$

so that $C(S) = C/S$.

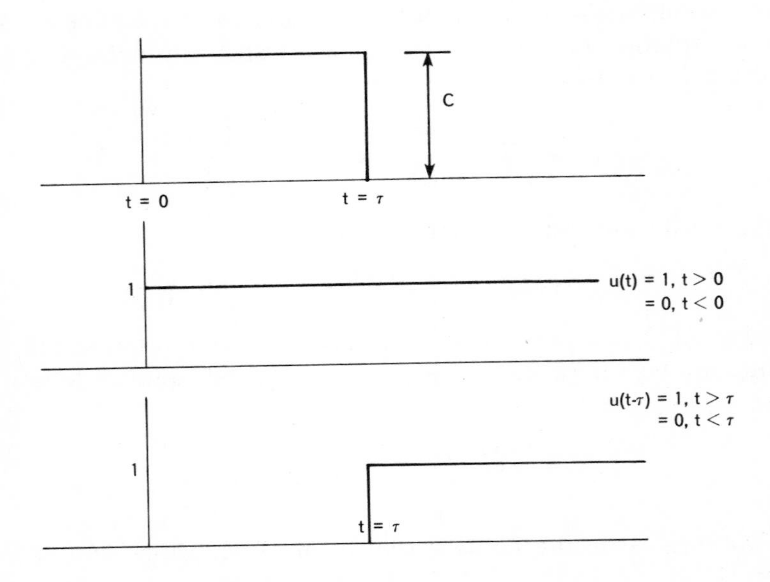

Figure 110. Representation of sudden heating.

For $t > \tau$

$$C(t) = C\ [u(t) - u(t-\tau)]$$

which yields

$$C(S) = C \left[\frac{1}{S} - \frac{e^{-S\tau}}{S} \right] \tag{14.73}$$

having used the familiar Laplace transform rules that

$$L\ F(t + b) = e^{-bS}\ f(S)$$

$$L\ F(t) \quad = f(S)$$

$$L\ \text{const} \quad = 1/S$$

With these, the complete solution during the heating phase, *i.e.*, $t < \tau$ can be written as

$$T(x,t) = C\sqrt{\alpha} \left[2 \sqrt{\frac{t}{\pi}}\ e^{-x^2/4\alpha t} - \frac{x}{\sqrt{\alpha}}\ \text{erfc} \left(\frac{x}{2\sqrt{\alpha t}} \right) \right] \tag{14.74}$$

where Laplace inversion tables or the convolution theorem can be used to obtain this result. In this expression the function erfc(x) is given by

$$\text{erfc}(x) = \frac{2}{\sqrt{\pi}} \int_x^{\infty} e^{-\xi^2}\ d\xi \tag{14.75}$$

which is related to the familiar error function

$$\text{erf}(x) = \frac{2}{\sqrt{\pi}} \int_0^{x} e^{-\xi^2}\ d\xi$$

by the fact that $\text{erf}(x) + \text{erfc}(x) = \text{erf}(\infty) = 1$. At $x = 0$ equation (14.74) reduces to

$$T(o,t) = C \sqrt{\frac{4\alpha t}{\pi}} = \frac{E}{A\tau} \sqrt{\frac{4}{\pi \rho k C_p}}\ \sqrt{t} \tag{14.76}$$

and for $t = \tau$ it gives the maximum temperature increase of the surface of the solid due to dumping of energy E per area A in time τ, *i.e.*,

$$(\Delta T)_{max} = \frac{E}{A} \frac{1}{\sqrt{\tau}} \sqrt{\frac{4}{\pi \rho k C_p}} = T(o,\tau) = T_{max} - T_o \tag{14.77}$$

During the cooling phase, *i.e.*, $t > \tau$ the solution is given by equations (14.71) and (14.73) or

$$T)\ T(x,S) = \sqrt{\alpha} \frac{C}{S\sqrt{S}} \quad e^{-x\sqrt{S/\alpha}} \ [1 - e^{-S\tau}] \tag{14.78}$$

and if once again we use the inverse transform

$$\frac{e^{-a\sqrt{S}}}{S\sqrt{S}} \rightarrow \frac{2\sqrt{t}\ e^{-a^2/4t}}{\sqrt{\pi}} - a \text{ erfc } (a/2\sqrt{t})$$

then the spatial and temporal distribution of the temperature in the solid becomes

$$T(x,t) = C\sqrt{\alpha} \left[2\sqrt{\frac{t}{\pi}}\ e^{-x^2/4\alpha t} - \frac{2}{\sqrt{\pi}} \sqrt{t-\tau} \quad e^{-\frac{x^2}{4\alpha(t-\tau)}} \right.$$

$$\left. - \frac{x}{\sqrt{\alpha}} \text{ erfc } (x/2\sqrt{\alpha t}) + \frac{x}{\sqrt{\alpha}} \text{ erfc } (x/2\sqrt{\alpha(t-\tau)}) \right] \tag{14.79}$$

At the surface, *i.e.*, at $x = 0$ this expression simplifies to

$$T(o,t) = C\sqrt{4\alpha/\pi}\ [\sqrt{t} - \sqrt{t \cdot \tau}\ ] \tag{14.80}$$

The time history of the temperature during both phases is shown in Figure 111, and if we choose to focus our attention on the cooling phase we can shift our origin by letting $t - \tau \rightarrow t$ so that equation (14.80) becomes

$$T(o,t) = C\sqrt{4\alpha/\pi}\ [\sqrt{t+\tau} - \sqrt{t}\ ]$$

or

$$T(o,t) = C\sqrt{4\alpha\tau/\pi}\ [\sqrt{1 + t/\tau} - \sqrt{t/\tau}\ ]$$

If we now utilize equation (14.77) the above result further becomes

$$T(o,t) = T_{max}\ [\sqrt{1 + t/\tau} - \sqrt{t/\tau}\ ] \tag{14.81}$$

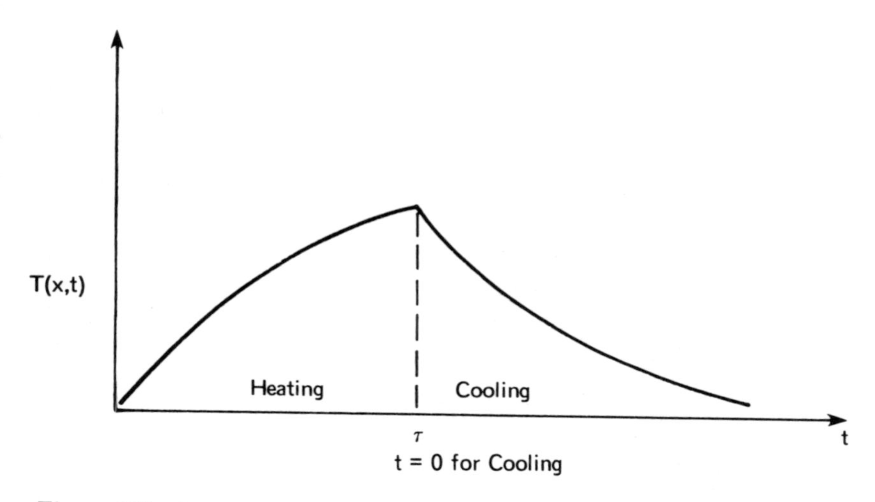

Figure 111. Variation of temperature with time during heating and cooling phases.

Of special interest is the surface temperature shortly after the end of heating, *i.e.*, for $t \leqslant \tau$, a condition which allows us to approximate the above relation by

$$T(o,t) = T_{max} \exp [- \sqrt{t/\tau}] \tag{14.82}$$

This result along with equation (14.76) will be used to calculate the rate of evaporation from the surface of the first wall. Before we do that let us note that the temperature profile as a function of depth x in the semi-infinite solid at the end of heating, *i.e.*, at $t = \tau$ is given by equation (14.74) evaluated at $t = \tau$ and can be written with the aid of equation (14.77) in terms of the universal unit

$$u = \frac{x}{\sqrt{\tau}} \sqrt{\frac{\rho C_p}{4k}} = \frac{x}{2\sqrt{\alpha\tau}} \tag{14.83}$$

as

$$T(x) = T(o,\tau) [e^{-u^2} - 2u \int_u^\infty e^{-\xi^2} d\xi] \tag{14.84}$$

For several materials of interest equations (14.77) and (14.84) are shown respectively in Figure 112 and 113 using the values of Table 37.

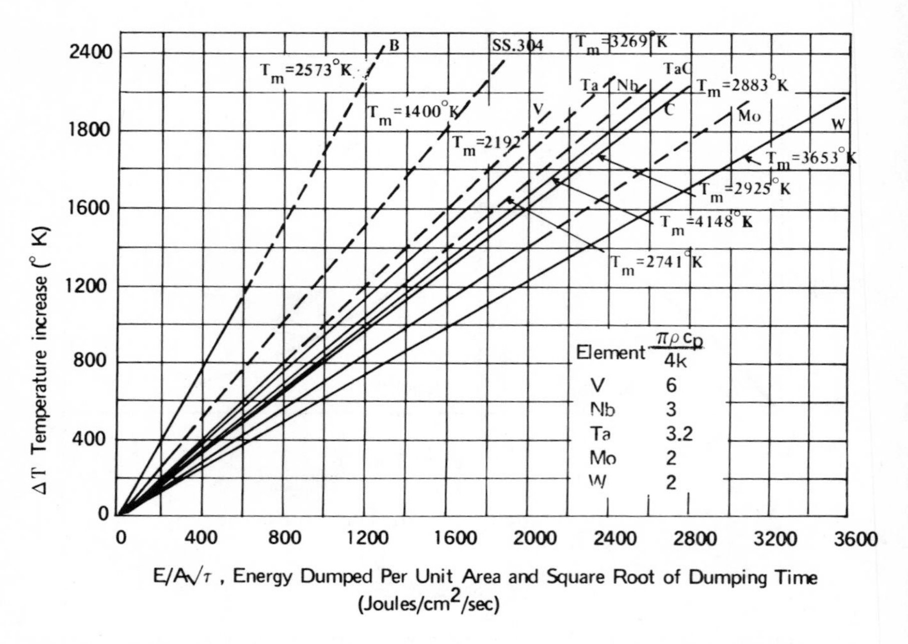

Figure 112. Temperature increase of surface due to sudden heating.

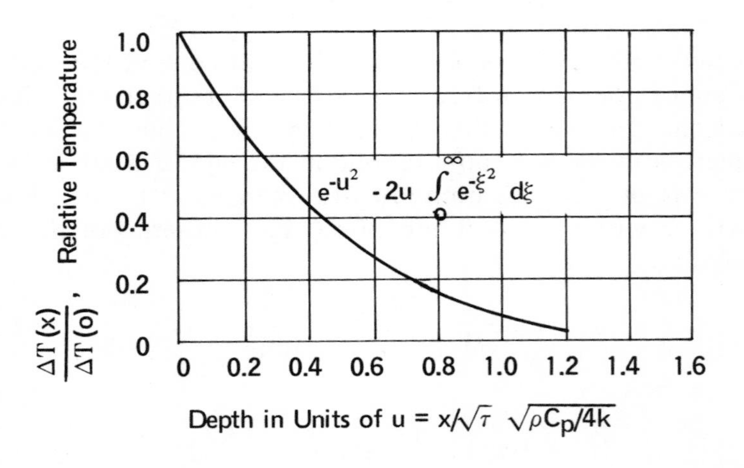

Figure 113. Depth distribution of temperature after sudden heating.

Table 37

Thermodynamic Data of Various Elements (Behrisch 1972)

Element		*V*	*Nb*	*Ta*	*Mo*	*W*	*C*Graphite*	*B*	*S.S.403*	*TaC**
Atomic No. Z		23	41	73	42	74	6	5	(18% Cr	
Mass No. M		50.94	92.9	180.95	95.94	183.85	12	10.82	8% Ni)	
Density	gm/cm^3	5.87	8.57	16.6	9.01	19.3	2.25	2.34	7.9	13.9-14.5
$(atoms/cm^3)\ 10^{-22}$		6.93	5.56	5.52	5.65	6.31	11.3	13		
Melting Pt. °K		2192	2688	3269	2883	3653	Subl.3925	2573	1400	4148
Heat of sublimation	300°K	5.33	7.5	8.1	6.83	8.8	8.2-10.4	5.6		
	500	5.27	7.43	8.05	6.78	8.78				
	1000	5.16	7.28	7.95	6.63	8.64				
	1500	4.96	7.14	7.77	6.48	8.5				
	2000	4.75	6.97	7.62	6.3	8.34				
ΔH(eV/atom)										
Thermal Conductivity	500°K	0.331	0.567	0.582	1.3	1.49	0.8-1	1.41	0.19	0.4
k[J/cm s °K]	1000	0.386	0.644	0.602	1.12	1.21	.49-.64	0.063		0.4
	1500	0.447	0.721	0.622	0.97	1.09	0.4-0.5			0.4
	2000	0.509	0.791	0.64	0.88	1.00	0.3-0.4			0.4
Specific Heat	500°K	0.5	0.28	0.145	0.254	0.135	1.2	1.69	0.5	0.25
C_p[J/gr °K]	1000	0.64	0.3	0.155	0.29	0.15	1.67	2.4		0.29
	1500	0.7	0.33	0.165	0.33	0.16	1.8			0.32
	2000	0.85	0.37	0.17	0.38	1.17	2			0.33

*Chemical sputtering under hydrogen bombardment expected.

The semi-infinite approach used in obtaining the above results is valid provided u > 1, which in the case of a first wall with thickness "d" means

$$u = \frac{d}{\sqrt{\tau}} \sqrt{\frac{\rho C_p}{4k}} > 1$$

or that the dumping time is smaller than

$$\tau < \frac{\pi \rho C_p}{4k} \; d^2 \tag{14.85}$$

For larger τ the temperature increase is given by equation (14.77) with (14.85) substituted for τ, *i.e.*,

$$\Delta T = \frac{E}{A\tau} \; \frac{d}{k} \tag{14.86}$$

which is the same result as would be obtained in the steady-state situation.

We turn now to the calculation of the number of atoms evaporated during one heat pulse and for that we assume uniform deposition during τ, constant heat of sublimation ΔH, constant thermal conductivity k, and constant specific heat C_p in the temperature range of interest. We recall from Figure 11 1 that the temperature rises with $\sqrt{t}$ during the dumping and drops exponentially afterward so that most of the evaporation occurs during the heating time τ. The number of atoms evaporated during the heating phase is obtained by substituting equation (14.76) along with equation (14.64) into equation (14.63) and integrating over the heating period. The result is

$$n = \frac{3.5 \times 10^{22}}{\sqrt{M}} \; P_o \int_0^t \frac{dt}{\sqrt{T_o + \frac{E}{A} \frac{\sqrt{t}}{\tau} \sqrt{\frac{4}{4\pi \rho k C_p}}}} \; \times$$

$$\times \; \exp \left\{ - \frac{\Delta H}{k_B \sqrt{\left(T_o + \frac{E}{A} \frac{\sqrt{t}}{\tau} \sqrt{\frac{4}{4\pi \rho k C_p}} \right)^2}} \right\} \tag{14.87}$$

The corresponding number during the cooling phase is obtained by using equation (14.82) along with equation (14.64) and (14.63) to give

$$n = \frac{3.5 \times 10^{22}}{\sqrt{M}} P_o \int_0^{\infty} \frac{dt \exp\left[-\frac{\Delta H}{k_B T_{max} \exp(-\sqrt{t/\tau})}\right]}{\sqrt{T_{max} \exp(-\sqrt{t/\tau})}} \tag{14.88}$$

These integrals can be readily carried out analytically if the various thermodynamic parameters are independent of temperature. With these assumptions and for $\Delta H/k_B T_{max} \ll 1$ the total number of atoms evaporated per heat pulse can be calculated using asymptotic expansions to yield

$$n = \dot{n}(T_{max}) (\tau_{h\,eff} + \tau_{c\,eff}) \tag{14.89}$$

where

$$\tau_{h\,eff} \approx \tau \frac{k_B T_{max}}{\Delta H} \frac{T_{max}}{T_{max} - T_o} \left(2 - \frac{5T_{max} - 3T_o}{T_{max} - T_o} \frac{k_B T_{max}}{\Delta H} \cdot \right)$$

$$\tau_{c\,eff} \approx 2\tau \left(\frac{k_B T_{max}}{\Delta H} \right)^2$$

where $\dot{n}(T_{max})$ is equation (14.63) evaluated at T_{max}. In expression (14.89) $\dot{n}(T_{max})$ is the dominating factor while $\tau_{h\,eff}$ and $\tau_{c\,eff}$ are relatively slowly varying functions of T_{max} and ΔH. Generally $\tau_{c\,eff}/\tau_{h\,eff} \ll 1$ and approaches ½ for large T_{max}. For Nb at $T_0 = 1400°K$ with the other parameters of interest for the first wall of fusion reactors it is found that

$$0.134\, \tau \leqslant \tau_{c\,eff} + \tau_{h\,eff} \leqslant 0.24\, \tau$$

and a reasonable mean value of 0.2 τ. With this equation (14.89) becomes

$$n = 0.2\, \tau\, \dot{n}\, (T_{max}) \tag{14.90}$$

and the results, using the data in Table 37, are plotted in Figure 114. It might be noted that equation (14.85) is reasonably accurate up to the temperature where the surface melts as indicated in Figure 114. At the melting point the temperature will remain constant for a time giving a slower increase of evaporation with the thermal load $E/A\sqrt{\tau}$. But due to the lower ΔH of the liquid, evaporation will increase faster with $E/A\sqrt{\tau}$ after the surface has

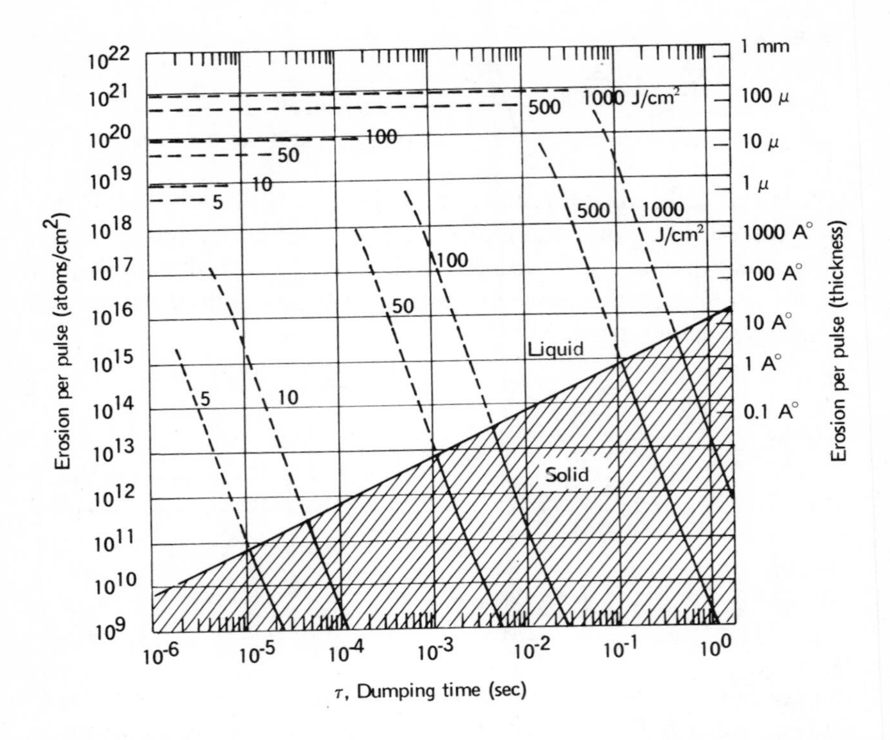

Figure 114. Number of evaporated Nb-atoms versus dumping time for different energy loads.

melted. The total number of atoms which can evaporate is limited by the total amount of energy available; these limits are shown as dashed lines in Figure 114. They will be reached only within a factor of 2 or less because of cooling by radiation and because part of the energy will always be conducted into the solid.

14.4 WALL DAMAGE BY BLISTERING

Energetic projectiles impinging on a solid can penetrate through the medium and displace lattice particles from their sites; and when they have slowed down sufficiently they can be trapped in the lattice. Near the end of the range of these projectiles a region of intense radiation damage occurs where the vacancies formed can coalesce and form voids. If the solubility of the gaseous projectiles trapped in the solid is very small (as it is for inert gases in metals) a fraction of the gas can precipitate out of solution and convert voids into gas-filled bubbles. Such bubbles can grow in regions of lower surface energy (*e.g.*, regions of high defect concentration) and ultimately explode.

In a controlled fusion reactor (CTR) blistering can lead to (1) serious damage and erosion of bombarded wall surfaces thus affecting its life time, and (2) the release of gases which will contaminate the plasma and thereby cool it below the temperature required for fusion reaction. Metals of interest for CTR first wall can be divided into two groups so far as their interaction with hydrogen is concerned. The first group containing V, Nb, Ta, Ti, and Zr has a very high solubility or positive heat of solution for hydrogen while the second group which contains Mo, W, Cu, and Be has a negative heat of solution and the solubility of hydrogen in it is extremely small. Helium on the other hand is not soluble in any metal to a large extent, and it is expected to diffuse very well in a metal lattice as long as it is not trapped at a lattice defect like a vacancy. This means that helium can be trapped in a metal only if the incoming ions have an energy high enough ($>$ 250 eV for He in W) to form defects or if lattice defects are introduced by other means. It has been concluded (Kornelsen, 1971) that one vacancy in W is able to trap up to 3 atoms and that implanting beyond 3×10^{13} per cm^3 of He atoms results in larger clustering of trapped He atoms with higher binding energies. At a total dose of 10^{15}-10^{16} ions/cm^2 the helium gas forms relatively large bubbles in the solid which are very stable and cannot be annealed out at temperatures up to near the melting point.

Recent results (Das and Kaminsky, 1972 and 1973) show that Nb irradiated with helium ions will blister at room temperature with an average size blister of 400 μm. The gas pressure in the bubbles can be estimated from measurements of the diameter and skin thickness. For bubbles with radii less than 1 μm in the bulk material, surface tension σ provides the restraining force and the pressure can be estimated from $P = \sigma/r$ where r is the radius of the bubble. For larger bubbles formed by plastic deformation near the free surface the restraint is mostly by the strength of the metal. In this case the pressure is estimated by

$$P = 4\,h\,t\,Y\,/\,(h^2 + a^2) \tag{14.91}$$

where h is the height of the spherical shell illustrated in Figure 115, t the thickness of the shell, 2a the chord, and Y is the yield modulus which may be related to Young's modulus; or if the blister shatters, it may be a rupture modulus. As an example,

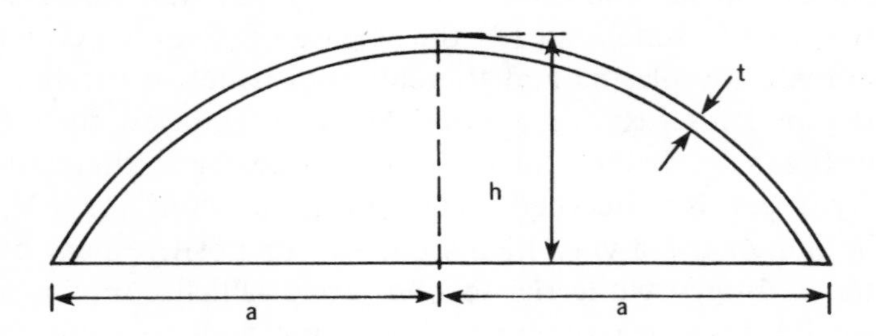

Figure 115. Schematic of a blister.

consider a blister with h = 10 μm, t = 1.2 μm, a = 60 μm, and Y (Nb) = 10.4×10^{11} dyne/cm^2; then the pressure is P = 1.35×10^{10} dyne/cm^2 = 1330 atm which appears to be too large. If on the other hand one considers the critical pressure P_0 needed to cause the onset of buckling in a circular membrane with fixed edges as the pressure in the blister, one obtains

$$P_o = 4\,\sigma_o\,t^2/3a^2 \tag{14.92}$$

where σ_o is the yield strength of the metal and the remaining symbols have the same meaning as before. For Nb, σ_o = 500 kg/cm^2 and the above equation gives $P_0 = 2.6 \times 10^5$ dyne/cm^2 = 0.25 atm, a much lower estimate. However it must be kept in mind that P_0 is the critical pressure needed to initiate buckling, and h, the height of the blister, does not enter equation (14.92); hence the actual pressure may be higher than P_0.

Another estimate may be obtained by calculating the deformation of a circular metal diaphragm rigidly clamped around its perimeter. When pressure is applied on one side, bulging by plastic deformation occurs and if we use the stress-strain relation given by

$$\sigma = \sigma_o\,(1 + H\epsilon) \tag{14.93}$$

then the pressure is given by

$$P = \sigma_o \left(\frac{4ht}{a^2}\right) \left[1 + \tfrac{1}{2}(3H - 5)(h^2/a^2)\right] \tag{14.94}$$

In these equations σ_o is the yield stress as before, ϵ is the strain, H is a nondimensional constant, and the remaining symbols are as defined earlier. For a spherical deformation in a metal whose stress-strain relation is

$$\sigma = \sigma_o \exp \epsilon \tag{14.95}$$

the pressure takes on the form

$$P = 4\,\sigma_o\, h\, t/(h^2 + a^2) \tag{14.96}$$

We note that this equation is similar to equation (14.91) except that Y is taken to be the yield stress rather than Young's modulus. Taking a typical value H = 3 and other parameters as in the previous examples, we find that p as given by equation (14.94) is 7.26×10^6 dyne/cm^2 $\approx$ 7 atm, and by (14.96) is 6.27 atm. For large plastic deformation, *i.e.*, for large h, the general solution (14.94) must be used.

From these estimated pressures and the known volume of the blisters one can calculate the fraction of the total dose of the incident (bombarding) gas that is present in the blisters and may be released when they rupture. For example for a cold-worked polycrystalline niobium irradiated to a total dose of 0.1 Coulombs per cm^2 of helium a very crude approximation can be obtained by considering only one type of large blister for which h = 10 μm and a = 60 μm have been observed. These results indicate that 14% of the ions striking the area are trapped in it and presumably can be ejected back into the plasma upon rupture.

REFERENCES

1. Behrisch, R. "First-Wall Erosion in Fusion Reactors," *Nuclear Fusion 12*, 695 (1972).
2. Behrisch, R. and W. Heiland. Sixth Symposium on Fusion Technology, Aschen, Sept. (1970).
3. Bethe, H. and J. Ashkin. in *Experimental Nuclear Physics,* O. E. Segre, Ed. (New York: John Wiley, 1953), p. 166.
4. Carruthers, R. *BNES Nuclear Fusion Reactors* (Proc. Int. Conf. Culham) (1969), p. 337.

5. Das, S. K. and M. Kaminsky. *J. Appl. Phys., 44,* 25 (1973).
6. Fusion Reactor First Wall Materials, USAEC Report, Wash. 1206 (1972).
7. Honig, R. E. and D. A. Kramer. "Vapor Pressure Data for the Solid and Liquid Elements," *RCA Review, 30,* 285 (1969).
8. Kaminsky, M. *Atomic and Ionic Impact Phenomena on Metal Surfaces,* (Berlin: Springer, Verlag, 1965).
9. Kaminsky, M. "Surface Phenomena Leading to Plasma Contamination and Wall Erosion in Thermonuclear Reactors and Devices," Proc. Int. Working Sessions on Fusion Reactor Technology, USAEC CONF-710624 (1971).
10. Kaminsky, M. and S. K. Das. *Appl. Phys. Lett., 21,* 443 (1972).
11. Lindhard, J., M. Scharff, and H. E. Schiott. Kgl. Danske Videnskab Selskab, Mat-Fys. Medd, *33*, N. 14 (1963).
12. Robinson, M. T. *BNES Nuclear Fusion Reactors* (Proc. Int. Conf. Culham), p364 (1969).
13. Robinson, M. T. *Phil. Mag., 12,* 741 (1965), *17*, 639 (1968).
14. Sigmund, P. *Physical Review, 184,* 383 (1969).
15. Vernickel, H. *Nuclear Fusion, 12,* 386 (1972).

CHAPTER 15

HEAT REMOVAL AND THERMAL CONSIDERATIONS IN FUSION REACTOR BLANKETS

In previous chapters, especially Chapter 13, we examined those aspects of fusion reactor design that deal primarily with the source of the fusion energy, *i.e.*, the plasma. We have seen how to calculate the various parameters of a reactor intended to produce a certain amount of a power assuming certain efficiencies for the various components. One of the major considerations of any power-producing device is that dealing with the heat removal system since it plays a significant role in the overall efficiency. We recall that in a D-T reactor most of the fusion energy is carried by the 14 MeV neutrons whose energy must ultimately be extracted from a blanket which is placed around the first wall to moderate them. In addition to being a good moderator the blanket material must be a breeder of tritium with sufficiently good heat transfer characteristics for efficient heat removal.

We have noted that lithium is one such material, and recovery of heat from a lithium-bearing blanket at sufficiently high temperature results in an efficient generation of electrical power from a fusion reactor. While in principle a range of lithium-bearing molten salts such as flibe ($2LiF.BeF_2$) can be used, these provide a marginal tritium breeding ratio due to parasitic neutron capture whereas the use of lithium metal itself would allow a significantly higher breeding ratio. As in liquid-metal fast breeder fission reactors, the lithium can also be considered as the coolant circulating through the blanket giving outlet temperatures of up to 650°C. Thus, conceptually, the blanket could be a simple structure with the lithium in it serving all the functions of a neutron moderator, tritium breeding, and heat removal medium. Such a scheme does have a major disadvantage in that the blanket must be placed inside the windings which generate the confining magnetic field, and the lithium must therefore be pumped across the magnetic field. This may result in an unacceptably high pumping power.

In this chapter we shall first derive the equations which govern the flow of a conducting fluid in a duct in the presence of a magnetic field. The physical principles of such a flow will be examined particularly with regard to the ways in which the magnetic field changes the pressure drop along the duct and the heat transfer from the duct walls to the fluid. Pressure drop calculations will then be applied to the reactor design of Chapter 13 for which the pumping power of the coolant will be estimated. We shall follow quite closely the work of Hunt and Hancox (1971).

15.1 THE EQUATIONS OF MAGNETOHYDRODYNAMIC (MHD) FLOW IN DUCTS

The basic equations for incompressible, steady MHD flows are the familiar Navier-Stokes equations of hydrodynamics modified to account for the interaction of a conducting fluid with electromagnetic forces. They are the continuity equation

$$\frac{\partial \rho}{\partial t} + \nabla \cdot (\rho \vec{v}) = 0 \tag{15.1}$$

and the momentum equation

$$\rho \frac{\partial \vec{v}}{\partial t} + \rho(\vec{v} \cdot \nabla)\, \vec{v} = -\nabla p + \vec{J} \times \vec{B} + \eta\, \nabla^2 \vec{v} \tag{15.2}$$

where ρ is the fluid density, v is its velocity, p is the pressure, J is the current density, B is the magnetic field strength, and η is the viscosity. In the case of an incompressible fluid where the density is constant and a steady flow for which $\partial v/\partial t = 0$, the above conservation of mass and momentum equations become

$$\nabla \cdot \vec{v} = 0 \tag{15.3}$$

$$\rho(\vec{v} \cdot \nabla)\, \vec{v} = -\nabla p + \vec{J} \times \vec{B} + \eta\, \nabla^2 \vec{v} \tag{15.4}$$

The electric and magnetic forces in these equations must satisfy Maxwell's equations which in mks units can be written as

$$\nabla \times \vec{E} + \frac{\partial \vec{B}}{\partial t} = 0 \tag{15.5}$$

$$\nabla \cdot \vec{B} = 0 \tag{15.6}$$

$$\nabla \times \vec{B} = \mu \vec{J} \tag{15.7}$$

$$\vec{J} = \sigma(\vec{E} + \vec{v} \times \vec{B}) \tag{15.8}$$

where the displacement current (usually small in many cases of interest) has been ignored in equation (15.5). The fluid electrical conductivity σ, magnetic permeability μ, and viscosity η are assumed constant, and for an infinitely conducting medium equation (15.8) is replaced by $\vec{E} + \vec{v} \times \vec{B} = 0$.

We apply these equations to the flow in the duct shown in Figure 116 whose length and radius are respectively given by ℓ, a, and which is situated in a uniform magnetic field B_o pointing

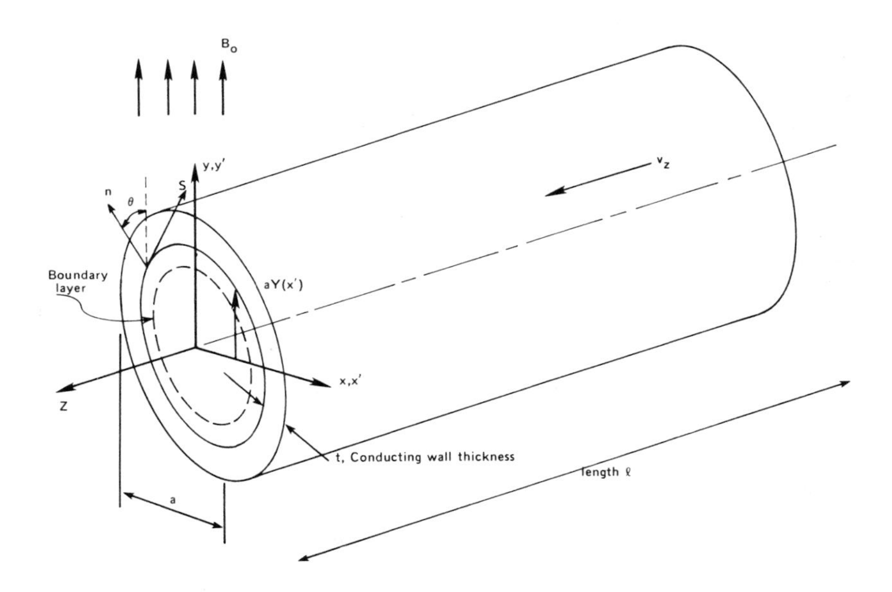

Figure 116. Geometry of a coolant duct.

in the y direction. We let H be the induced magnetic field caused by the induced currents in the system and write $\vec{B} = \vec{B}_o + \mu\vec{H}$. We assume that $|\mu H| \ll B_o$ and if we take the curl of equations (15.7) and (15.8) and combine the results we get

$$\nabla^2 \vec{B} = - \mu \sigma [(\vec{B} \cdot \nabla) \vec{v} - (\vec{v} \cdot \nabla) \vec{B}]$$

having taken advantage of some vector identities.

If we now substitute for B in terms of B_o and H and note that B_o is static and uniform we obtain after rearranging

$$\mu [(\vec{v} \cdot \nabla) \vec{H} - (\vec{H} \cdot \nabla) \vec{v}] = \frac{1}{\sigma} \nabla^2 \vec{H} + (\vec{B}_o \cdot \nabla) \vec{v} \tag{15.9}$$

It is convenient at this point to introduce the magnetic Reynolds Number R_m which is defined as the ratio of the induced magnetic field to the applied magnetic field. We note from Ohm's law that for a fluid moving with some average velocity v the induced current (with E = 0) is given by $J \sim \sigma v B_o$, and if we substitute this in equation (15.7) and utilize Stokes' theorem we get for the induced field B_i

$$B_i = \mu \sigma v B_o a$$

If we now divide through by B_o we get

$$R_m = \frac{B_i}{B_o} = \mu \sigma v a \tag{15.10}$$

or

$$R_m = \frac{\mu H}{B_o}; \quad H = \sigma v B_o a$$

The assumption made earlier that $\mu H \ll B_o$ implies that $R_m \ll 1$ and in that case equation (15.9) reduces to

$$(\vec{B}_o \cdot \nabla) \vec{v} + \frac{1}{\sigma} \nabla^2 \vec{H} = 0 \tag{15.11}$$

In the absence of a magnetic field a basic assumption of the theory of "laminar" duct flow is that if the duct is sufficiently long the flow eventually reaches a steady state, *i.e.*, $\partial v/\partial z = 0$. It can then be shown that $v_x = v_y = 0$, so that the flow has only one component, namely v_z. Given similar assumptions under

MHD conditions, *i.e.*, that the flow eventually becomes fully developed then it can also be shown that no secondary flow can exist and $v = (0,0,v_z)$. With this, we proceed by first substituting equation (15.7) into (15.4) noting that the induced current has a component only in the x-direction, *i.e.*, $J = J_x$, $J_y = J_z = 0$, which in turn implies that $H_x = 0$. The result after invoking the assumption of small R_m is

$$\frac{\partial p}{\partial z} = B_o \frac{\partial H_z}{\partial y} + \mu H_z \frac{\partial H_z}{\partial z} + \eta \nabla^2 v_z$$

From equation (15.6) upon substituting $\vec{B} = \vec{B}_o + \mu\vec{H}$ we find that $\partial/\partial z\ (\mu H_z) = 0$ so that the equation for the pressure gradient along the duct finally becomes

$$\frac{\partial p}{\partial z} = \eta \nabla^2 v_z + B_o \frac{\partial H_z}{\partial y} \tag{15.12}$$

The x and y components of equation (15.4) readily reveal that $\partial p/\partial x = \partial p/\partial y = 0$ so that the pressure is only a function of z under the above assumptions. If we now let $x' = x/a$ and $y' = y/a$ we observe that

$$\nabla'^2 = \frac{\partial^2}{\partial x'^2} + \frac{\partial^2}{\partial y'^2} = a^2 \nabla^2 \tag{15.13}$$

which if introduced into equation (15.12) and divided through by $-\partial p/\partial z$, we obtain

$$-1 = \frac{\eta}{a^2} \nabla'^2 v_z \left(1/ - \frac{\partial p}{\partial z}\right) + \frac{B_o}{a(-\partial p/\partial z)} \frac{\partial H_z}{\partial y'} \tag{15.14}$$

This result can be simplified by further letting

$$v = \frac{v_z}{(a^2/\eta)(-\partial p/\partial z)}\ ; \quad h = \frac{H_z}{a^2 (\sigma/\eta)^{1/2} (-\partial p/\partial z)}$$

and introducing the Hartmann number, $M = a\,B_o\,(\sigma/\eta)^{1/2}$, which represents the ratio of the electromagnetic stress to the viscous stress. The result is

$$\nabla'^2 v + M \frac{\partial h}{\partial y'} = -1 \tag{15.15}$$

Equation (15.11) has only a z-component which when normalized in a manner similar to the above equation becomes

$$M \frac{\partial v}{\partial y'} + \nabla'^2 h = 0 \tag{15.16}$$

When solved subject to appropriate boundary conditions equations (15.15) and (15.16) will yield the flow characteristics in the duct. To achieve this we must establish the boundary conditions and for that we need to consider the distribution of current in conducting walls where they exist. Noting that v = 0 at the wall and recalling that the induced magnetic field is much smaller than the applied equations (15.5 - 15.8) yield

$$\vec{J} = \sigma_W \vec{E}; \quad \nabla \times \vec{E} = 0 \tag{15.17}$$

$$\vec{J} = \nabla \times \vec{H}; \quad \nabla \cdot \vec{H} = 0 \tag{15.18}$$

Taking the curl of the last equation and utilizing (15.17) we readily note that H satisfies the equation

$$\nabla^2 \vec{H} = 0 \tag{15.19}$$

The boundary conditions at the fluid-wall interface can be deduced as follows. From (15.17) we see upon applying Stokes' theorem that the tangential component of the electric field must be continuous across the interface, *i.e.*, $E_{sf} = E_{sw}$ where the subscripts f and w denote fluid and wall, respectively. Moreover from the curl of equation (15.18) along with the Gauss's theorem we observe that the normal component of the current must also be continuous at the fluid-wall interface, *i.e.*, $J_{nf} = J_{nw}$, so that the boundary conditions at that surface may be written as

$$E_{sf} = E_{sw}; \quad J_{nf} = J_{nw}; \quad v = 0 \tag{15.20}$$

On the outer boundary of the duct walls, *i.e.*, at n = t, $J_{nw} = 0$ where the local coordinates s,n are those shown in Figure 116. From equations (15.11) and (15.19) we note that $\nabla^2 H_z = 0$ which when normalized and expressed in terms of the local coordinates s,n becomes

$$\frac{\partial^2 h}{\partial n^2} \equiv \frac{\partial^2 h}{\partial x'^2} + \frac{\partial^2 h}{\partial y'^2} = 0 \tag{15.21}$$

This result shows that h in the duct wall varies linearly with space and since it is constant at n = t we take this constant to be zero. This implies that the continuity of h along the normal at the fluid-wall interface demands that

$$\left.\frac{\partial h}{\partial n}\right|_{w} = -\frac{a}{t} \left. h \right|_{f} \tag{15.22}$$

Moreover from equations (15.17) and (15.20) we see that $J_{sw}/\sigma_w = J_{sf}/\sigma_f$ which, upon substituting in (15.18) and employing the previously used normalization, we can write

$$\frac{1}{\sigma_w} \left.\frac{\partial h}{\partial n}\right|_{w} = \frac{1}{\sigma_f} \left.\frac{\partial h}{\partial n}\right|_{f} = -\frac{a}{t\,\sigma_w} h_f \tag{15.23}$$

having utilized equation (15.22). Introducing the ratio of the conductances $\varphi = t\sigma_w/a\sigma_f$ into the above equation the conditions on h and v at n = 0 can finally be written as

$$\partial h/\partial n + h/\varphi = 0; \qquad v = 0 \tag{15.24}$$

The mathematical problem has therefore been reduced to finding the solution to equations (15.15) and (15.16) subject to the boundary conditions (15.24).

Since for all problems of relevance to heat removal in magnetic-type fusion reactors we expect the Hartmann number to be large, *i.e.*, $M \gg 1$, we proceed with the analysis by first recognizing that in this case the flow can be divided into two regions. The first region, the "core," occupies most of the duct away from the walls where the gradients are small so that (15.15) and (15.16) become

$$M \frac{\partial H}{\partial y'} = -1$$

$$M \frac{\partial v}{\partial y'} = 0 \tag{15.25}$$

The second region is the "boundary layer" region on the walls (see Figure 116) where gradients normal to the wall are large. In terms of the coordinate n, normal to the wall, given by $n = x' \sin\theta = y' \cos\theta$, we see that $\partial h/\partial y' = (\partial h/\partial n) \cos\theta$ and

$\nabla^2 = \partial^2/\partial n^2$ so that equations (15.15) and (15.16) assume the form

$$M' \frac{\partial h}{\partial n} + \frac{\partial^2 v}{\partial n^2} = -1$$

$$M' \frac{\partial v}{\partial n} + \frac{\partial^2 h}{\partial n^2} = 0 \tag{15.26}$$

where $M' = M \cos\theta$. We note that $M' < 0$ on the lower wall.

The second essential step in the analysis is to consider the combined variables $z_1 = (v + h)$ and $z_2 = (v - h)$ which transform equation (15.25) into

$$\frac{\partial z_1}{\partial y'} = -\frac{1}{M}; \quad \frac{\partial z_2}{\partial y'} = \frac{1}{M} \tag{15.27}$$

moreover, it is useful to introduce the value of h on the upper wall, $h_w(x')$ which is at this stage an unknown function. Then by symmetry it follows that $h = -h_w(x')$ on the lower wall. Also if the equation $y' = Y(x')$ defines the duct wall, then again by symmetry the upper wall is given by $y' = Y(x')$ and the lower wall by $y' = -Y(x')$. Turning now to equation (15.27) we note that the solutions in the core are given by

$$z_1 = -\frac{1}{M} y' + C_1(x')$$

$$z_2 = \frac{1}{M} y' + C_2(x') \tag{15.28}$$

where C_1 and C_2 are constants of integration. If we apply the second condition in (15.24) and note that $h = h_w(x')$ on $y' = y(x')$ and $h = -h_w(x')$ on $y' = -Y(x')$ we get

$$C_1 = C_2 = \frac{Y(x')}{M} + h_w(x')$$

so that (15.28) finally becomes

$$z_1 = [Y(x') - y']/M + h_w(x')$$

$$z_2 = [Y(x') + y']/M + h_w(x') \tag{15.29}$$

The solutions to equation (15.26) have been examined in detail by Chang and Lundgren (1961) and they are in the boundary layers at $y' = Y(x')$ given by

$$z_1 = h_w(x') - n \cos \theta / M$$

$$z_2 = 2\, Y(x')/M + h_w(x') + n \cos \theta / M - 2[Y(x')/M + h_w(x')]\, e^{M'n} \tag{15.30}$$

whereas at $y' = -Y(x')$ they are

$$z_2 = h_w(x') - n \cos \theta / M$$

$$z_1 = 2\, Y(x')/M + h_w(x') + n \cos \theta / M - 2[Y(x')/M + h_w(x')]\, e^{M'n} \tag{15.31}$$

Noting that $h = \frac{1}{2}(z_1 - z_2)$, we readily obtain from (15.30) after differentiation with respect to n

$$\frac{\partial h}{\partial n} = -\frac{\cos \theta}{M} + M' \,[Y(x')/M + h_w(x')]\, e^{M'n}$$

which, when substituted in the first condition of (15.24) at $n = 0$, we get the value of h at the fluid-wall interface, *i.e.*,

$$h_w(x') = \frac{\varphi \cos \theta \,[1 - MY(x')]}{M\,(1 + M \varphi \cos \theta]} \tag{15.32}$$

If on the other hand we combine z_1 and z_2 in (15.29) to construct v, we find that the velocity in the core can first be expressed as

$$v = \frac{Y(x') + \varphi \cos \theta}{M\,[\,1 + M\varphi \cos \theta\,]}$$

From Figure 116 we observe that $\cos \theta = dx'/ds$ or simply

$$\cos \theta = [1 + Y'^2(x')]^{-1}$$

where $Y'(x')$ is the derivative with respect to x', so that upon insertion in the above expression for v it finally becomes

$$v = \frac{Y(x') + \varphi/\sqrt{1 + Y'^2(x')}}{M[1 + \varphi M/\sqrt{1 + Y'^2(x')}]} \tag{15.33}$$

Two interesting limits of this result emerge when the walls of the duct are poorly conducting and when they are purely resistive. In the first case $\varphi \ll 1$ but with $M\varphi \gg 1$ equation (15.33) reduces to, after reverting to the physical variables,

$$v_z = \frac{Y(x')\,(-\,\partial p/\partial z)\,\sqrt{1 + Y'^2(x')}}{\sigma\,\varphi\,B_o^{\,2}} \tag{15.34}$$

and for the second case where $\varphi = 0$, it reduces to

$$v_z = \left(-\frac{\partial p}{\partial z}\right)\frac{Y(x')\,a^2}{\eta\,M} \tag{15.35}$$

It is interesting, by way of example, to apply these calculations to a duct with a diamond-shaped cross section as illustrated in Figure 117.

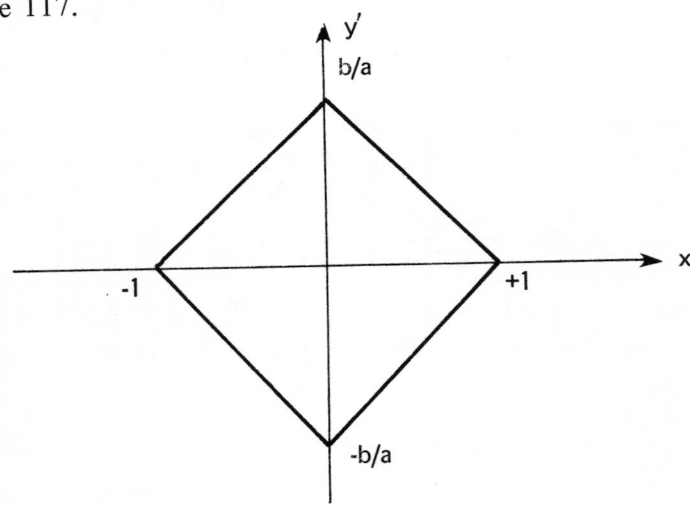

Figure 117. Diamond-shaped duct.

This configuration possesses the symmetry about the x'-axis demanded by the above analysis. The equation for the duct is

$$Y(x') = \pm\,(1 - |x'|)\,b/a \tag{15.36}$$

and the average flow velocity is expressed by

$$\bar{v} = \int_A v_z \, dA \Big/ \int_A dA = Q/A$$

where Q, as before, is the flow volume. Substituting for v_z from equation (15.34) we see that Q can be written as

$$Q = \frac{4(-\partial p/\partial z)(b/a)}{\sigma \varphi B_o^{\,2}} \int_{x'=0}^{1} \int_{y'=0}^{Y(x')} Y(x') \sqrt{1 + Y'^2(x')} \; dx' \, dy'$$

Moreover, from (15.36) we observe that $Y'(x') = \mp\, b/a$ so that upon insertion into the above equation and carrying out the integration we get

$$Q = \frac{4(-\partial p/\partial z)(b/a)^2 \sqrt{1 + (b/a)^2}}{\sigma B_o^{\,2} \varphi} \int_0^1 (1\text{-}x')^2 \, dx'$$

which, upon carrying out the last integration, further becomes

$$Q = \frac{4/3\ (-\partial p/\partial z)(b/a)^2 \sqrt{1 + b^2/a^2}}{\sigma B_o^{\,2} \varphi}$$

The total area is simply four times that of each quadrant or

$$A = \int dA = 4\ (\tfrac{1}{2}\ b/a)$$

so that finally $\bar{v}$ becomes

$$\bar{v} = \frac{Q}{A} = \frac{(-\partial p/\partial z)}{\sigma B_o^{\,2} \varphi} \quad (2/3)\ (b/a) \sqrt{1 + b^2/a^2} \qquad (15.37)$$

This quantity represents a relevant parameter which demonstrates the effect of the duct's shape. We note that if $b/a = 2$ the mean velocity is increased by a factor of 3 and if the diamond shape is replaced by an elliptical shape with $b/a = 2$ the increase would only be a factor of 2.

Before concluding this section it might be useful to estimate the boundary layer thickness in an MHD flow. In this kind of flow the boundary layer represents a region in which the viscous force of the fluid is balanced by the magnetic force. From equation (15.2) we note that the viscous force is given by $\eta\, \nabla^2\, v$, so that for a one-dimensional flow this force can be written as $\eta\, v_z/\delta^2$. The balancing force is simply the J x B, and from Ohm's law with E = 0, J = σ v x B, so that with B and v_z as given in Figure 116 the current density becomes J = $\sigma\, v_z\, B^2$. Equating the two forces we obtain

$$\eta\, v_z/\delta \approx \sigma\, v_z\, B^2$$

or

$$\delta \approx \frac{1}{B}(\eta/\sigma)^{1/2} = \frac{a}{M} \qquad (15.38)$$

where we have utilized the definition of the Hartmann number. If as shown in Figure 116 the normal to the surface makes an angle θ relative to the applied magnetic field, then B in the above equation would be replaced by B cos θ so that thickness of the boundary layer becomes $\delta \approx a/M'$, where $M' = M \cos\theta$.

15.2 SOME PHYSICAL ASPECTS OF MHD FLOW IN DUCTS

When a conducting fluid is pumped along a duct situated in a magnetic field considerable changes can take place in the properties of the flow and in the velocity distribution across the duct. Of special concern to flow in cooling ducts in fusion reactors is the effect of the magnetic field on the pressure drop along the duct and the heat transfer from the duct walls to the fluid.

In the absence of a magnetic field when the fluid mean velocity v is quite low the flow is quite steady, any disturbances being damped by viscous forces, and the flow can be regarded as thin layers of fluid sliding over each other at different speeds. Such a flow is referred to as *laminar* and the velocity distribution is parabolic with the center moving faster than at the walls where the velocity is zero. When the mean velocity is increased beyond a critical value, small disturbances are amplified and distorted by inertial effects until the velocity becomes random throughout the duct. This "turbulent" flow differs from the laminar flow not only because it contains random fluctuations but also because the velocity distribution changes, the parabolic profile being replaced by an approximately constant velocity in the center of the duct and narrow boundary layers [see equation (15.38)] close to the walls where the velocity decreases to zero. This difference is caused by eddies transferring momentum from the center to the walls and back more effectively than the viscous effects in the laminar flow. By analogy, therefore, turbulent flow is more effective in the process of heat transfer than laminar flow.

The dividing line between the two regimes of flow is specified by the Reynolds number, $R_e = a v \rho / \eta$, which gives the order of magnitude of the ratio of the inertial to viscous forces in the fluid. Typically, flows characterized by a Reynolds number of 1000 or more are considered turbulent. The practical significance of the difference between the two regimes and the effect of the velocity distribution can be appreciated from equation (15.33) where we note the dependence of the pressure drop on the shear stresses at the wall which depend on the local velocity gradient. By the same token, the heat transfer depends on the temperature gradient and, in view of the earlier statements, we conclude that the pressure drop as well as the heat transfer are greater in a turbulent flow than in a laminar flow at the same flow rate. It follows therefore that any true assessment and calculation of coolant flow and heat transfer characteristics in a fusion (magnetically-confined) reactor blanket should incorporate understanding and examination of the magnetic field effects on the velocity distribution as well as its effects, in the case of turbulent flow, on the damping or amplification of fluctuations.

The effect of a magnetic field on the flow of a conducting fluid in a duct in a fusion reactor will depend very much on the orientation of the field relative to the duct. In the simple scheme for cooling the blanket of a toroidal reactor illustrated in Figure 118 we note the two extreme situations where the field is either parallel or perpendicular to the duct. Clearly, in any realistic situation the orientation will vary between these two limits and we can easily estimate the effects in these cases by knowing the effects in the two extremes.

We first examine the case where the magnetic field is at right angles to the duct simply illustrated in Figure 119. We recall from equation (15.8) that when a conducting fluid flows at a velocity v in the presence of a magnetic field B, a current J is induced in the direction of v x B. If the walls of the ducts are nonconducting (electrical insulators), then the currents generated in the core return

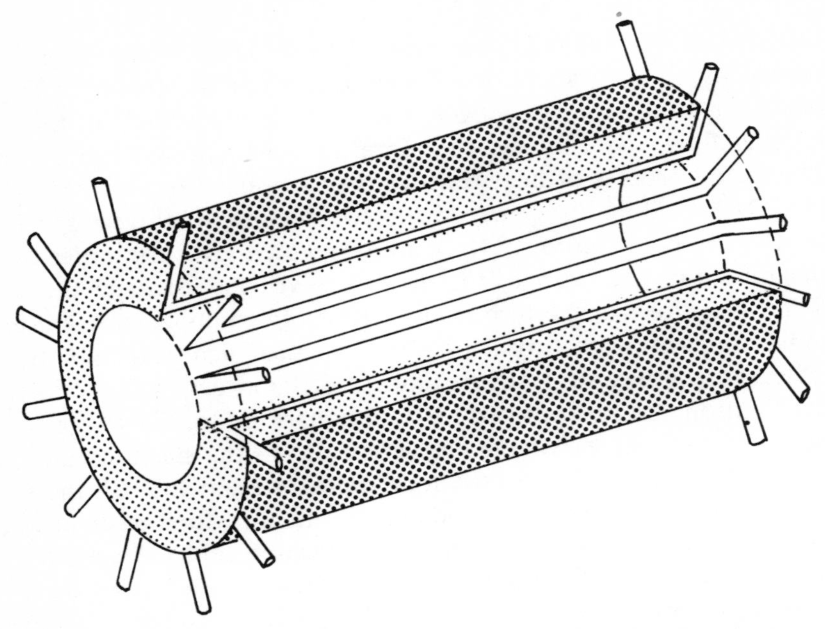

Figure 118. Cooling scheme of a toroidal fusion reactor (Hunt and Hancox, 1971).

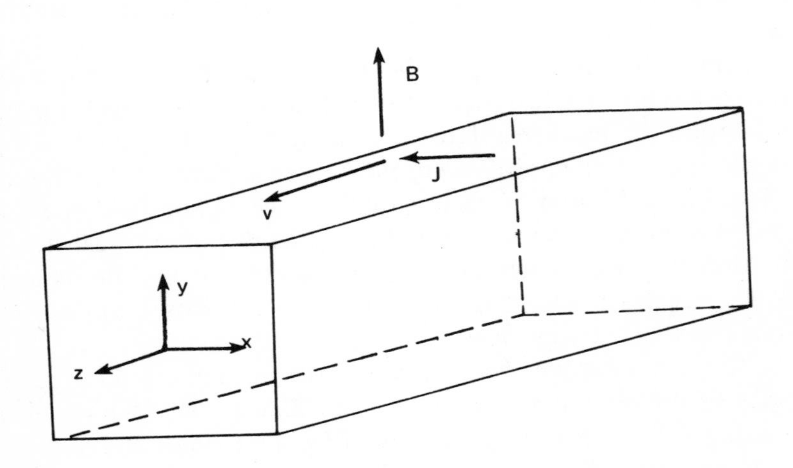

Figure 119. Induced current in a duct.

through the boundary layers on the side walls as shown in Figure 120a. If on the other hand the walls are electrically-conducting, then the currents return through the walls as in Figure 120b, especially if the conductivity of the walls is high compared with that of the fluid and the boundary layer thickness is small compared to the thickness of the walls. These currents give rise to

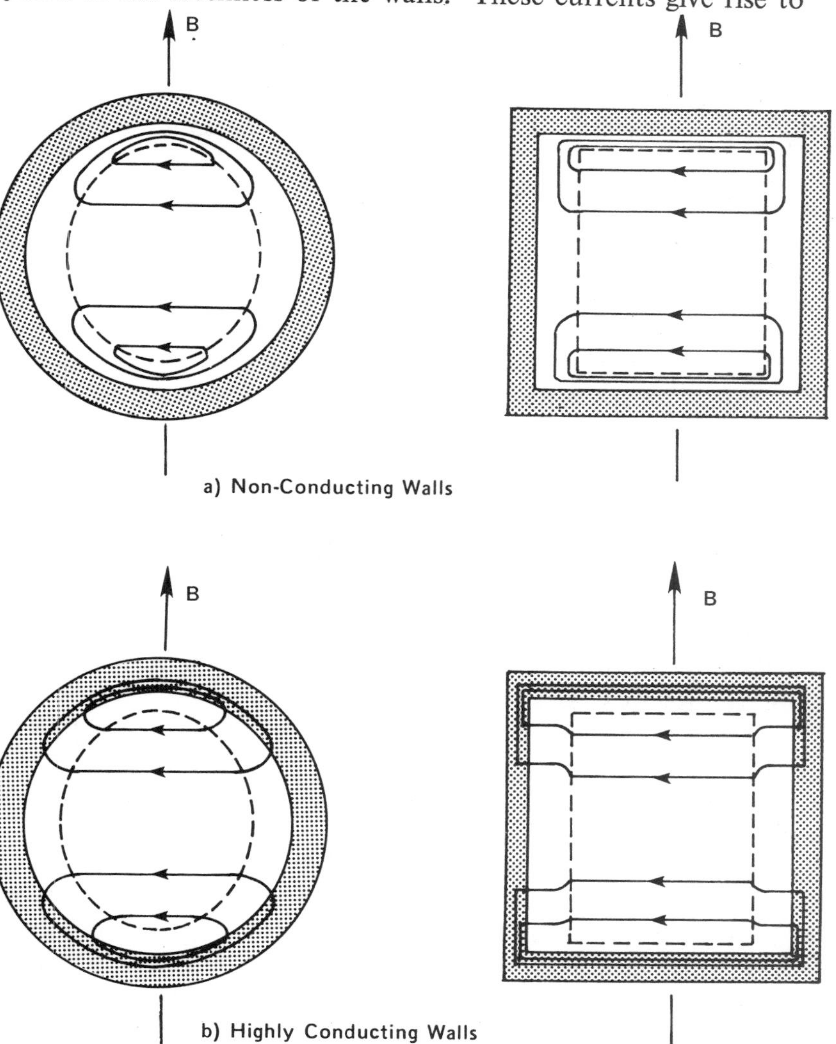

Figure 120. Current paths in ducts due to a strong transverse magnetic field (Hunt and Hancox, 1971).

their own magnetic field which will be directed mainly along the duct, and the ratio of the induced magnetic field to the applied field, known as the magnetic Reynolds number R_m, is usually small in most practical situations.

The effects these currents have on the velocity distribution can be assessed from the J x B force term in equation (15.4). If this force does not vary across the duct then it would simply represent an additional contribution to the pressure gradient. For a duct with nonconducting walls the current distribution (proportional to v x B) follows closely the velocity distribution (parabolic profile when B = 0) which has a maximum at the core. It follows therefore that the electromagnetic force tends to brake the fluid in the core region of the flow but speeds up the flow near the walls. The situation is quite similar in the case of a duct with highly conducting walls. In this case the J x B force also has a large change between the core and the boundary layer approaching that of the nonconducting walls when the boundary layers are very thin. These changes in the velocity profile across the ducts parallel and perpendicular to the magnetic field as well as the effects on the profile of the duct shape are illustrated in Figure 121. We note that the velocity profile parallel to the magnetic field $v_z(y)/v$ is similar for both the circular and square ducts while substantially different in the two shapes perpendicular to the magnetic field. The extent to which the velocity distribution is affected by the magnetic field is indicated by the value of the Hartmann number M, which as we recall represents the ratio of electromagnetic to viscous stresses.

Since the magnetic field appears to reduce the differences in the time-averaged velocity, it is reasonable to assume that it will have the same effect on fluctuating velocities, *i.e.*, the magnetic field tends to suppress turbulence. It has been suggested by Branover (1967) that a sufficient condition for the turbulence to be completely suppressed by a magnetic field is that $R_e/M < 130$. In some cases, however, as pointed out by Hunt and Shercliff (1971) when the conductivity of the wall varies around the perimeter, turbulence could be promoted by the magnetic field because it produces very unstable boundary layers.

The magnetic field has a direct and a nondirect effect on the pressure gradient through the electromagnetic force J x B. In a

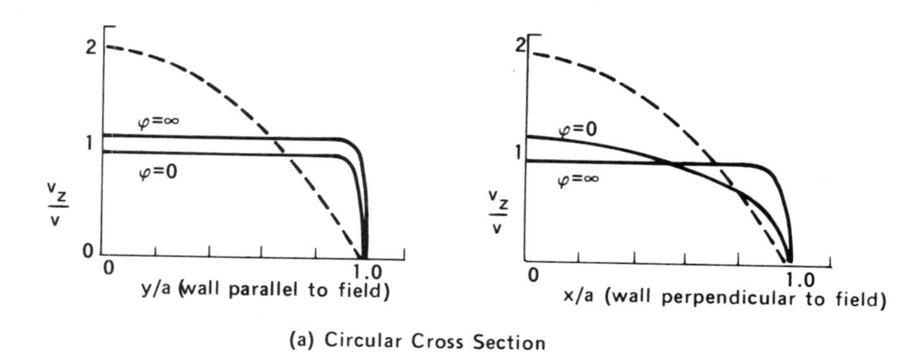

(a) Circular Cross Section

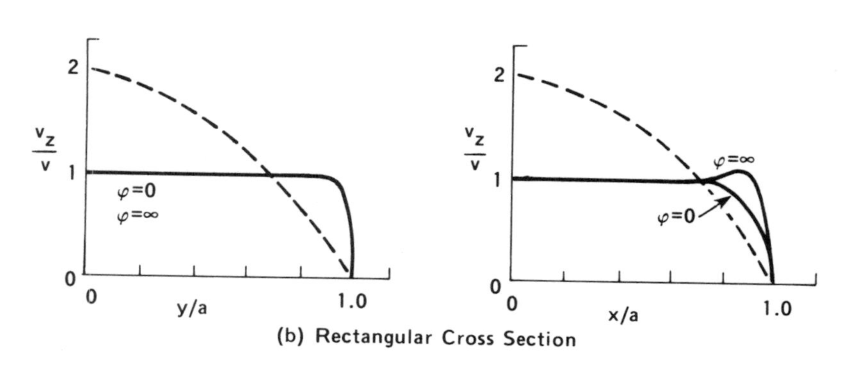

(b) Rectangular Cross Section

Figure 121. Velocity profiles in ducts with zero (- - - -) and strong (———) mag field, and walls which are nonconducting (φ=0) and highly conducting (φ=∞) (Hunt and Hancox 1971).

duct with nonconducting walls the currents form closed paths in the fluid (as illustrated in Figure 120a) so that

$$\int_{-a}^{+a} J_x \, dy = 0$$

correspondingly

$$\int_A J_x B_y \, dx \, dy = 0$$

so that there can be no net body force. However, we noted earlier that the effect of J x B forces is to increase the shear stresses at the wall, and this indirectly increases the frictional force and hence the pressure gradient. We have also seen that these electromagnetic forces reduce turbulence and thereby indirectly decrease the pressure gradient. These two conflicting effects have been seen in numerous experiments whose results were correlated by Branover and Tsinober (1970). These results show that if the flow is turbulent with zero magnetic field, then at low flow rates as the magnetic field is increased the pressure gradient first decreases slightly and then increases; but at higher flow rates the pressure gradient increases with increasing magnetic field. If the initial flow is laminar the pressure gradient always increases as the magnetic field increases.

In the case of a duct with conducting walls the currents circulate through the walls and the body force

$$\int_A J_x B_y \, dx \, dy$$

is not zero; hence it makes a net contribution to the pressure gradient. In general this effect dominates over the other two indirect effects, namely the effect of increasing wall shear stresses and reducing turbulence, that even for turbulent flows the pressure gradient always increases as the magnetic field increases.

We turn now to the case where the applied magnetic field is parallel to the duct, *i.e.,* in the z-direction of Figure 119. If the flow is steady there will be no induced currents and, in turn, no induced magnetic fields to affect the velocity profiles so that in such flows a parallel magnetic field has no effect on the pressure gradient or heat transfer. If on the other hand the flow is turbulent, then there exist motions across the magnetic field and, as in the case of a duct at right angles to the magnetic field, the latter tends to damp the turbulence. But since the field does not distort the time-averaged velocity distribution which provides the turbulent energy, it is found that a much larger field is needed to suppress turbulence than in the case of a perpendicular magnetic field. We conclude therefore that a parallel magnetic field will always reduce the pressure gradient in a turbulent flow.

Of equal importance in the heat removal from reactor blankets is the effect of magnetic fields on the heat transfer from the duct walls to the fluid. This also depends on the orientation of the magnetic field as well as on the nature of the flow after the field is applied. If the flow is laminar in the absence of a magnetic field then, as we have seen, the effect of the perpendicular field is to maintain the flow laminar and to reduce the thickness of the boundary layers. This leads to an increase in heat transfer which is of the order of 50% as demonstrated in Figure 122.

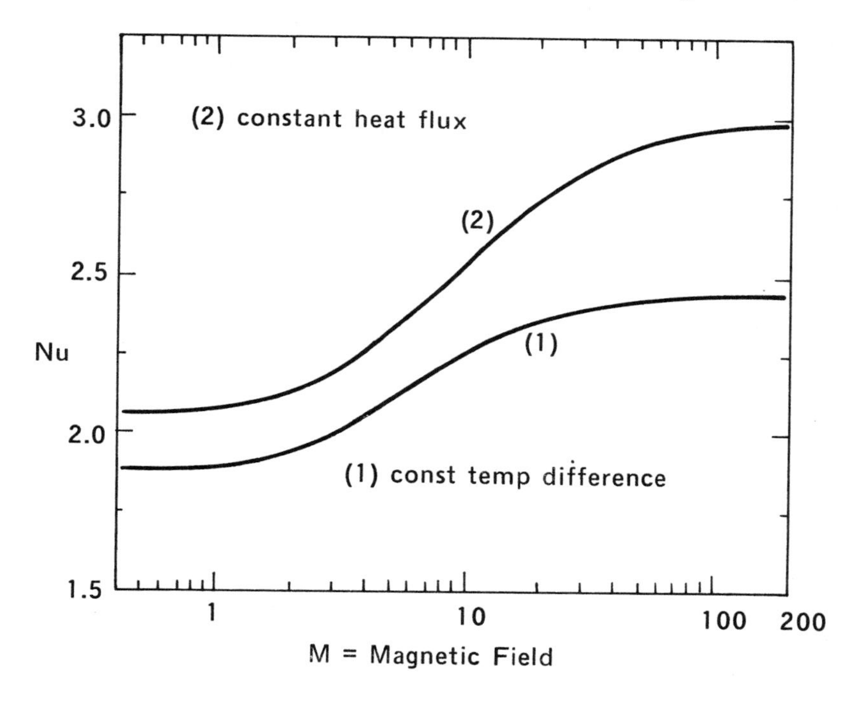

Figure 122. Effect of transverse magnetic field on the heat transfer (Hunt and Hancox 1971).

This change is reflected in the increase of the Nusselt number $N_u = a\,h/K$, which is defined in terms of the heat transfer coefficient h and the thermal conductivity of the fluid K. If, however, an initially turbulent flow crosses a magnetic field sufficiently strong for its turbulence to be suppressed, then the heat can no longer be transferred from the wall to the fluid by the eddies and the heat transfer will be substantially reduced from its initial value. If the flow remains turbulent even in the presence of the field, then the field must be relatively weak; but even in this case there still will be some damping of turbulence and a corresponding reduction in the heat transfer. A parallel magnetic field does not affect the heat transfer in a laminar but will damp the eddies and reduce the heat transfer in a turbulent flow. Because we have only considered long straight ducts the effects of uniform magnetic fields on the pressure gradients and heat transfer have been readily deduced by examining only the interaction of the viscous, electromagnetic pressure and turbulent stresses. In a reactor, however, the ducts will have curved and variable cross sections passing through nonuniform magnetic fields and an assessment of these effects requires the understanding of the interaction of inertial forces and electromagnetic forces. This is a much more difficult problem but fortunately these effects are so small in the fusion reactor situation that they can be safely neglected.

15.3 CALCULATION OF THE PRESSURE DROP AND PUMPING POWER

We turn now to the calculation of the pressure drop in a duct carrying a coolant in a fusion reactor blanket. We assume that the magnetic field is transverse to the duct and that the following conditions are satisfied: (a) the duct is straight and its dimensions remain constant along its length; and (b) the magnetic field does not change appreciably over a distance of the order of the duct size. These conditions may not be untypical in the heat removal system of a toroidal fusion reactor. Moreover, in such a system the Hartmann number, Reynolds number, and the magnetic Reynolds number which characterize the fluid flow have the values, $M \gg 1$, $R_e \sim M$ and $R_m \ll 1$. With these assumptions the flow can be considered laminar, according to Branover, since $R_e/M \ll 130$, and it is fully developed as long as $\ell/a \gg R_eM$ where ℓ is the length of the duct. Both these conditions are satisfied by a lower value of the Reynolds number or a higher value of the Hartmann number if the duct walls are highly conducting. In this case it can be assumed that the inertial forces in the fluid are negligible and that the mathematical analysis presented above concerning the balance between viscous, pressure, and electromagnetic forces will be applicable. We recall that in such a flow thin boundary layers are created on the walls of the duct with thickness $\delta \sim a/M \cos\theta$ where θ is the angle between the normal to the wall and the magnetic field. Hunt and Shercliff (1971) have shown that for a rectangular duct the boundary layer on the walls, which are parallel to the magnetic field, has a thickness $\delta \sim (aM^{-1/2})$; and in the case of a round pipe whose walls are also parallel there is a small region with $\delta \sim (aM^{-2/3})$.

In the core of the duct inertial and viscous forces are neglected and from equation (15.4) we see that the pressure forces are balanced by the electromagnetic forces so that

$$\frac{\partial p}{\partial z} = (\vec{J} \times \vec{B})_z \tag{15.39}$$

The current density in this case is in the x-direction and from equation (15.8) it has the value given by $J_x = \sigma(E_x - v_z B_y)$ which when substituted in the above equation becomes

$$\frac{\partial p}{\partial z} = \sigma B_y (E_x - v_z B_y) \tag{15.40}$$

If the conductivity of the walls is very high, *i.e.*, $\varphi \gg 1$, the electric field is zero and equation (15.40) reduces to

$$\frac{\partial p}{\partial z} = -\sigma v_z B_o^2 \tag{15.41}$$

where we have replaced B_y by B_o to indicate that it is the applied magnetic field. The total volume flow rate is $Q = Av = Av_z$ where A is the cross sectional area of the duct, and in terms of this quantity the pressure drop assumes the form

$$\frac{\partial p}{\partial z} = -Q \sigma B_o^2 / A \tag{15.42}$$

If, however, the conductivity of the walls is small and finite so that $0 < \varphi \ll 1$, but the magnetic field is so large that $M\varphi \gg 1$, then the currents will still circulate through the walls because of the high resistance of the boundary layers. At the mid-plane (x=0) of a duct whose cross section is characterized by the dimensions "a" and "b" the total current flowing through the walls is equal to the total current in the fluid, *i.e.*, $I_w = I_f$ or $tJ_w = -bJ_f$. In this case the electric field in the wall E_x may be expressed by

$$E_x \sim \frac{J_w}{\sigma_w} \sim \frac{bJ_f}{t\sigma_w} \sim \frac{J_f}{\varphi \sigma_f}(b/a) \tag{15.43}$$

where, as we recall, the conductance ratio φ is given by $\varphi = \sigma_w t/\sigma a$. We now define a wall effective current density in the x-direction by $J_{eff} = \varphi(a/b)\sigma E_x = \varphi(a/b)J_x$, which allows us to write equation

(15.39) in the form

$$\frac{\partial p}{\partial z} = \varphi(a/b)J_x$$

which upon combining with Ohm's law, equation (15.8), it further becomes upon neglecting E_x,

$$\frac{\partial p}{\partial z} = -\varphi(a/b)\sigma\, v_z\, B_o^{\,2} \qquad (15.44)$$

This order of magnitude result agrees with the exact solution represented by equation (15.34). To see this, we apply that result to the rectangular cross section shown in Figure 123 whose wall is expressed by the function Y(x′) = 1.

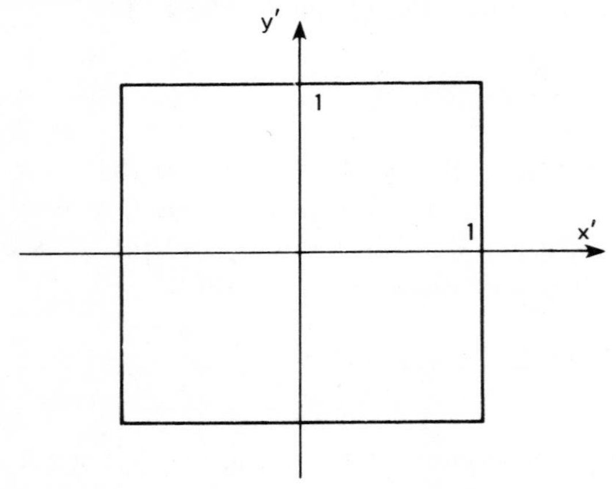

Figure 123. Cross section of a rectangular duct.

The axial velocity is given by

$$v_z = \frac{(-\partial p/\partial z)\, Y(x')\sqrt{1 + Y'^2(x')}}{\sigma\, \varphi\, B_o^{\,2}}$$

which leads to the following average speed of flow

$$\bar{v} = \frac{Q}{A} = \frac{(-\partial p/\partial z) \iint dA\; Y(x')\sqrt{1 + Y'^2}}{A\, \sigma\, \varphi\, B_o^{2}}$$

or

$$\bar{v} = \frac{-\partial p/\partial z}{\sigma\varphi B_o^2} \quad \frac{\int_0^1\int_0^1 dx'\,dy'}{1} = -\frac{\partial p/\partial z}{\sigma\varphi B_o^2}$$

From this result we see that the pressure drop is given by

$$\frac{\partial p}{\partial z} = -\frac{Q}{A}\sigma\,\varphi\,B_o^2 \tag{15.45}$$

which is quite comparable to (15.44).

In the case of a nonconducting wall with $M\varphi = 0$, equation (15.35) is used and for the rectangular duct in question we observe that

$$v = \frac{Q}{A} = \frac{-\partial p/\partial z}{\eta M}\,a^2$$

Multiplying the numerator and denominator by M and utilizing the definition of the Hartmann number we find that

$$v = \frac{Q}{A} = \frac{(-\partial p/\partial z)M}{\sigma\,B_o^2}$$

or

$$\frac{\partial p}{\partial z} = -\frac{Q}{A}\,\frac{\sigma B_o^2}{M} \tag{15.46}$$

Comparison of this result with (15.42) reveals that the pressure gradient has decreased by a factor of M from the case of a perfectly conducting wall.

The above results can now be used to calculate the pumping power for the toroidal reactor discussed in Chapter 13. The main parameters of that reactor are shown in Table 38.

Table 38

Reactor Parameters (Carruthers, *et al.*, 1967)

Parameter	*Value*
Power output	5000 Mw(Th)
Magnetic field	10 tesla = 100 kG
Plasma radius	1.25 meters
Inner radius of blanket	1.75 meters
Outer radius of blanket	3.00 meters
Outer radius of mag winding	3.50 meters
Major radius of torus	5.6 meters
Power entering blanket	13 MW/m^2

For the purposes of our calculation we shall assume that the torus is divided into N sections along its circumference, each of which has the form shown in Figure 118. Liquid lithium is pumped into the blanket across the magnetic field through several feeder tubes, passes along the ducts in the parallel to the magnetic field and leaves through a second group of feeder tubes. We shall assume that the ducts are situated near the inner surface of the blanket where most of the neutron and radiation energy is deposited. Assuming 10% of the total surface area of the blanket is available for the tubes and that the lithium temperature rise in passing through the blanket can be 300°C, Hunt and Hancox (1971) deduce the flow parameters shown in Table 39.

Table 39

Flow Parameters

Heat removed in each section	5000/N	MW
Length of each section	35/N	meters
Area of feeder tubes	19.3/N	m^2
Volume flow of lithium	8.4/N	m^3/sec
Mass flow of lithium	4000/N	kg/sec
Velocity of lithium flow	0.43	m/sec

The physical properties of the materials concerned, in the temperature range 300-600°C, are taken by those authors to be those shown in Table 40.

Table 40

Physical Properties of Coolant and Duct

Density of lithium	475 kg/m^3
Specific heat of lithium	4200 joules/kg C
Viscosity of lithium	4.5×10^{-4} kg/m sec
Electrical conductivity of lithium	2.2×10^{6} mho/m
Electrical conductivity of duct walls	5.0×10^{6} mho/m

If we further assume that the feeder tubes have a radius of 0.05 m and a ratio of wall thickness to radius of 0.05 the non-dimensional parameters that characterize the flow assume the values shown in Table 41, and they satisfy the conditions for

Table 41

Flow Conditions		
Hartmann number	M	35×10^3
Reynolds number	R_e	28×10^3
Magnetic Reynolds number	R_m	0.07
Conductance ratio	φ	0.1

fully developed laminar flow. From these values we note that $0 < \varphi \ll 1$, and $M\varphi \gg 1$, and from Table 40 we note that $Q = 8.4/n\ m^3/sec$, $A = 19.3/N\ m^2$, $v = 0.43$ m/sec and $\sigma = 2.2 \times 10^6$ mho/m. For these conditions the pressure drop is given by equation (15.45) to be

$$\frac{\partial p}{\partial z} = -\frac{Q}{A}\sigma\varphi B_o^2$$

$$= -\frac{8.4/N}{19.3/N}(2.2 \times 10^6)(0.1)(10^2)$$

$$= 9.5 \times 10^6 \frac{\text{Newtons}}{m^3} = 9.5\ MN/m^3$$

(Note that ($MN/m^2 = 10^6\ N/m^2 = 145$ psi). Assuming an effective length of the duct to be 2.8 m the total pressure drop is

$$\Delta p = (\ell)\partial p/\partial z = 2.8 \times 9.5 \times 10^6$$

$$= 27\ MN/m^2$$

$$= 4000\ \text{psi}$$

This result can now be used to calculate the pumping power per section. It is $\Delta p \times Q = 225/N$ megawatts, so that the total pumping power for the reactor is 225 MW. Since the reactor power output is 5000 MW(Th) then we see that the pumping power represents 4.5% of the thermal output. Clearly a number of assumptions have gone into the calculation of this ratio, but it represents a pumping power which is of the order of 10% of the electrical power output of the reactor and that is unacceptably high. It might be noted that the pressure drop in the absence of the magnetic field is about six orders of magnitude smaller.

15.4 STRESS CONSIDERATIONS IN COOLANT DUCTS

We have seen in the previous section that the use of lithium might be prohibitive due to the high pumping power requirements. We will now examine its heat transfer characteristics and since it is fertile breeding material, we shall see that the possibility also exists for it as a heat transfer fluid. As before, we shall focus our attention on a steady-state toroidal reactor consisting of a single toroidal magnetic field of the type discussed in Chapter 13. The pressure required to circulate the lithium will be calculated together with the mechanical stresses in the duct to show the limitations that these impose on the reactor parameters. These limitations will then be compared with those of the plasma physics so as to define a possible operating regime.

In order to minimize the adverse effects on breeding in the front sections of the blanket and to accommodate the high pressure of lithium, circular ducts with relatively thin walls will be considered. The tensile stress in the wall is related to the pressure "p" by

$$S = p\, a/t \tag{15.47}$$

where "a" is the duct radius and "t" is the wall thickness. We recall from equation (15.45) that the pressure gradient is given by

$$\frac{\partial p}{\partial z} = -\frac{Q}{A}\, \sigma_f\, \varphi\, B_o^2$$

where σ_f is the fluid's electrical conductivity, and φ is the ratio of the conductances, *i.e.*, $\varphi = t\sigma_w/a\sigma_f$. The volume flow rate, Q, is determined by the permissible temperature rise in the coolant, ΔT, the area of the blanket served by each duct, A_w, and the power loading of the first wall of the blanket, P_w, defined (in Chapter 13) as the thermal power deposited in the blanket per unit area facing the plasma. The volume flow is therefore given by

$$Q = P_w\, A_w/C\,\rho\Delta T \tag{15.48}$$

where C and ρ are the specific heat and density of the coolant, respectively. If ℓ is the effective length of the duct perpendicular to the magnetic field, then using the expression for the pressure drop with φ replaced by its value we calculate the pressure p to be

$$p = \ell Q B_o^2 \sigma_w t/a A \qquad (15.49)$$

Substitution of this result into (15.47) and utilizing (15.48) we get for the stress

$$S = \sigma_w \ell B_o^2 P_w A_w/C \rho \Delta T A \qquad (15.50)$$

The stress in the duct wall can therefore be readily determined since the parameters in the above equation are usually known or limited. If $\ell \approx 5$ meters, and using the previously suggested ratio $A/A_w \sim 0.05$, then for lithium flowing in stainless steel ducts at an average temperature of 450°C and ΔT of 300°C the stress is only a function of the magnetic field B_o and the wall loading P_w. This is demonstrated in Figure 124 where we note that the use of lithium as a coolant appears feasible at low magnetic fields or low wall loading. Moreover we observe from equation (15.50) that the stress is independent of the wall thickness or the ratio (t/a). In order to fully appreciate the information in Figure 124

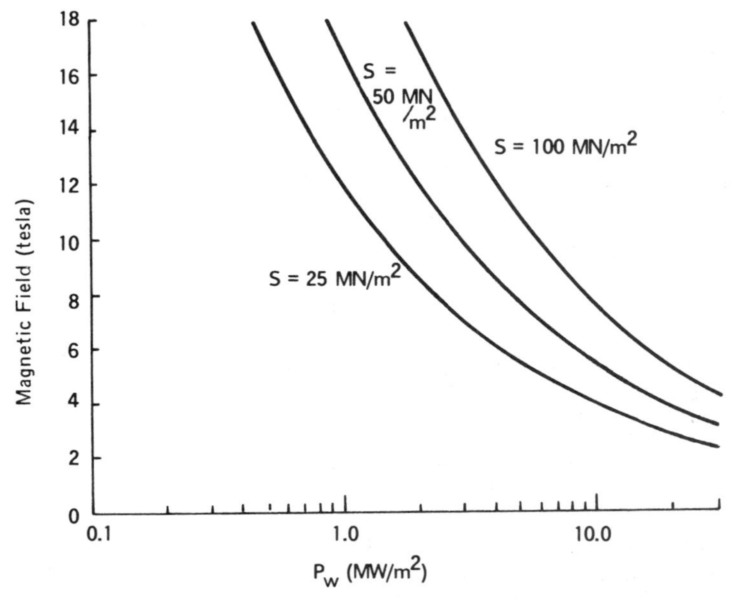

Figure 124. Magnetic field versus wall loading for different duct wall stresses (Hancox and Booth 1971).

we must first determine the permissible stresses in the materials of interest. Because of the high operating temperatures, the limitation is set by the long-time creep allowed in the material. It has been found reasonable to set the working stress as half the stress required to produce 0.2% creep in 100,000 hours. From the typical creep curves for stainless steel and nimonic alloy shown in Figure 125 the data can be extrapolated to yield the suggested working stresses given in Table 42. If we accept in the case of stainless steel the working stress of 50 MN/m^2 (7250 psi) then together with the other parameters equation (15.50) reduces to

$$P_w B_o^2 < 3 \times 10^{18} \tag{15.51}$$

For a magnetic field of 10 tesla this shows that the maximum wall loading is 3 MW/m^2 which is much smaller than (13 MW/m^2) that was suggested in Chapter 13.

We proceed now to assess the effect of the stress limitation on the pumping power. We recall that the pumping power P_p can be written as $P_p = Qp$, and if we substitute for the pressure from equation (15.47) we find that

$$P_p = \frac{St}{a} Q \tag{15.52}$$

Moreover we have seen in Chapter 13 that the thermal power output of the reactor can be expressed in terms of the product of the wall loading P_w and the area of the first wall A_w, *i.e.*, $P_t = A_w P_w$, and if we replace this product by its equivalent from equation (15.48) we obtain

$$P_t = Q C \rho \Delta T \tag{15.53}$$

In terms of this power, the pumping power required to circulate the lithium coolant can therefore be written as

$$\frac{P_p}{P_t} = \frac{s\,t}{a\,C\,\rho\,\Delta T} \tag{15.54}$$

Since the working stress S has already been fixed this ratio depends only on the ratio (t/a) and if this quantity is less than 0.1, the pumping power will be less than 2% of the reactor thermal power for a stress of 50 MN/m^2 (7250 psi). In practice (t/a) may be

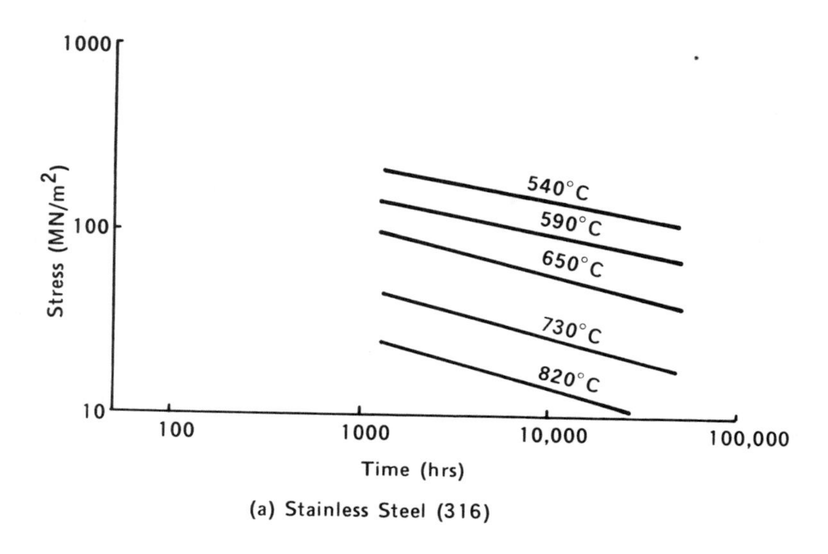

(a) Stainless Steel (316)

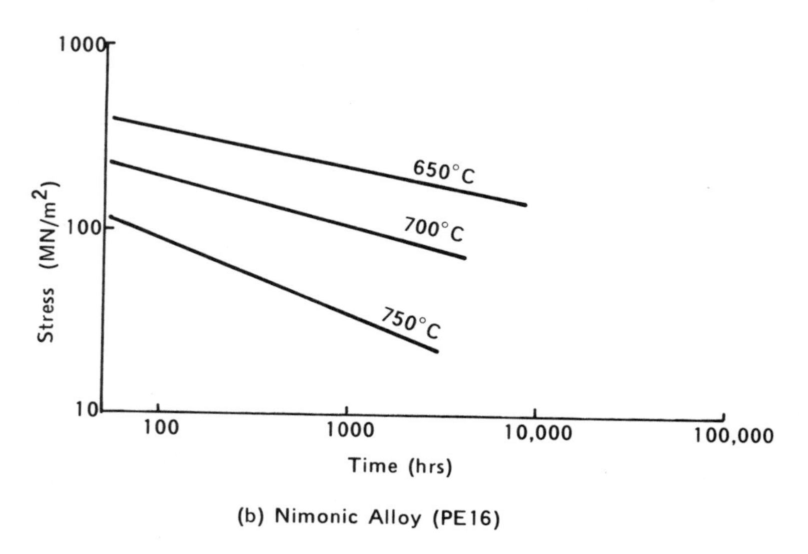

(b) Nimonic Alloy (PE16)

Figure 125. Stress-time curves for 0.2% creep (Hancox and Booth 1971).

Table 42

Working Stress for Various Duct Materials
for 0.2% Creep in 100,000 Hours (Hancox and Booth 1971)

Material		*Creep Stress*	*Working Stress*
Stainless Steel (316)	at 550°C	100	50
	at 600°C	80	40
Nimonic Alloy (PE16)	at 600°C	150	75
	at 650°C	100	50
Molybdenum Alloy (TM)	at 980°C	120	60

more severely limited by the need to reduce the structural component of the blanket to optimize tritium breeding and hence the pumping limitation may be less severe than the stress limitation. This situation is only reversed when very high stresses ($\sim$ 100 MN/m^2) can be tolerated or the temperature difference ΔT of the lithium must be reduced—as may occur in a low energy reactor.

We surmise from the above discussion that the most serious limitation to the use of lithium as coolant in a toroidal reactor is the stress in the pumping ducts so that either low magnetic fields or low wall loadings are required. These are not independent and a more explicit limitation may be deduced from the physics of confinement in toroidal systems. Once again, we recall from Chapter 13 that the reactor thermal power can be written as

$$P_t = A_w P_w = 4\pi^2 r_p R_o P_w/y \tag{15.55}$$

where r_p is the plasma radius, R_o is the major radius of the torus and y is the ratio of plasma radius to wall radius. If the energy deposited in the blanket per fusion reaction is Q_b then the rate of energy deposition in the blanket from a (50-50) D-T plasma is proportional to $n^2 \langle\sigma v\rangle Q_b$ and the reactor thermal power may also be expressed as

$$P_t = (2\pi R_o)(\pi r_p^2) \frac{n^2}{4} \langle\sigma v\rangle Q_b$$

$$= \pi^2 r_p^2 R_o \frac{n^2}{2} \langle\sigma v\rangle Q_b \tag{15.56}$$

We now introduce the reactor safety factor

$$q = \frac{B_T r_p}{B_p R_o}$$

where B_T is the toroidal magnetic field (B_o in this case), and B_p is the poloidal magnetic field (see Chapter 1) generated by the ohmic heating current. This factor must be larger than unity (usually between 1 and 3) to insure hydromagnetic stability of the plasma. In general the poloidal magnetic field is substantially smaller than the toroidal field, and its effect on the thermal aspects examined in this chapter has been and can be ignored. In a steady-state Tokamak type reactor the limit on the plasma pressure ratio is

$$\beta_o = [(r_p/R_o)^2 q^2]^{-1} \beta_p \qquad (15.57)$$

and if we take an aspect ratio R_o/r_p of 6, plasma temperature of 20 keV, a ratio of plasma radius to wall radius "y" of 0.8, we find from equations (15.55) - (15.57) that the average magnetic field required to confine the plasma is

$$B_o = 0.177\, q\, P_w^{3/8}\, P_t^{-1/8} \qquad (15.58)$$

A plot of the magnetic field as a function of P_w given by the above is shown in Figure 126 for three reactors with different electrical power output assuming a thermal conversion efficiency of 43%.

The limitations illustrated in Figures 126 and 124 can now be compared to establish the possibility of using lithium as a reactor coolant. A straightforward comparison is misleading in that Figure 126 gives the average value of the magnetic field whereas the stress limitation is most severe at the point in the toroidal reactor where the local field is highest. This correction factor relating the maximum field in the reactor blanket to the average field in the plasma has been computed by Booth and Hancox (1971) and is shown in Figure 127. Utilizing these correction factors we can now combine the limitations of Figures 124 and 126 to establish the operating regimes of toroidal reactors using lithium. These two criteria are illustrated in Figures 128 and 129 for a 2500 MW(e) and a 125 MW(e) reactor where the

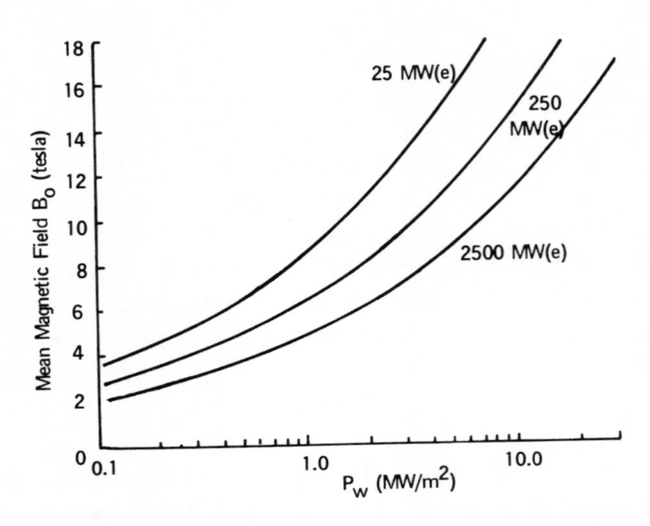

Figure 126. Average magnetic field for different size Tokamak reactors (Hancox and Booth 1971).

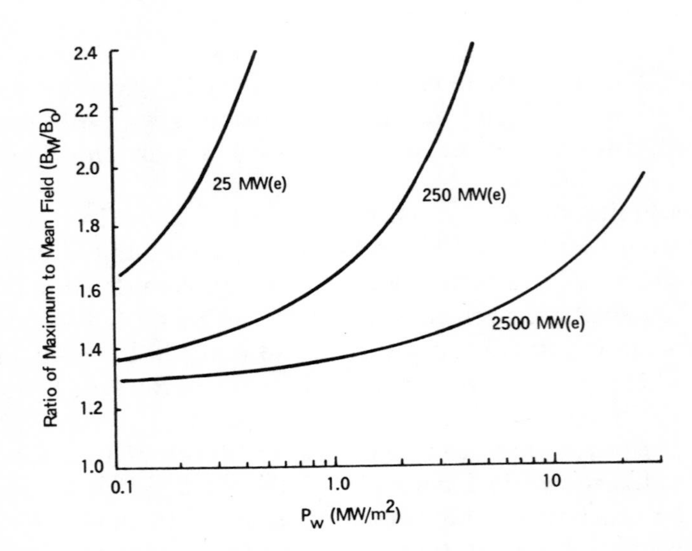

Figure 127. The ratio B_M/B_o versus wall loading for different size reactors (Hancox and Booth 1971).

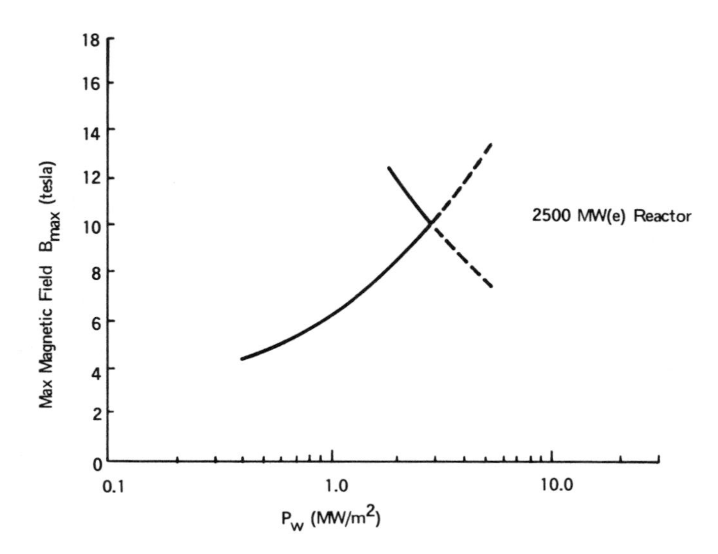

Figure 128. Combined limitations of magnetic field and wall loading in a 2500 MW(e) reactor (Hancox and Booth 1971).

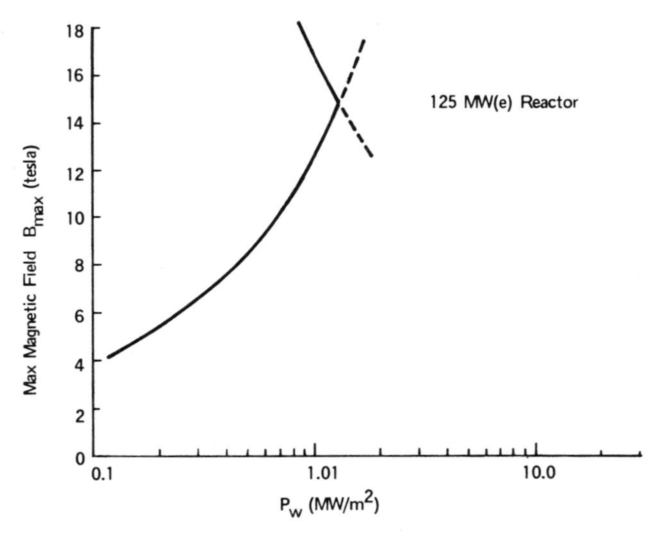

Figure 129. Maximum magnetic field versus wall loading for a 125 MW(e) reactor (Hancox and Booth 1971).

magnetic field is expressed in terms of the maximum magnetic field in the system. The first reactor may be viewed as a commercial reactor while the second is a prototype reactor due to its relatively small power output.

In the case of the power reactor Figure 128 shows that the optimum magnetic field is about 10 tesla (100 kilogauss), and the wall loading is then 2.9 MW/m^2 which is much less than the 13 MW/m^2 mentioned in Chapter 13. Therefore it appears that the use of liquid lithium forces the acceptance of a system which is physically larger (hence more expensive) than generally appears possible from other considerations. For the low-power prototype reactor it is seen from Figure 129 that the optimum maximum magnetic field is about 15 tesla giving a wall loading of 1.3 MW/m^2. This small wall loading is consistent with the small power output of such a reactor where the economic argument for high power density may not be relevant. In this case the use of liquid lithium cooling appears feasible.

REFERENCES

1. Branover, G. G. "Suppression of Turbulence in Pipes with Transverse and Longitudinal Magnetic Fields," *Magnetohydrodynamics, 47,* 107 (1967).
2. Branover, G. G. and A. B. Tsinober. *Magnetohydrodynamics of Incompressible Fluids* (in Russian) (Moscow, 1970).
3. Carlson, G. A. and M. A. Hoffman. "Effect of Magnetic Fields on Heat Pipes," Lawrence Livermore Laboratory Report UCRL-72060 (1969).
4. Chang, C. C. and T. S. Lundgren. "Duct Flow in MHD," *ZAMP, 12,* 100 (1961).
5. Hancox, R. and J. A. Booth. "The Use of Liquid Lithium as Coolant in a Toroidal Fusion Reactor," Part II. Culham Laboratory Report CLM-R116 (1971).
6. Hoffman, M. A. and G. A. Carlson. "Calculation Techniques for Estimating the Pressure Losses for Conducting Fluid Flows in Magnetic Fields," Lawrence Livermore Laboratory Report UCRL-51010 (1971).
7. Hughes, W. F. and F. J. Young. *The Electromagnetodynamics of Fluids*. (New York: John Wiley and Sons, 1966).
8. Hunt, J. C. R. and R. Hancox. "The Use of Liquid Lithium as Coolant in a Toroidal Fusion Reactor," Part 1 Culham Laboratory Report CLM-R115 (1971).
9. Hunt, J. C. R. and J. A. Shercliff. "High Hartmann Number Magnetohydrodynamics," *Ann. Rev. Fluid Mech., 3,* 37 (1971).

10. Shercliff, J. A. *A Text Book of Magnetohydrodynamics.* (Pergamon Press, 1965).
11. Shercliff, J. A. *J. Fluid Mech., 13,* 513 (1962).
12. Sutton, G. W. and A. Sherman. *Engineering Magnetohydrodynamics.* (New York: McGraw-Hill, 1965).

CHAPTER 16

A COMPARATIVE STUDY OF THE APPROACHES TO FUSION POWER

The lengthy and somewhat detailed examination in the previous chapters of the characteristic and design parameters of the various fusion reactors may be classified in the category of "specific" studies since in each case the analysis was particularly responsive to the confinement physics appropriate to the device. Although some general comparisons between open and closed-ended systems as well as between steady-state and pulsed systems were alluded to, no meaningful comparison of the various concepts was made since the specific studies have not been carried out on a common basis. For economic assessment as well as for critical evaluation of each concept it would be desirable to have some sort of a general formalism which will provide a common basis for comparison. Such formalism can be extracted from a complete formulation of a general power balance equation which can be readily applied to the various reactor concepts in a detailed manner. In this chapter we will follow the work of Nozawa and Steiner (1974) and deduce a general energy balance equation for the fusion plant which we will subsequently apply to the four concepts: laser-ignited, mirror, theta-pinch and Tokamaks. These various reactor concepts will then be compared on the basis of required plasma properties for power break-even and for net power production.

16.1 GENERAL FORMULATION OF THE POWER BALANCE EQUATION

We consider a hypothetical fusion reactor power plant schematically illustrated in Figures 130 and 131. The first figure represents such devices as the mirror machine where the energy of escaping charged particles can be directly converted to electrical energy, while the latter represents those systems with direct conversion recovery through the magnetic system such as the theta-pinch reactor. We assume a pulsed mode of operation for which the energy balance can be written as

$$W_n = W_s E \eta_d \left\{ \alpha Q \eta_a + \ell \eta_a + m F \eta_a + \delta(1 - \eta_a) \right\}$$

$$+ W_s E \eta_{t1} \left\{ (1 - \alpha) Q \eta_a + (1 - \ell) \eta_a + (1 - m) F \eta_a + \right.$$

$$\left. + (1 - \delta)(1 - \eta_a) \right\} + W_s E \eta_{t2} \left(\frac{1}{\eta_h} - 1 \right) + W_s E F \eta_a \eta_{t3} \left(\frac{1}{\eta_m} - 1 \right)$$

$$- W_s (1 + a) \tag{16.1}$$

where

W_n = net electrical energy output from the plant,

W_h = the heating energy supplied to the plasma with η_h efficiency,

W_m = the pulsed magnetic field energy produced with η_m efficiency,

and W_s = the sum of the electrical energies to the plasma heating subsystem and the pulsed magnetic field supply subsystem, *i.e.*, it is given by

$$W_s = \frac{W_h}{\eta_h} + \frac{W_m}{\eta_m} \tag{16.2}$$

We let

$$F = W_m / W_p \tag{16.3}$$

be the ratio of the pulsed magnetic field energy to the plasma energy, and if

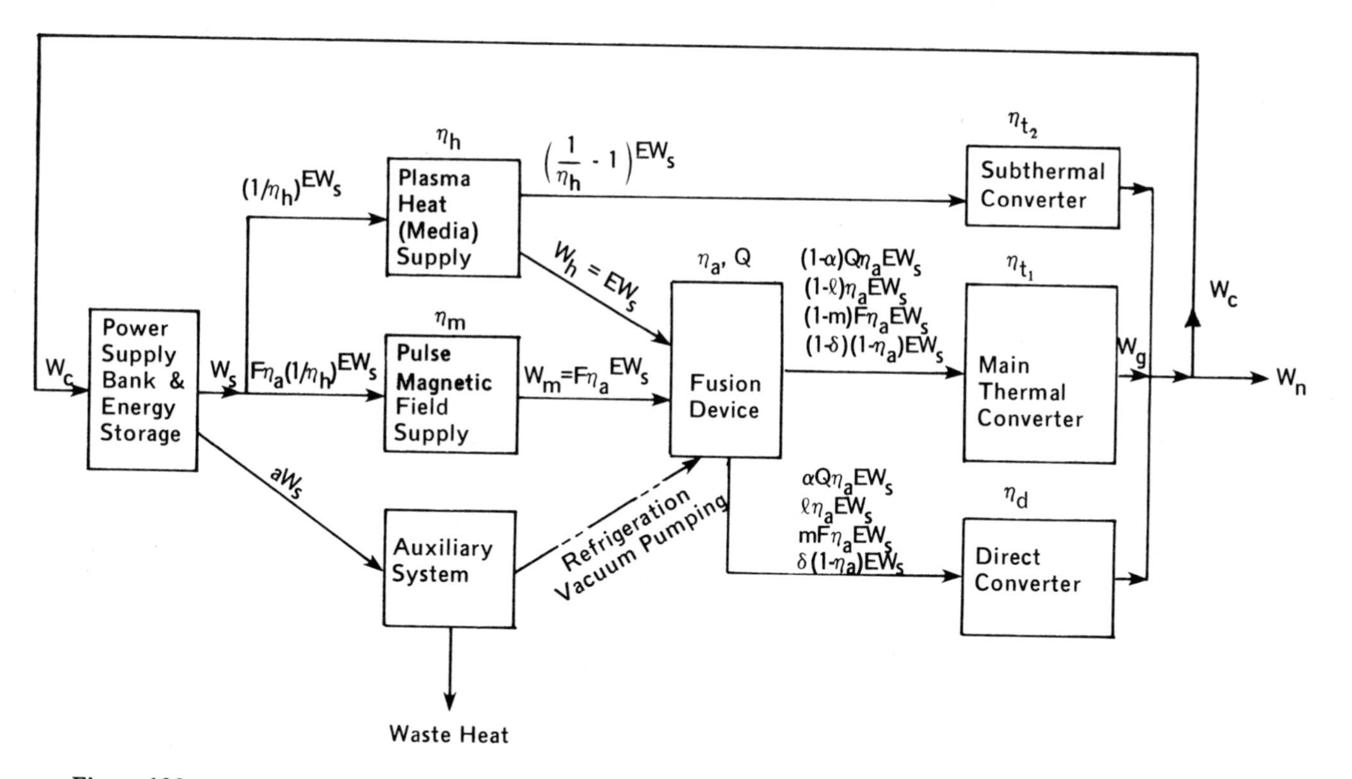

Figure 130. Schematic diagram of a hypothetical fusion reactor power plant with Direct Converter (Nozawa and Steiner, 1974).

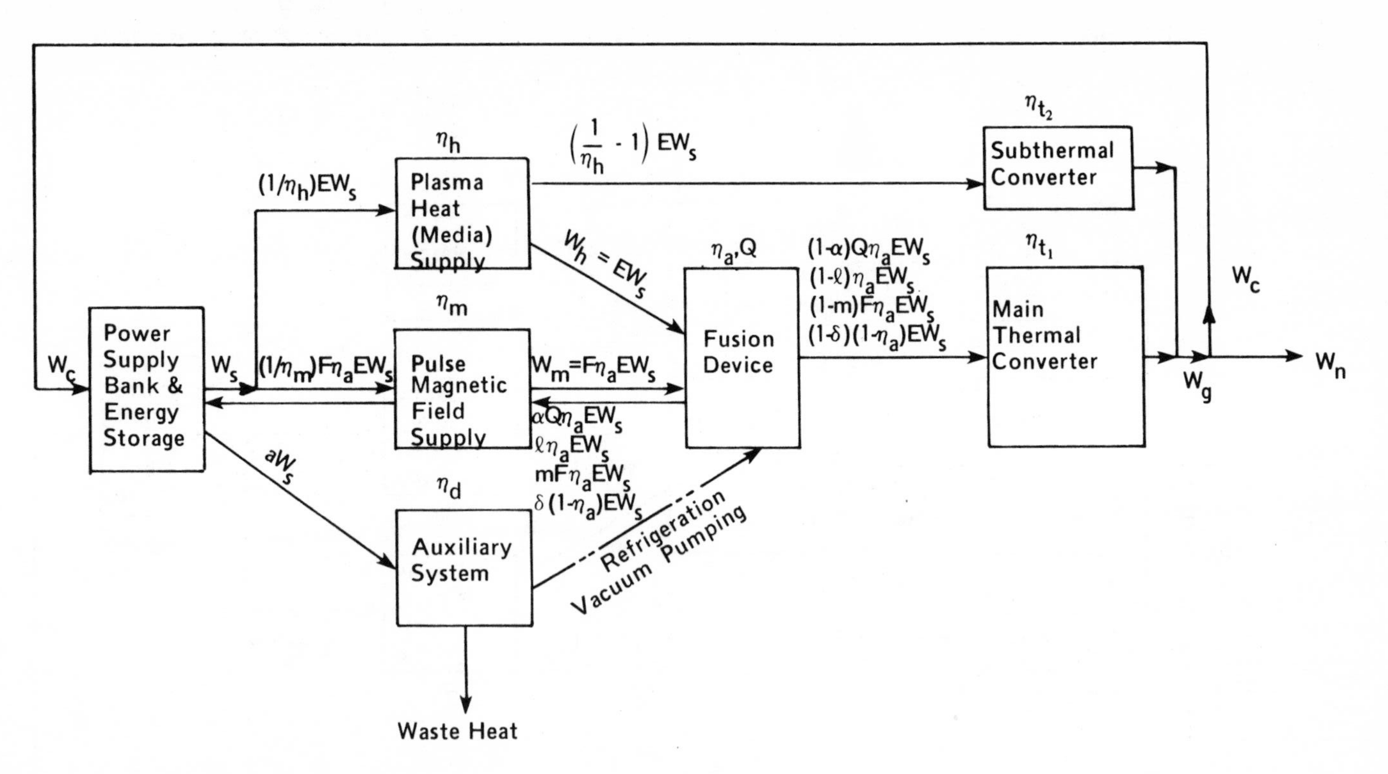

Figure 131. Schematic diagram of a fusion reactor power plant without a direct converter (Nozawa and Steiner, 1974).

$$\eta_a = W_p/W_h \tag{16.4}$$

represents the efficiency with which the heating energy to the plasma is absorbed by the plasma, then we can write for the fraction E of W_s which appears as heating energy to the plasma

$$E = \frac{W_h}{W_s} = \frac{W_h}{W_h/\eta_h + W_m/\eta_m} = \frac{\eta_h \, \eta_m}{\eta_m + \eta_h \, F \, \eta_a} \tag{16.5}$$

In view of these relations we can rewrite equation (16.2) as

$$W_s = \frac{W_h}{\eta_h} + \frac{F \, \eta_m \eta_h}{\eta_m} \tag{16.6}$$

so that W_h can be further expressed as

$$W_h = W_s \frac{\eta_h \, \eta_m}{\eta_m + F \, \eta_a \, \eta_h} = E \, W_s \tag{16.7}$$

We now introduce the familiar figure of merit, Q, which we recall has the definition

$$Q = \frac{\text{fusion energy produced in the plasma and blanket}}{\text{heating energy absorbed by the plasma}}$$

and if we represent the numerator by W_f, then

$$Q = W_f/W_p \tag{16.8}$$

Returning to equation (16.1), we find that the first term represents the energy recovered by direct conversion with efficiency η_d. The quantity α denotes that portion of the fusion energy that is recovered by direct convertion, while ℓ, m, and δ denote those portions recovered by direct conversion of the plasma energy supplied, the pulsed magnetic energy supplied, and the heating energy not absorbed by the plasma, respectively. This can be seen by in incorporating the definitions presented earlier in the first term, *i.e.*,

$$W_s\, E\, \eta_d \left\{ \alpha\, Q\, \eta_a + \ell \eta_a + m\, F\, \eta_a + \delta(1 - \eta_a) \right\} =$$

$$W_s \left(\frac{W_h}{W_s}\right) \eta_d \left\{ \alpha \frac{W_f}{W_p} \frac{W_p}{W_h} + \ell \frac{W_p}{W_h} + m \frac{W_m}{W_p} \frac{W_p}{W_h} \right.$$

$$\left. + \delta(1 - W_p/W_h) \right\} = \eta_d \left\{ \alpha\, W_f + \ell\, W_p + m\, W_m \right.$$

$$\left. + \delta(W_h - W_p) \right\}$$

In a similar manner, we see that the second term, *i.e.*,

$$W_s\, E\, \eta_{t1} \left\{ (1 - \alpha)\, Q\, \eta_a + (1 - \ell)\, \eta_a + (1 - m) F\, \eta_a + (1 - \delta)(1 - \eta_a) \right\}$$

$$= \eta_{t1} \left\{ (1 - \alpha) W_f + (1 - \ell) W_p + (1 - m) W_m + (1 - \delta)(W_h - W_p) \right\}$$

represents the energy recovered by the main thermal converter with efficiency η_{t1}. The third term represents heat loss recovered from the plasma heating subsystem by a secondary thermal converter with efficiency η_{t2}, *i.e.*,

$$W_s\, E\, \eta_{t2}\, (1/\eta_h - 1) = W_s\, (W_h/W_s)\, \eta_{t2} (1/\eta_h - 1)$$

$$= \eta_{t2}\, (W_h/\eta_h - W_h)$$

The fourth term, which will be neglected, represents the heat loss recovered from the pulsed magnetic supply subsystem by a thermal converter with efficiency η_{t3}. This may be seen as follows

$$W_s\, E\, F\, \eta_a\, \eta_{t3} \left(\frac{1}{\eta_m} - 1\right) = W_s\, (W_h/W_s)(W_m/W_p)(W_p/W_h) \eta_{t3} \left(\frac{1}{\eta_m} - 1\right)$$

$$= \eta_{t3} \left(\frac{W_m}{\eta_m} - W_m \right)$$

The total input energy term is given by the last term in equation (16.1), *i.e.*, $- W_s(1 + a)$ where “a” indicates the portion of energy consumption due to the auxiliary system in relation to W_s. If we call "f" the repetition rate (in cycles per second) then the relations between energies and powers are

$$
\begin{aligned}
P_n &= f\,W_n \\
P_s &= f\,W_s \\
P_a &= a\,f\,W_s
\end{aligned}
\tag{16.9}
$$

with the last relation representing the continuous power P_a consumed by the auxiliary system.

In the case of a steady-state fusion reactor, the energy balance equation is obtained from (16.1) by simply replacing each energy term W_i by a corresponding power term P_i. For Q of a steady-state system the definition simply becomes

$$Q = \frac{\text{rate of fusion energy produced}}{\text{rate of heating energy trapped by plasma}}$$

When the plant's net electrical energy output W_n is set equal to zero, equation (16.1) yields a "critical" value Q_c given by

$$Q_c = \frac{1}{\eta_e}\left[\frac{1+a}{E\eta_a} - \eta_d(\ell + Fm) - \eta_{t1}\left\{(1-\ell) + F(1-m)\right\} - \left\{\delta\eta_d + (1-\delta)\eta_{t1}\right\}\left(\frac{1}{\eta_a} - 1\right) - \eta_{t2}\,\frac{1-\eta_h}{\eta_a\,\eta_h}\right] \tag{16.10}$$

where

$$\eta_e = \alpha\,\eta_d + (1-\alpha)\,\eta_{t1} \tag{16.11}$$

represents the efficiency with which fusion energy is converted into electrical energy. We note that if all the efficiencies were unity (a = 0) then Q_c would be equal to zero. Moreover, by combining equations (16.1) and (16.10) we find that

$$W_n = (Q - Q_c)\,E\,\eta_a\,\eta_e\,W_s \tag{16.12}$$

where the term $Q_c E \eta_a \eta_e W_s$ corresponds to the energy necessary to sustain the power plant without the production of net output energy.

For an assessment of the power cost of a fusion plant it is necessary to relate the energy capacity of the subsystems to the plant net output. We therefore define the quantity S to be the ratio of the total supplied energy, $(1 + a)\, W_s$, to the net output energy W_n, and define another quantity C to be the ratio of the circulating power W_c to W_n. Then, utilizing equation (16.12) we get

$$S = \frac{(1 + a)\, W_s}{W_n} = \frac{(1 + a)}{(Q - Q_c)\, E\, \eta_a\, \eta_e} \tag{16.13}$$

and note that S^{-1} is an energy multiplication factor which is applied to the supplied energy $(1 + a)\, W_s$ to yield an output energy of W_n. When all the directly converted energies contribute to the gross output W_g, as is the case for the plant in Figure 130 then

$$C = \frac{W_c}{W_n} = \frac{(1 + a)\, W_s}{(Q - Q_c)\, E\, \eta_a\, \eta_e\, W_s} = S$$

but when these energies go to the input side of the plant as in Figure 131, C becomes

$$C = \frac{W_c}{W_n} = \frac{[(1 + a) - \eta_e\, \eta_d\, E\, \{\alpha Q + \ell m + \delta(1/\eta_a - 1)\}]\, W_s}{(Q - Q_c)\, E\, \eta_a\, \eta_e\, W_s} \tag{16.14}$$

The above expression reduces to (16.13) when $\eta_d = 0$, and if there are two kinds of direct conversion mechanisms then equation (16.14) must be modified accordingly.

We recall from equation (9.5) of Chapter 9 that the overall plant efficiency is defined as the ratio of the net electrical output energy to the total fusion energy, and if we call it here η then we can write it as

$$\eta = \frac{W_n}{W_f} = \frac{W_n}{Q\, E\, \eta_a\, W_s} \tag{16.15}$$

having used equations (16.4), (16.7) and (16.9). If in addition we utilize equation (16.12) then η could further be written as

$$\eta = \frac{Q - Q_c}{Q}\, \eta_e \tag{16.16}$$

which allows us to write Q in the form

$$Q = \frac{\eta_e}{\eta_e - \eta} Q_c = Y Q_c \tag{16.17}$$

This equation defines the value of Q which must be attained for a given set of plant conditions expressed by Q_c and the efficiencies η and η_e. It might be noted in concluding this section (as we did in Chapter 9 for the mirror reactor) that the larger the η the better the plant performance and the less the thermal discharge. Moreover, small values of S and C are obviously desirable from the viewpoint of capital cost.

16.2 CALCULATION OF Q FOR PULSED AND STEADY-STATE SYSTEMS

Perhaps more than any other parameter, the Q value is the most important characteristic parameter of a fusion power-producing system. In this section we deduce the equation for this parameter in terms of relevant plasma and confinement parameters for both pulsed and steady-state reactors.

If we let $n(\vec{r})$ be the ion (or electron) density, $T_e(\vec{r})$ and $T_i(\vec{r})$ be the electron and ion temperatures, respectively, and V_1 be the plasma volume to be heated by the heating subsystem, then the plasma energy W_p can be written as

$$W_p = \int_{V_1} \frac{3}{2} n(\vec{r}) \left\{ T_e(\vec{r}) + T_i(\vec{r}) \right\} d^3r \tag{16.18}$$

where the temperatures are in electron-volt units. We focus our attention on a 50-50 D-T mixture, and if Q_f denotes the energy released per D-T fusion reaction including the energy produced in the blanket we get for the fusion energy W_f produced in the reactor

$$W_f = \int_{V_2} \frac{1}{2} n(\vec{r}) f_b Q_f d^3r \tag{16.19}$$

where V_2 is the plasma volume which undergoes burning, and f_b is the fuel fractional burn-up given by (see equation 2.38 of Chapter 2)

$$f_b = \left[1 + \frac{2}{n\tau\ \langle\sigma v\rangle}\right]^{-1} \approx \tfrac{1}{2} n\tau\ \langle\sigma v\rangle \qquad (16.20)$$

We note also that equation (16.19) can be put in the form

$$W_f = \int_{t_1}^{t_f}\int_{V_2} \tfrac{1}{4} n^2(\vec{r},t)\ \langle\sigma v\rangle\ [T_i(\vec{r},t)]\ Q_f\ d^3r\ dt \qquad (16.21)$$

if the ion density and temperature are functions of time as well as space. The limits on the time integral represent the time at which plasma burning begins and ceases, respectively. For spatially uniform plasma density and temperature, equations (16.18) and (16.19) can be combined to give the value of Q in accordance with equation (16.8). The result is

$$Q = f_b \frac{Q_f}{3(T_e + T_i)} \frac{V_2}{V_1} \qquad (16.22)$$

which reveals that Q may be increased by as much as V_2/V_1 when a portion, V_1 of V_2 is ignited and then nuclear heating takes the whole volume V_2 to the ignition point. When $V_2 = V_1$ and $T_e = T_i = T$ then equation (16.22) reduces to

$$Q = \frac{f_b\ Q_f}{6T} \qquad (16.23)$$

Equations (16.18) and (16.21) can be combined to give yet another expression for Q, namely

$$Q = \frac{\int_{t_i}^{t_f} n^2(t)\ \langle\sigma v\rangle\ [T_i(t)]\ Q_f\ dt}{6\ n(t_i)\ (T_e + T_i)_{t=t_i}} \ \frac{V_2}{V_1} \qquad (16.24)$$

assuming n, T_e and T_i to be spatially uniform. If in addition the plasma burn-up is small or is kept constant in time by refueling, then n stays constant in time. Assuming $V_2 = V_1$ and that the plasma operating temperature is brought to T_2 after ignition, then equation (16.24) assumes the form

$$Q = \frac{\langle\sigma v\rangle\ (T_2)\ Q_f}{6(T_e + T_i)_{t=t_i}}\ n\tau \qquad (16.25)$$

where $\tau = t_f - t_i$.

We proceed now to apply the above expressions to a pulsed reactor where we assume that the initial plasma temperature before heating is negligible; that heating energy is supplied up to the temperature at which the plasma is ignited (above this point the plasma will be heated by nuclear heating); that heating is required for each pulse; and that the temperature and density of the plasma are uniform over the plasma volume. If $T_e = T_i = T_1$ where T_1 is the ignition temperature then from equations (16.23) and (16.25) we can write for the pulsed reactor

$$Q = \frac{f_b\, Q_f}{6\, T_1} \tag{16.26}$$

or

$$Q = \frac{\langle\sigma v\rangle\, [T_i(t)]\, Q_f}{12\, T_1}\; n\tau \tag{16.27}$$

where $\langle\sigma v\rangle$ is still a function of time. If we let T_2 be the representative temperature of the plasma burn, then the above expression can be put in the form

$$Q = g\, q(T_2)\, n\tau \tag{16.28}$$

with

$$g = \frac{T_2}{T_1};\quad q(T_2) = \frac{\langle\sigma v\rangle\, (T_2)\, Q_f}{12\, T_2} \tag{16.29}$$

The factor g can be viewed as an energy multiplication factor, while $q(T_2)$ represents the rate of energy multiplication per fusion event as a function of the burning temperature. Moreover, we note from equation (16.26) that Q has a maximum value, Q_{max}, for a given ignition temperature when $f_b = 1$. This is in addition to the increase that can be achieved through the ratio V_2/V_1 as indicated by equation (16.22).

For a steady-state reactor, a continuous supply of plasma particles with appropriate energy is needed for its operation. The energy balance equation in this case assumes the form

$$\frac{d\, W_p}{dt} = I\, E_o\, \eta_a - \frac{W_p}{\tau_c} - \frac{W_p}{\tau_b} \tag{16.30}$$

where I is the rate of particle injection, E_o is the energy of injected particle so that the energy $I\,E_o$ is injected at an efficiency of

η_a. The quantity τ_c represents the geometric confinement time whereas τ_b denotes the plasma life time due to burn-up. For the stationary state, equation (16.30) becomes

$$I\, E_o\, \eta_a = W_p \left(\frac{1}{\tau_c} + \frac{1}{\tau_b}\right) \tag{16.31}$$

which upon substitution in the definition of Q for steady-state systems gives

$$Q = \frac{\frac{1}{4} n^2 \langle \sigma v \rangle (T_2)\, Q_f}{W_p \left(\frac{1}{\tau_c} + \frac{1}{\tau_b}\right)} \tag{16.32}$$

We note, however, from the definition of the burn-up fraction f_b that it is equal to $f_b = \tau_c / \tau_b$, and if we combine this with W_p as given by (16.18) the above expression for Q reduces to

$$Q = \frac{\langle \sigma v \rangle (T_2)\, Q_f}{6\,[T_e(1) + T_i(1)]\,(1 + f_b)}\, n\tau_c \tag{16.33}$$

where we have labeled the electron and ion temperatures at injection by $T_e(1)$ and $T_i(1)$. It is interesting to observe that (16.33) has the same form as (16.25) except for the factor $(1 + f_b)^{-1}$. The corresponding relationship between Q and f_b can be readily obtained through the use of (16.20); the result is

$$Q = \frac{f_b\, Q_f}{3[T_e(1) + T_i(1)]\,(1 + f_b)} \tag{16.34}$$

16.3 APPLICATION TO THE VARIOUS FUSION CONCEPTS

The Laser-Ignited Fusion Reactor

We recall from Chapter 11 that fusion energy is produced in this inertial-confinement scheme as a result of irradiating a deuterium-tritium pellet by intense laser beams. We recall also that the incident laser energy could, in addition to heating the plasma, lead to compression of the pellet to solid state or greater densities, thereby increasing the fusion reaction rate. A power-producing reactor based on this concept is, clearly, a pulsed system and an energy flow diagram for such a reactor is schematically illustrated in Figure 132.

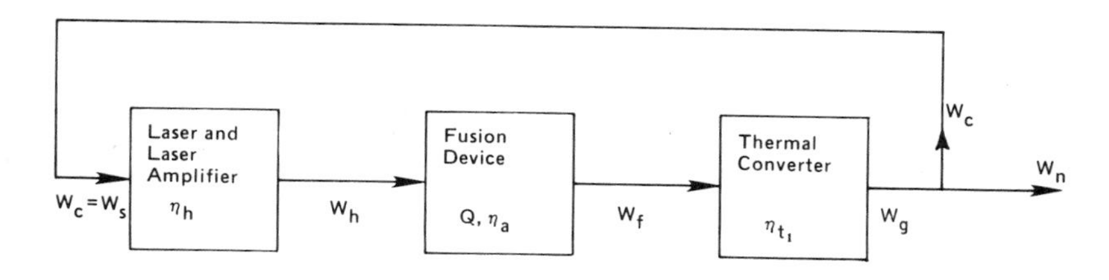

Figure 132. Schematic of a laser fusion reactor power plant.

In its simplest form, the fusion device consists of a blast containment vessel having windows through which the laser light is introduced and a blanket to utilize the fusion neutrons. The laser and laser amplifier in Figure 132 correspond to the plasma heat subsystem of Figure 130 and η_h represents the efficiency of producing and introducing the laser energy into the fusion device. The efficiency of coupling the laser energy to the fuel pellet is denoted by η_a, which clearly represents the absorption efficiency of the laser energy. As in other systems, we assume that the fusion energy is converted with efficiency η_{t_1} by a thermal converter, while the energy losses in the laser subsystem may be recovered by an auxiliary thermal converter with efficiency η_{t_2}. Since no pulsed magnetic field is used in this system, then F = 0 and E = η_h in equation (16.1). Moreover, η_d = o and $\alpha = \ell = m = \delta = 0$ since there is no direct recovery of either plasma or fusion energies; thus equation (16.1) becomes for the laser-fusion system

$$W_n = W_s \eta_h \eta_{t_1} (Q\eta_a + 1) + W_s \eta_h \eta_{t_2} \left(\frac{1}{\eta_h} - 1\right) - (1 + a) W_s \tag{16.35}$$

If we now assume that the energy losses in the laser subsystem are not recoverable, *i.e.*, $\eta_{t_2} = 0$, and that the auxiliary power requirements are also negligible, *i.e.*, a = 0, then the above equation reduces to

$$W_n = W_s \left\{ \eta_h \eta_{t_1} (Q\eta_a + 1) - 1 \right\}$$

$$= \frac{W_h}{\eta_h} \left\{ \eta_h \eta_{t_1} \left(\frac{W_f}{W_p} \frac{W_p}{W_h} + 1 \right) - 1 \right\} \qquad (16.36)$$

$$= \eta_{t_1} (W_f + W_h) - W_h/\eta_h$$

with the last form showing that W_n is simply the difference between the output and input energies. Further examination of equation (16.36) reveals the following. W_s is the energy required to produce the laser energy W_h, and η_h is the efficiency of the laser subsystem. The quantity $W_s \eta_h \eta_a$ represents the thermal energy given to the pellet and corresponds to the plasma energy W_p. That portion of the laser energy not absorbed by the pellet appears as heat in the blanket and amounts to $W_s \eta_h(1 - \eta_a)$. Thus, as shown by the intermediate and final forms of (16.36) the fusion energy produced is $W_s Q\eta_a \eta_h$ and the total energy delivered to the thermal converter is $W_s Q\eta_a \eta_h + W_s \eta_h$. The gross electrical output, W_g, is then seen to be

$$W_g = \eta_{t_1} W_s \eta_h (Q\eta_a + 1) = \eta_{t_1}(W_f + W_h) \qquad (16.37)$$

The critical value of Q for the system, *i.e.*, Q_c, is obtained by setting $W_n = 0$ in equation (16.35), *i.e.*,

$$Q_c = \frac{1}{\eta_a} \left(\frac{1 + a}{\eta_h \eta_{t_1}} - \frac{1 - \eta_h}{\eta_h} \frac{\eta_{t_2}}{\eta_{t_1}} - 1 \right) \qquad (16.38)$$

and if, once again, we set $a = \eta_{t_2} = 0$ we find that

$$Q_c = \frac{1}{\eta_a} \left(\frac{1}{\eta_h \eta_{t_1}} - 1 \right) \qquad (16.39)$$

It can be seen from the above equations that $a > 0$ tends to increase the value of Q_c whereas $\eta_{t_2} > 0$ tends to decrease it. Moreover, we note from equations (16.13), (16.14) and (16.15) that that the power balance parameters for the laser-fusion reactor have the forms

$$S = \frac{1 + a}{(Q - Q_c)\, \eta_a\, \eta_h\, \eta_{t_1}} \tag{16.40}$$

$$C = S \tag{16.41}$$

and

$$\eta = \frac{Q - Q_c}{Q}\, \eta_{t_1} \tag{16.42}$$

The equivalence of C and S as indicated by (16.41) is due to the fact that there is no direct recovery of energy from the fusion device to the power storage subsystem.

According to some of the references cited at the end of Chapter 11, the efficiency of the laser subsystem η_h will be about 25% for CO_2 lasers which are currently considered the most efficient lasers. We recall also from the same chapter that the laser coupling efficiency ϵ_L which corresponds to η_a in the present notation has a value between 5% and 10%. The quantity $Q = W_f/W_p$ corresponds to M/ϵ_L when it is equal to Q_c, and we recall that it can be as large as 200. For comparison, we take $\eta_a = 5\%$, $\eta_h = 25\%$, and $Q_c \approx 180$ to represent the results given in Chapter 11 and compare them with some reference cases shown in Table 43.

Table 43

Laser Fusion Reactor (Nozawa and Steiner 1974)

	Reference Cases			*Results from*
Parameter	*Case 1*	*Case 2*	*Case 3*	*Chapter 10*
η_{t_1}	0.38	0.57	0.38	0.40
η_h	0.20	0.20	0.20	0.25
η_a	0.10	0.10	0.10	0.05
a	0	0	0.05	0
η_{t_2}	0	0	0.20	0
Q_c	121.6	77.7	107.1	180

For a D-T fuel cycle, equation (16.26) yields for the system under consideration

$$Q = \frac{3733}{T_1}\, f_b \tag{16.43}$$

where it has been assumed that $T_e = T_i = T_1$ with T_1, in keV, being the ignition temperature. After ignition, the plasma temperature will rise due to alpha-particle heating and the temperature at which the bulk of fusion energy will be released is substantially higher than T_1 and may in fact exceed 100 keV. For purposes of comparison, we assume an ignition temperature of 5 keV and a mean burning temperature of 50 keV. The values of Q, f_b, and $n\tau$ required to attain a plant efficiency, η, of 0.30, 0.40, and 0.50 are shown in Table 44. We observe from this table that high burn-up fractions are necessary for most cases. The quantity Y is calculated from equation (16.17), while the Lawson parameter $n\tau$ is obtained from (16.28).

We may conclude from the above discussion that improvement of the laser efficiency is perhaps the most important technological problem for the realization of laser-ignited fusion power reactors. It also appears that efficient power production by laser fusion (*e.g.*, η = 50%) will require relatively high burn-up of the fuel. The pulsed nature of the system may nevertheless allow for some direct conversion mechanisms to improve the overall plant efficiency.

Table 44

Q, f_b, and $n\tau$ for Given Values of η of Laser Fusion Reactor (Nozawa and Steiner 1974)

Given η	*Quantity*	*Case 1*	*Case 2*	*Chapter 4 Values*
0	Q_c	121.6	77.7	180
	η_{t_1}	0.38	0.57	0.40
0.30	Y	4.75	2.11	4.0
	Q	578	164	720
	f_b	0.77	0.22	0.96
	$n\tau(10^{14})$	17.8	5.1	22.2
0.40	Y		3.35	
	Q		260	
	f_b		0.35	
	$n\tau(10^{14})$		8.0	
0.50	Y		8.14	
	Q		633	
	f_b		0.85	
	$n\tau(10^{14})$		19.5	

The Mirror Fusion Reactor

The mirror fusion power plant, shown schematically in Figure 133, is a system which is envisaged to operate in a steady-state mode. We recall from previous chapters that this reactor is expected to be heated and fueled by neutral beam injection and the Q for the device is given by equations (16.33) or (16.34). We also recall that direct conversion to electricity of the energy of the escaping charged particle is almost an inherent part of the system. In order to apply the power balance equation derived earlier to this fusion concept, a number of assumptions will be made. They are: (1) the confining magnetic field is supplied by "superconducting" coils so that the power required for the steady-state field is negligibly small; (2) the neutral beam injection is 300 keV; (3) the fuel consists of 50-50 D-T mixture; (4) a fraction η_a of the injected power $P_s \eta_h$ will be trapped in the confinement region (it has been shown by Futch, *et al.* (1971) that ten per cent of the trapped power will go to the electrons, and that the plasma ion temperature is about the same as that of the injection energy]; and finally (5) that the escaping D and T ions carry out 90% of the trapped energy while the electrons carry out the remaining 10%.

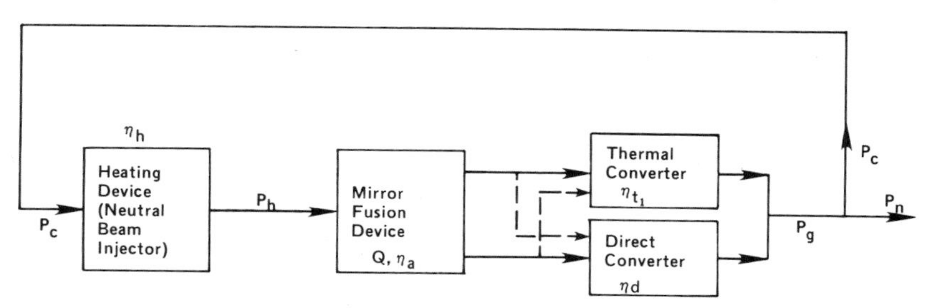

Figure 133. Schematic of a mirror reactor power plant.

On the basis of these assumptions and the general features of mirror machines the following facts can be utilized in equation (16.1): (1) no pulsed magnetic field is supplied, hence F = 0 and E = η_h; (2) no heat loss is recovered from the neutral beam injection subsystem, thus η_{t_2} = 0; and (3) the injected energy which is not trapped will go to the direct converter, thus δ = 1. With these conditions, equation (16.1) reads

$$P_n = P_s \eta_h \eta_d \left\{ \alpha Q \eta_a + \ell_e \eta_a + (1 - \eta_a) \right\} + P_s \eta_h \eta_{t_1}$$

$$\left\{ (1 - \alpha) Q \eta_a + (1 - \ell_e) \eta_a \right\} - P_s (1 + a) \quad (16.44)$$

where

$$\ell_e = \alpha f_b + \ell_o (1 - i f_b) \quad (16.45)$$

and ℓ_e is the effective fraction of the injected energy which is directly recoverable, ℓ_o is the fraction of the escaping energy which is directly recoverable in the absence of burn-up, and "i" is the fraction of the injected energy carried out of the plasma by the escaping ions. The first term on the right-hand side of equation (16.45) represents the amount of injected kinetic energy which is added to the charged-particle portion (*e.g.*, the 3.5 meV alpha) of the fusion energy as a result of the burn-up. The last term in (16.45) shows that the number of injected ions escaping from the plasma is reduced by f_b and, as a result, that the recoverable injected energy is reduced by i f_b where "i" has a value of 90% by assumption (5) mentioned earlier. Equation (16.45), therefore, accounts for the effects of burn-up on the available charged-particle energy to be recovered by direct conversion.

From equation (16.34) we see that, if we neglect $(1 + f_b)$, the burn-up fraction is related to Q by

$$f_b = \frac{3[T_e(1) + T_i(1)]}{Q_f} Q = B Q \quad (16.46)$$

where the electron and ion temperatures are those at injection. Combining equations (16.44), (16.45) and (16.56), we get for the net power output of the reactor

$$P_n = P_s \eta_h \eta_d [\{ \alpha(1 + B) - i \ell_o B \} Q \eta_a + \ell_o \eta_a + (1 - \eta_a)]$$

$$+ P_s \eta_h \eta_{t_1} [\{ (1 - \alpha)(1 + B) - (1 - i \ell_o) B \} Q \eta_a$$

$$+ (1 - \ell_o) \eta_a] - P_s(1 + a) \quad (16.47)$$

which upon setting $P_n = 0$ yields the critical value of Q, namely Q_c, *i.e.*,

$$Q_c = \frac{(1 + a) - \eta_d \eta_h + (1 - \ell_o)(\eta_d - \eta_{t_1}) \eta_a \eta_h}{\eta_a \eta_h \eta_{em}} \tag{16.48}$$

where

$$\eta_{em} = \left\{ \alpha(1 + B) - i\, \ell_o\, B \right\} \eta_d + \left\{ (1 - \alpha)(1 + B) - (1 - i\, \ell_o)\, B \right\} \eta_{t_1} \tag{16.49}$$

If we consider the limiting case of $\eta_a = \ell_o = 1$, and $a = B = 0$ then equation (16.48) reduces to

$$Q_c = \frac{1 - \eta_d \eta_h}{\eta_h \{ \alpha \eta_d + (1 - \alpha)\, \eta_{t_1} \}} \tag{16.50}$$

which is a result quoted by many workers in the field. The above results can now be inserted in the expressions for the power balance parameters, S, C and η to yield

$$S = C = \frac{1 + a}{(Q - Q_c)\, \eta_a \eta_h \eta_{em}} \tag{16.51}$$

and

$$\eta = \frac{Q - Q_c}{Q}\, \eta_{em} \tag{16.52}$$

Returning to (16.49), we note that η_{em} defines an effective conversion efficiency for mirror reactors. The physical meaning of the terms containing B can be best appreciated by first introducing an effective α which we call α^* defined by

$$\alpha^* \equiv \alpha(1 + B) - i\, \ell_o\, B \tag{16.53}$$

which makes η_{em} assume the form

$$\eta_{em} = \alpha^* \eta_d + (1 - \alpha^*)\, \eta_{t_1} \tag{16.54}$$

Before examining the significance of each of these terms, we first note that for $T_e = 1/10\ T_i$ equation (16.46) yields

$$f_b = \frac{3.3\ T_i}{Q_f}\, Q = B\, Q \tag{16.55}$$

where $T_i(1) = T_i(2) = T_i$, *i.e.*, the operating temperature of the ions is equal to the injected energy. Equation (16.54) indicates that the energy directly recoverable is increased by $B(1 - i\ell_o)$ while the thermal energy is decreased by the same amount. For the D-T fuel cycle, α varies between 0.16 and 0.20, depending on the degree of contribution from the D-D and D-He^3 reactions. The quantity "i" is taken to be 90% and ℓ_o must be larger than 90%, thus leading to the fact that $B(\alpha - i\ell_o)$ is negative and proportional to B. We see, therefore, that the overall effect of burn-up is to shift the recovery of the injected energy from the direct converter to the thermal converter.

We turn our attention now to an examination of the effects on Q_c of the efficiencies η_h and η_d for which the following will be adopted as reference values for a, B, i, ℓ_o, α, and η_a:

a	B	i	ℓ_o	α	α^*	η_a
0	0.03	0.90	0.98	0.20	0.18	0.98

The choice of a = 0 implies that the auxiliary power requirement is neglected; the value of B is obtained from (16.55) assuming an injection energy of 300 keV; i = 0.90 means that 90% of the plasma energy is carried out by the escaping ions; $\ell_o = 0.98$ means that in the absence of burn-up 98% of the total escaping power from the plasma enters the direct converter; $\alpha = 0.20$ means that 20% of the total fusion energy will appear as charged particle energy and thus be available for direct conversion; α^* is obtained from (16.54), and $\eta_a = 0.98$ means that only 2% of the injected neutral beam will not be trapped but will go to the direct converter. The variation of Q_c with η_d for various values of η_h and the two reference values of η_{t_1} is shown in Figure 134.

Some recent experimental and analytical work by Moir and Barr (1973) and by Fries (1973) has shown that η_d falls in the range 60-80%, and we choose $\eta_d = 0.70$ as the reference value for the present study. Moreover, the neutral beam power efficiency η_h may be decomposed into two factors, namely

$$\eta_h = \eta_{ac} \cdot \eta_{ne} \tag{16.56}$$

where η_{ac} is the acceleration efficiency, *i.e.*, the accelerated ion power per input power, and η_{ne} is the neutralization efficiency,

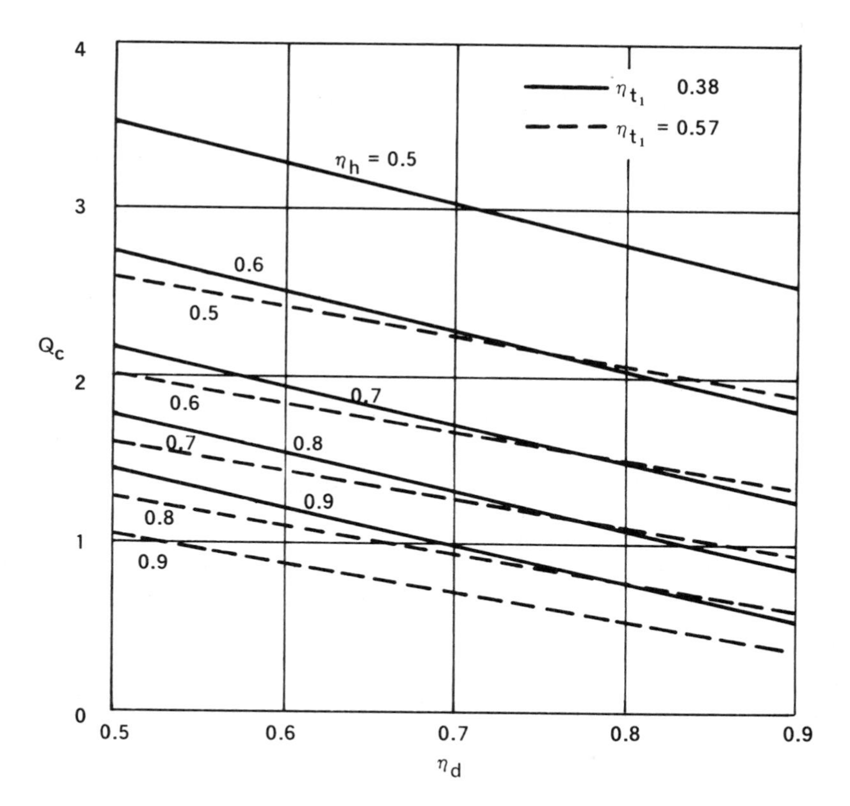

Figure 134. Q_c versus η_d for the mirror reactor.

i.e., the neutralized beam power per accelerated ion power. It has been suggested that η_{ac} can be in the range 90-95% for positive ions, and η_{ne} can be 76% for D_2^+ in the energy range of 200-800 keV. With D^{-1}, η_e, can be as high as 80%, so η_h will be taken to be 0.70 for the present discussion. Selecting η_d = 0.70 and η_h = 0.70 as the reference values, the values of Q_c for these reference values and for cases where a more optimistic value of 0.8 is assumed for both η_h and η_d are shown in Table 45.

The ratio C of circulating power P_c to net electrical power P_n and the overall plant efficiency η are shown as functions of Q in Figure 135 for the reference parameters. We note as we did in Chapter 9 that even at Q = 5 the plant efficiency is less than 30% for case 1, in which the thermal conversion efficiency

Table 45

Q_c for Mirror Fusion Reactor
(Nozawa and Steiner 1974)

	Reference Cases			
Parameter	*Case 1*	*Case 2*	*Case 3*	*Case 4*
η_{t_1}	0.38	0.57	0.38	0.57
η_h	0.70	0.70	0.80	0.80
η_d	0.70	0.70	0.80	0.80
η_{em}	0.438	0.593	0.456	0.611
Q_c	0.712	1.258	1.025	0.759

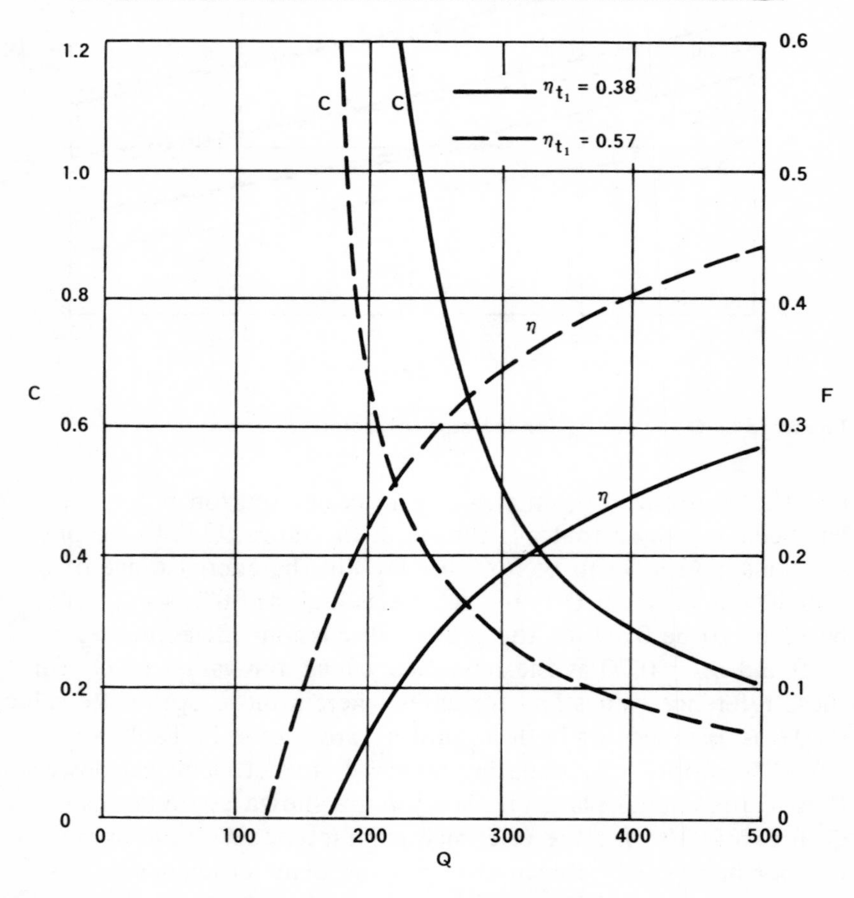

Figure 135. C and η versus Q for the mirror reactor.

η_{t1} is taken as 38%. We note also the improvement in the value of η—to 44%—as a result of increasing η_{t1}, to 57% at the same Q.

Table 46 gives the values of Q, f_b, and $n\tau$ to achieve plant efficiencies of 0.30, 0.40, and 0.50, respectively. These quantities are obtained using the previously mentioned assumptions of $T_e = 0.1\ T_i$, and an ion temperature of 200 keV corresponding to an injection energy of 300 keV (since $T = 2/3\ E_o$).

Table 46

Q, f_b, and $n\tau$ for Mirror Fusion Reactor (Nozawa and Steiner 1974)

η	*Quantity*	*Case 1*	*Reference Cases* *Case 2*	*Case 3*	*Case 4*
0	Q_c	1.712	1.258	1.025	0.759
	η_{em}	0.438	0.593	0.456	0.611
0.30	Y	3.174	2.024	2.923	1.965
	Q	5.43	2.55	3.00	1.49
	f_b	0.16	0.08	0.09	0.04
	$n\tau(10^{14})$	5.03	2.36	2.77	1.38
0.40	Y	11.526	3.073	8.143	2.896
	Q	19.73	3.87	8.35	2.20
	f_b	0.58	0.11	0.25	0.06
	$n\tau(10^{14})$	18.27	3.58	7.73	2.04
0.50	Y		6.376		5.505
	Q		8.02		4.18
	f_b		0.24		0.12
	$n\tau(10^{14})$		7.43		3.87

The above results show that a Q value of 2.55 is required to achieve a plant efficiency of η = 30% even with the advanced thermal converter of η_{t1} = 0.57. Increasing the efficiencies of the thermal converter and direct converter to 80% reduces the Q required for η = 30% to 1.49. Even greater Q values are needed to achieve overall plant efficiencies in excess of 30%. We may, therefore, conclude, as we did in Chapter 9, that direct conversion and injection efficiencies well in excess of 80% must be achieved if mirror reactors are to attain competitive plant efficiencies.

The Theta-Pinch Fusion Reactor

This reactor, as we noted in Chapter 13, is a pulsed system in which plasma heating is accomplished in two stages, an implosion heating stage followed by an adiabatic compression. The reference theta-pinch reactor detailed in the above mentioned chapter will form the basis of the present analysis. Figure 136 shows its schematic energy flow diagram. It would be a large

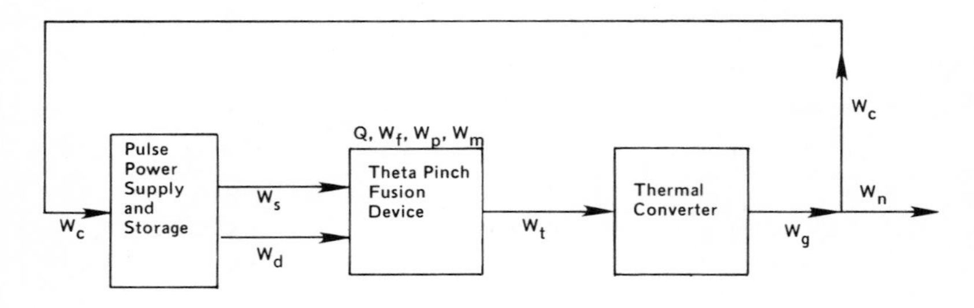

Figure 136. Schematic of theta-pinch reactor.

torus whose major radius is determined in part by the power rating of the reactor output. We also recall that the design calls for the region surrounding the plasma to consist of a tritium-breeding blanket surrounded by the implosion coil, the compression coil, and finally the shield. The compression coil not only supplies the plasma heating energy and confining magnetic field but also serves in the direct conversion scheme associated with the recovery of some of the fusion energy through expansion against the magnetic field. This energy must be transferred to the energy storage system with high efficiency.

Furthermore, according to the reactor model, eddy current losses in the blanket and the energy delivered by the implosion heating system are recovered by the thermal converter, while similar losses in the compression coil, at room temperature, are not recoverable. These considerations lead to the following conditions to be applied to the energy balance equation (16.1): (1) the supplied plasma energy is accounted for in the magnetic energy term and the plasma heating loss term, thus terms containing the parameter ℓ are dropped; (2) the supplied magnetic energy is recovered by induction only, hence m = 1; (3) there is no recovery of the ohmic heating losses incurred in the compression coil, thus $\eta_{t2} = 0$; and (4) there is no direct recovery of the energy losses due to plasma heating, thus $\delta = 0$. In view of these facts, equations (16.1) and (16.10) for this reactor become

$$W_n = W_s E \eta_d (\alpha Q \eta_a + F \eta_a) + W_s E \eta_{t_1} \left\{ (1 - \alpha) Q \eta_a + 1 - \eta_a \right\} - W_s (1 + a) \tag{16.57}$$

and

$$Q_c = \frac{1}{\eta_e} \left\{ \frac{1 + a}{E \eta_a} - F \eta_d - \eta_{t_1} \left(\frac{1}{\eta_a} - 1 \right) \right\} \tag{16.58}$$

$$= \frac{1}{\eta_e} \left\{ \frac{(1 + a) - \eta_h \eta_{t_1} (1 - \eta_a)}{\eta_a \eta_h} + F \frac{(1 + a) - \eta_d \eta_m}{\eta_m} \right\} \tag{16.59}$$

where

$$\eta_e = \alpha \eta_d + (1 - \alpha) \eta_{t_1} \tag{16.60}$$

We note from equation (16.59) that Q_c consists of two terms, one which represents losses in the heating process and another which is associated with losses in the pulsed magnetic energy supply and recovery system.

The power balance parameters, S, C and η, for the theta-pinch reactor take on the forms

$$S = \frac{(1 + a)}{(Q - Q_c) E \eta_a \eta_e} \tag{16.61}$$

$$C = \frac{(1 + a) - E \eta_a \eta_d (\alpha Q + F)}{(Q - Q_c) E \eta_a \eta_e} \tag{16.62}$$

$$\eta = \frac{W_n}{W_f} = \frac{(Q - Q_c) W_s E \eta_a \eta_e}{Q E \eta_a W_s} = \frac{Q - Q_c}{Q} \eta_e \tag{16.63}$$

which reveal that S is not equivalent to C in this case.

To calculate F, we begin by writing the energy of a plasma contained in a volume V_p, *i.e.*,

$$W_p = 3/2\ p\ V_p \tag{16.64}$$

where p = 2nT is the plasma pressure. In the theta-pinch reactor, this pressure is effectively equal to the magnetic pressure, *i.e.*, $\beta = 1$ so that we can write

$$W_p = \frac{3}{2}\frac{B_o^2}{8\pi} V_p \tag{16.65}$$

By the same token, the magnetic energy W_m in a volume V_m is given by

$$W_m = \frac{B_o^2}{8\pi} V_m \tag{16.66}$$

and upon combining these results we obtain

$$\begin{aligned} F &= \frac{W_m}{W_p} \\ &= \frac{(B_o^2/8\pi)\pi\ r_c^2\ L}{3/2(B_o^2/8\pi)\pi\ a_o^2\ L} \\ &= \frac{2}{3} r_c^2/a_o^2 \end{aligned} \tag{16.67}$$

where a_o, r_c, and L are the radius of the plasma at ignition, the effective radius of the compression coil, and the length of the reactor core, respectively.

We turn now to examination of the dependence of Q_c on the efficiencies of the theta-pinch reactor using the following reference parameters which are based on the analysis given in Chapter 13. The parameter "a" accounts for auxiliary power losses not including those associated with refrigeration losses in the magnetic energy transfer system which are included in η_m F = 61 indicates that the supplied magnetic energy is 61 times greater than the plasma energy at the ignition point. The quan tity α represents the portion of the fusion energy which is directly

a	F	α	η_h	η_a	η_m	η_d	m	η_{t_1}
0.01	61	0.09	0.61	0.57	0.97	1.00	1.00	0.38, 0.57

recoverable through induction, and η_h accounts for nonrecoverable energy losses, that is, losses associated with the various currents in the compression coil. The efficiency η_a accounts for thermally recoverable energy losses, *i.e.*, losses in the implosion heating process and in the blanket resulting from eddy currents during the transient in the compression coil. As we have already seen, η_m represents the energy losses in the magnetic energy transfer system and η_d and m in this case are unity.

It is perhaps more appropriate in the case of the theta-pinch reactor to study the dependence of Q_c on $\eta_m \cdot$ F, and α while holding the other parameters constant. The first is shown in Figure 137 where η_m is taken in the range 0.95 to 1.00 for the two thermal efficiencies η_{t1} = 0.38 and 0.57. We note that small changes in η_m produce large changes in Q_c, which points to the importance of minimizing the losses in the magnetic energy transfer systems. The dependence of Q_c on F can be clearly assessed

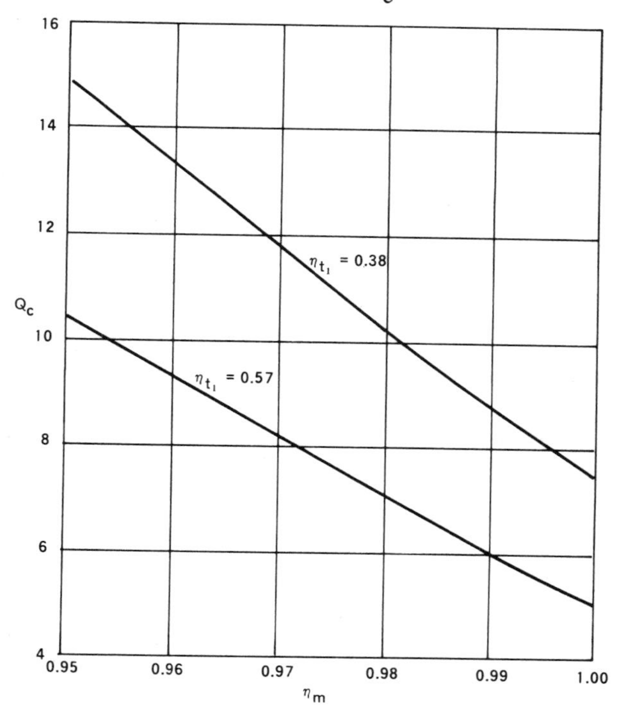

Figure 137. Q_c versus η_m for the theta-pinch reactor.

from equation (16.59) with the aid of (16.67). Large values of F resulting from large values of the ratio r_c/a_o lead to large values of Q_c, so that different values of Q_c can be obtained by altering the plasma radius. Although the effect of α on Q_c shown in Figure 138 is rather small, the quantity α enters into the effective conversion efficiency η_e [see equation (16.60)] and, therefore, plays an important role in the performance of the theta-pinch reactor. The values of Q_c for several cases are shown in Table 47 for the reference cases 1 and 2 and for cases 3 and 4 which correspond to the reference cases but with an increased value of α.

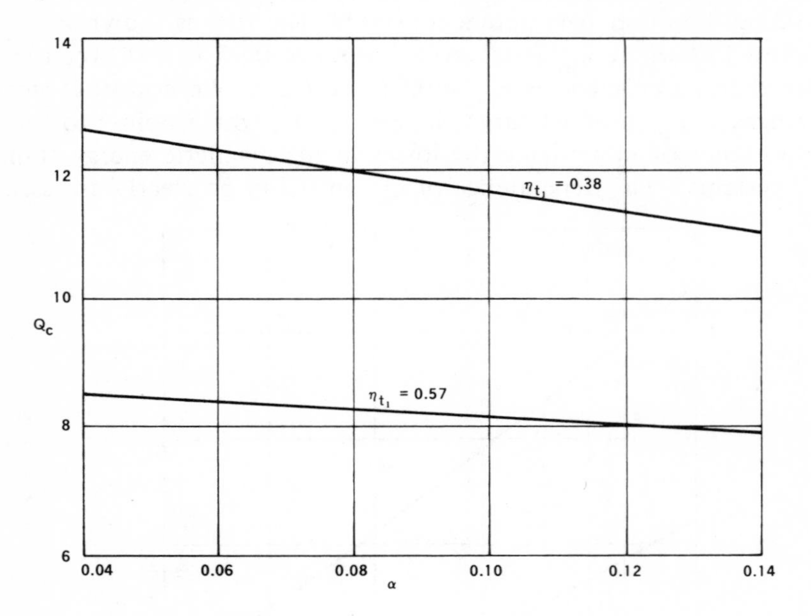

Figure 138. Q_c versus α for the theta-pinch reactor.

Assuming an ignition temperature of 6 keV and a mean operating temperature of 10 keV–although the final temperature may be around 15 keV–the values of Q, f_b and $n\tau$ for various plant efficiencies are given in Table 48. It is interesting to note that case 3 with α = 0.12 has rather small Q values compared to case 1 even though the improvement in Q_c is quite small. This is a reflection of the large effect on the value of Y as a result of the increases in η_e.

Table 47

Theta-Pinch Reactor (Nozawa and Steiner 1974)

Parameter	*Reference Cases*			
	Case 1	*Case 2*	*Case 3*	*Case 4*
η_{t_1}	0.38	0.57	0.38	0.57
α	0.09	0.09	0.12	0.12
F	61	61	61	61
η_m	0.97	0.97	0.97	0.97
η_e	0.436	0.609	0.454	0.622
Q_c	11.78	8.20	11.30	8.00

Table 48

Q, f_b, and $n\tau$ for Given η Theta Pinch Fusion Reactor (Nozawa and Steiner 1974)

η	*Quantity*	*Reference Cases*			
		Case 1	*Case 2*	*Case 3*	*Case 4*
0	Q_c	11.78	8.20	11.30	8.00
	η_{t_1}	0.38	0.57	0.38	0.57
0.30	Y	3.206	1.971	2.95	1.932
	Q	37.77	16.16	33.31	15.45
	f_b	0.061	0.026	0.054	0.025
	$n\tau(10^{14})$	10.56	4.51	9.31	4.32
0.40	Y	12.11	2.914	8.407	2.802
	Q	142.7	23.9	95.0	22.41
	f_b	0.229	0.038	0.152	0.036
	$n\tau(10^{14})$	39.9	6.71	26.5	6.27
0.50	Y		5.59		5.098
	Q		45.8		40.79
	f_b		0.074		0.066
	$n\tau(10^{14})$		12.79		11.39

Moreover, we see from the above analysis that variations in Q_c are very sensitive to η_m, the efficiency of the magnetic energy transfer system. In fact, the sensitivity to η_m is an order of magnitude greater than that to any other parameter. Since the parameters of the theta-pinch reactor are so strongly interrelated, they cannot be varied independently as was done in the other reactor concepts.

The Tokamak Fusion Reactor

The underlying principle of the Tokamak fusion reactor was examined in great detail in Chapter 13. We recall that in addition to providing plasma heating the large axial current induced in the system also provides the poloidal magnetic field required for stable confinement. We recall further that the ohmic (joule) heating associated with this current is not enough to bring the plasma to ignition and that a supplementary heating method such as the adiabatic compression of Chapter 7 or the energetic neutral beam injection of Chapter 4 will be required. Figure 139 shows schematically the energy flow in a Tokamak reactor power plant where the pulsed-power supply may consist of a superconducting electromagnetic energy storage system as proposed for the theta-pinch reactor. The magnetic induction subsystem corresponds to the primary winding for inducing the axial plasma current. The thermal converter would be the same as in any other reactor concept. Both the pulsed poloidal field and the steady toroidal magnetic field are assumed to be created by superconducting coils.

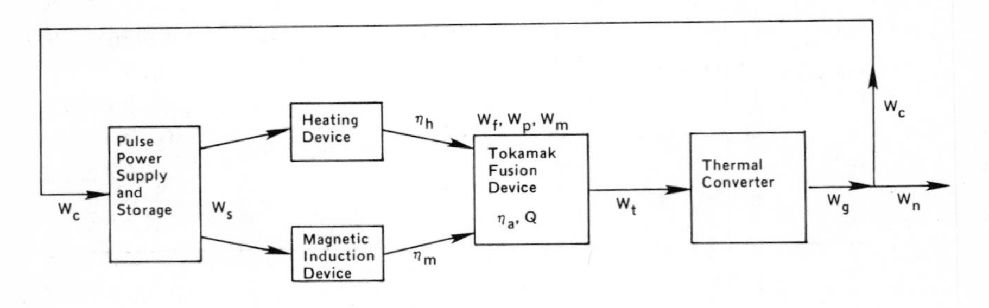

Figure 139. Schematic of the Tokamak fusion reactor power plant.

Three kinds of operational modes are being considered for Tokamak reactors. They are: (1) steady-state operation, (2) quasi-pulsed operation, and (3) pulsed operation. For the first kind of operation, it is assumed the plasma diffusion across the field lines (as explained in Chapter 1) gives rise to diffusion-driven currents known as "boot strap" currents which maintain the axial current. In the quasi-pulsed operation, the pulse time is considered long enough to allow refueling during the pulse, while in the pulsed operation the time is so short that no refueling is possible. Since the power balance in the steady-state operation is less stringent than that in the pulsed mode, the emphasis in this analysis will be focused on the pulsed operation. In this mode of operation, the plasma energy and poloidal magnetic field energy must be supplied at the start of each cycle; at the end of the pulse, portions of these energies will be recovered.

We will consider the following sequence of events to occur in the operation of the reactor. (1) Ionization of the fuel and ohmic heating to several hundred electron volts is accomplished by application of the poloidal magnetic field. (2) Neutral beam injection is applied to bring the plasma to ignition temperature at about 6 keV (see Chapter 7). (3) During the heating phase, only a fraction of the supplied energy η_a will be deposited in the plasma while the remainder will appear as heat in the first wall. (4) Once the plasma is ignited, alpha-particle heating will bring the plasma temperature to its operating (burning) value almost instantaneously. The operating temperature is about 25 keV and is determined by the conditions on plasma stability. The burning phase will proceed until the end of the pulse is reached. At this point, some of the plasma and poloidal magnetic field energy will be recovered and transferred by electromagnetic induction to the pulsed-power supply subsystem.

We shall further assume that for the Tokamak reactor: (1) there is no direct conversion of the fusion energy, hence $\alpha = 0$, $\eta_e = \eta_{t_1}$; (2) there is no recovery of the energy losses in the neutral beam injector except for direct recovery in the unneutralized beam, thus $\eta_{t_2} = 0$; (3) the energy losses from the plasma during heating will be recovered with the thermal converter, therefore $\delta = 0$.

The above conditions allow us to write equations (16.1) and (16.10) in the forms

$$W_n = W_s\, E\, \eta_d(\ell\, \eta_a + m\, F\, \eta_a) + W_s\, E\, \eta_{t_1} \Big\{ Q\eta_a + (1 - \ell)\,\eta_a + (1 - m)\, F\, \eta_a + (1 - \eta_a) \Big\} - W_s\,(1 + a) \tag{16.68}$$

and

$$Q_c = \frac{1}{\eta_{t_1}} \left\{ \frac{1 + a}{E\eta_a} - \eta_d\,(\ell + mF) \right\} - \left\{ \frac{1 - \eta_a}{\eta_a} + (1 - \ell) + (1 - m)F \right\}$$

$$= \left\{ \frac{(1 + a) - \eta_h\, \eta_{t_1}}{\eta_a\, \eta_h\, \eta_{t_1}} - \ell\, \frac{\eta_d - \eta_{t_1}}{\eta_{t_1}} \right\} + F \left\{ \frac{(1 + a) - \eta_m\, \eta_{t_1}}{\eta_m\, \eta_{t_1}} - m\, \frac{\eta_d - \eta_{t_1}}{\eta_{t_1}} \right\} \tag{16.69}$$

where F in this case is the ratio of the poloidal magnetic field energy to the plasma energy. The above equation for Q_c reveals that it consists of two parts: one which relates only to the efficiencies in the heating process and in recovery and another which relates only to efficiencies associated with magnetic field generation and with recovery.

The power balance parameters for the Tokamak reactor are

$$S = \frac{(1 + a)}{(Q - Q_c)\, E\, \eta_a\, \eta_{t_1}} \tag{16.70}$$

$$C = \frac{(1 + a) - E\, \eta_a\, \eta_d\, (\ell + mF)}{(Q - Q_c)\, E\, \eta_a\, \eta_{t_1}} \tag{16.71}$$

$$\eta = \frac{Q - Q_c}{Q}\, \eta_{t_1} \tag{16.72}$$

and we note that S and C are not equivalent in this case.

The parameter F referred to earlier can be expressed as

$$F = \frac{W_m\,(T_2, \beta_\theta)}{W_p(T_1)} \tag{16.73}$$

where the poloidal magnetic field energy W_m is a function of the poloidal beta (*i.e.*, the plasma pressure to the poloidal magnetic field pressure) and the plasma operating temperature T_2, while the plasma energy W_p is a function of the ignition temperature T_1.

If we assume that the plasma density and temperature are uniformly distributed over the plasma volume V_p, then F assumes the form

$$F = \frac{\frac{1}{2} L I^2 (T_2, \beta_\theta)}{3 nT_1 V_p} \tag{16.74}$$

where L is the inductance of the plasma loop, $I(T_2, \beta_\theta)$ is the axial plasma current, and n is the plasma density. Since the plasma pressure at the operating temperature p_2 is given by $p_2 = 2nT_2$, then β_θ can be written as

$$\beta_\theta = \frac{p_2}{\beta_\theta^2/8\pi} = \frac{16\, n\pi\, T_2}{\beta_\theta^2}$$

The poloidal field, B_θ, just outside the plasma is related to the current by $B_\theta = I/5a_o$, where a_o is the plasma radius and B_θ in gauss. Then I^2 can be expressed as

$$\begin{aligned} I^2 (T_2, \beta_\theta) &= (5\, a_o\, B_\theta)^2 \\ &= 400\, \pi\, a_o^2\, nT_2/\beta_\theta \end{aligned} \tag{16.75}$$

We recall from Chapter 7 that the inductance of the plasma column can be written as

$$L = 4\pi R \left[\ell n \frac{b}{a_o} + \frac{1}{4} \right] X\ 10^{-9} \quad \text{(Henry)} \tag{16.76}$$

where R and b are the major and minor radii of the torus, respectively. Inserting equations (16.76) and (16.75) into (16.74), we get

$$F(T_1, T_2, \beta_\theta) = \frac{T_2}{T_1} \frac{\nu}{\beta_\theta} \tag{16.77}$$

where $V_p = 2\pi^2 a_o^2 R$ has been utilized and the constant ν is given by

$$\nu = \frac{4}{3}\left[\ell n \frac{b}{a_o} + \frac{1}{4}\right] \tag{16.78}$$

In view of the definition of g in equation (16.29), the quantity F above can be expressed as

$$F = g \frac{\nu}{\beta_\theta} \tag{16.79}$$

and for $b/a_o = 2.6$, which lies in the range commensurate with the Tokamak design of Chapter 13, the constant ν has a value of 1.60. For the present analysis, we choose an aspect ratio $A = R/a_o = 4$ and assume that the maximum value of β_θ is given by $\sqrt{A} = 2$. Moreover, if we take the ignition temperature T_1 to be 6 keV and the operating temperature T_2 to be 18 keV, then $g = 3$. Combining these values of β_θ,· g and ν, we find $F = 2.4$, which will be used as the reference value.

The reference parameters for the Tokamak reactor to be used here are:

a	ℓ	m	F	η_a	η_d	η_h	η_m	η_{t_1}
0.05	0	0.8	2.4	0.8	1.00	0.70	0.97	0.38,0.57

It will be assumed that the auxiliary power, which includes the refrigeration power for the toroidal and poloidal superconducting coils and the vacuum pumping power, will incur losses which are 5% of the supply energy to the neutral beam injector and to the pulsed poloidal field supply unit. Setting $\ell = 0$ implies that there is no direct recovery of the plasma energy at the end of the pulse, which might be a conservative assumption. $m = 0.80$ means that 80% of the poloidal field energy is recoverable by induction and 20% appears as heat in the first wall. Eighty per cent of the injected energy is assumed to be deposited in the plasma while 20% is deposited in the first wall, hence $\eta_a = 0.80$. The choice of the value of F has been discussed above, but it is interesting to note at this juncture that this value is about 25 times smaller than that for the theta-pinch reactor. Also, as in the theta-pinch reactor case, the losses associated with the magnetic energy transfer system are accounted for in η_m, thus $\eta_d = 1.00$. We choose the efficiency of the neutral beam injector, η_h, to be the same as that for the mirror reactor, although the injected particle energy may be different in the two cases. Although plasma heating is totally provided by neutral injection in this analysis, it will be conservatively assumed that 5% of the plasma energy will come from ohmic heating.

We now examine the dependence of Q_c on the various efficiencies and note that its dependence on η_h and η_m is very instructive, as was the case in the other reactor systems. A plot of Q_c vs η_h for several values of η_m and η_{t_1} is shown in Figure 140, where we observe that Q_c varies slightly with η_m in contrast with the situation in the theta-pinch reactor. This difference is associated with the value of F in both reactor concepts.

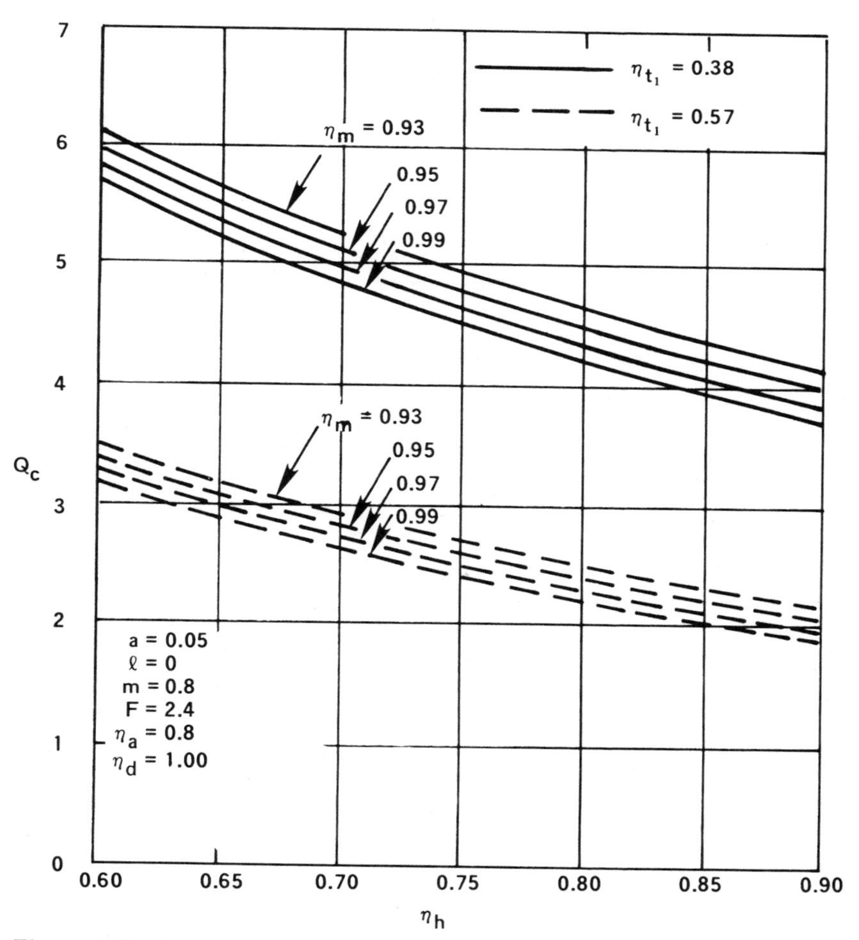

Figure 140. Q_c versus η_h for the Tokamak reactor.

For the reference cases (1 and 2) and for cases 3 and 4 in which an optimistic value of η_h = 0.80 is used, the values of Q_c and other efficiencies are given in Table 49. The values of Q and the related burn-up fraction f_b, and Lawson number $n\tau$ necessary to produce net electrical output with efficiencies of 0.30, 0.40, and 0.50 are given in Table 50 using T_1 = 6 keV, T_2 = 18 keV consistent with a toroidal magnetic field of 40-60 kilogauss at the center of the plasma and a density of about 10^{14} cm^{-3}.

Table 49

Q_c for the Tokamak Fusion Reactor
(Nozawa and Steiner 1974)

	Reference Cases			
Parameter	*Case 1*	*Case 2*	*Case 3*	*Case 4*
η_{t_1}	0.38	0.57	0.38	0.57
η_h	0.70	0.70	0.80	0.80
η_e	0.38	0.57	0.38	0.57
Q_c	4.99	2.75	4.37	2.34

Table 50

q, f_b, $n\tau$ for Given η, for Tokamak Fusion Reactor
(Nozawa and Steiner 1974)

		Reference Cases			
η η	*Quantity*	*Case 1*	*Case 2*	*Case 3*	*Case 4*
0	Q_c	4.99	2.75	4.37	2.34
	η_{t_1}	0.38	0.57	0.38	0.57
0.30	Y	4.75	2.111	4.75	2.111
	Q	23.70	5.81	20.76	4.94
	f_b	0.0381	0.0093	0.0336	0.0079
	$n\tau(10^{14})$	2.06	0.50	1.80	0.43
0.40	Y		3.353		3.353
	Q		9.22		7.85
	f_b		0.0148		0.0126
	$n\tau(10^{14})$		0.80		0.68
0.50	Y		8.143		8.143
	Q		22.39		19.05
	f_b		0.036		0.0306
	$n\tau(10^{14})$		1.95		1.66

The Tokamak power balance parameters, namely, the supply energy ratio S, the circulating energy ratio C, and the overall plant efficiency η, are plotted as functions of Q in Figures 141 and 142. We note that for values of $Q > 65$, S becomes less than 0.2 for both values η_{t_1}, while C becomes less than 0.1 for $Q > 75$. The overall plant efficiency η increases rapidly with Q and then for $Q > 20$ it asymptotically approaches the value of thermal converter efficiency η_{t_1}.

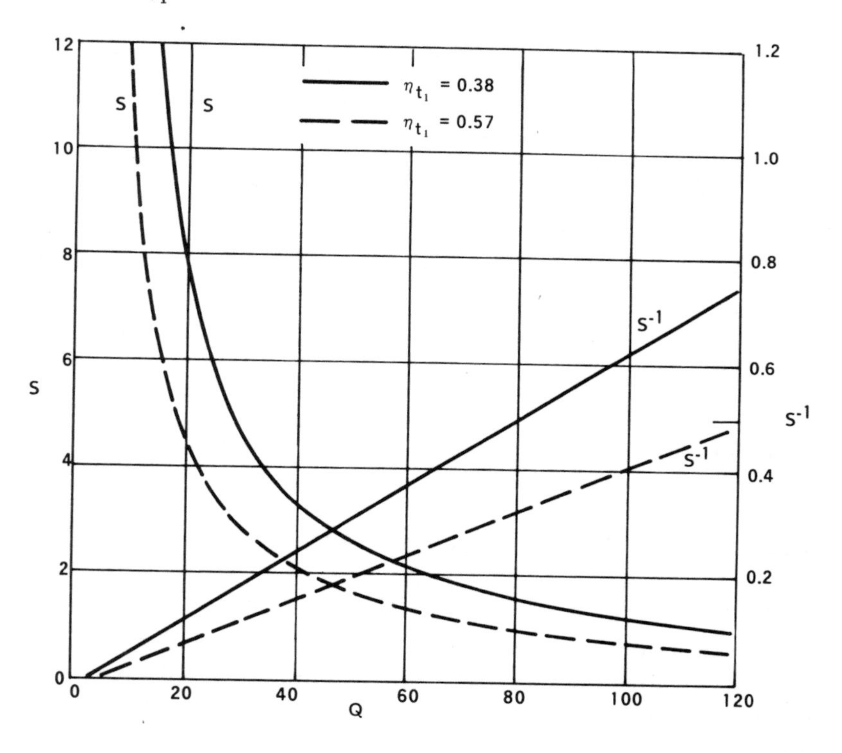

Figure 141. S and S^{-1} versus Q for the Tokamak reactor.

Although the above results have been obtained for a Tokamak reactor, their generalization to other types of toroidal diffuse pinch reactors should be straightforward. One striking feature of this reactor concept is the mild sensitivity of Q_c to the various parameters and the relatively low (less than 10^{14} cm^{-3}) $n\tau$ value for η = 0.40 with the advanced thermal converter.

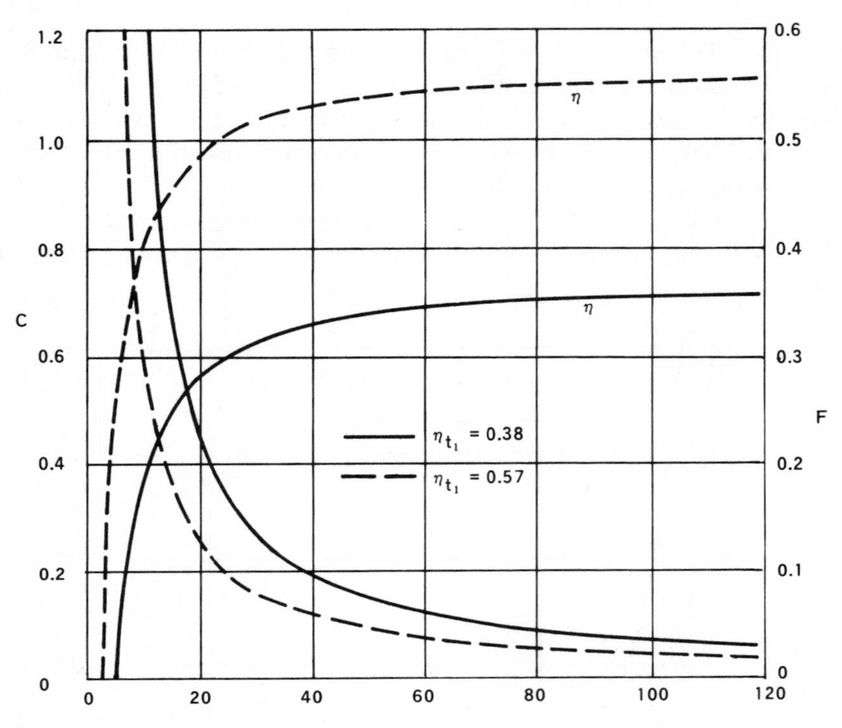

Figure 142. C and η as functions of Q for the Tokamak reactor.

16.4 INTERCOMPARISON AND CONCLUSIONS

The reference parameters and the resultant Q_c for the four D-T reactors considered above are summarized in Table 51, where it must be remembered that these parameters should be viewed as illustrative rather than definitive. The fractional burn-up, f_b and the Lawson number $n\tau$, as derived from the assumed ignition and operating temperatures and the power breakeven parameter Q_c, are given in Table 52. The following observations are readily made from these tables:

1. The mirror reactor exhibits the smallest value of Q_c for a given value of the thermal conversion efficiency η_{t_1}. It is followed by the Tokamak, theta-pinch, and laser-fusion reactors in that order. We note that the Q_c values for the laser-fusion reactor are 60-70 times greater than those for the mirror reactor. Since this quantity is a measure of the

Table 51

Reference Parameters for the Various D-T Fusion Reactors
(Nozawa and Steiner 1974)

Parameter		*Laser Fusion Reactor*	*Mirror Fusion Reactor*	*Theta-Pinch Fusion Reactor*	*Tokamak Fusion Reactor*
Heating Device Efficiency	η_h	0.20	0.70	0.61	0.70
Heat Deposition Efficiency	η_a	0.10	0.98	0.57	0.80
Pulse Magnetic Energy Supply Efficiency	η_m	-	-	0.97	0.97
Ratio of Pulse Magnetic Energy to Plasma Energy	F	-	-	61	2.4
Auxiliary Power Ratio	a	0.00	0.00	0.01	0.05
Direct Conversion Efficiency	η_d	-	0.70	1.00	1.00
Direct Conversion Fraction (Fusion Energy)	α	-	0.20	0.09	0.00
(Plasma Energy)	ℓ	-	0.98	-	0.00
(Magnetic Energy)	m	-	-	1.00	0.80
Thermal Conversion Efficiency	η_{t_1}	0.38 0.57	0.38 0.57	0.38 0.57	0.38 0.57
Effective Conversion Efficiency	η_e	0.38 0.57	0.438 0.593	0.436 0.609	0.38 0.57
Resultant Power Break-Even Parameter	Q_c	121.6 77.7	1.71 1.26	11.78 8.20	4.99 2.75

Table 52

Power Break-Even Parameters for the D-T Fusion Reactors
(Nozawa and Steiner 1974)

Parameter		*Laser Fusion Reactor*		*Mirror Fusion Reactor*		*Theta-Pinch Fusion Reactor*		*Tokamak Fusion Reactor*	
Power Break-Even Critical Q	Q_c	121.6	77.7	1.71	1.26	11.78	8.20	4.99	2.75
Fusion Ignition Temperature	T_1(keV)	5	5	200	200	6	6	6	6
Fusion Operation Mean Temperature	T_2(keV)	50	50	200	200	10	10	18	18
Fractional Burn-up	f_b(%)	16.3	10.4	5.04	3.72	1.89	1.32	0.80	0.44
Lawson Number	$n\tau(10^{14}$ cm^{-3} sec)	3.75	2.39	1.58	1.17	3.30	2.26	0.43	0.24
Assumed Plasma Density[a]	$n(10^{14}$ cm^{-3})	4.6×10^{10}[b]	4.6×10^{10}[b]	1	1	200	200	1	1
Critical Burning Time[c]	(sec (sec)	8.2×10^{-11}	5.2×10^{-11}	1.58	1.17	0.0165	0.0113	0.43	0.24

[a]For illustrative purposes, a tentative plasma density is assigned for each reactor.

[b]100 times the D-T frozen solid density.

[c]Lawson number divided by the plasma density assumed.

inherent energy handling and conversion efficiency of the plant, we see that the mirror displays a high inherent energy efficiency.

2. The low Q_c values exhibited by the mirror reactor are somewhat offset by the relatively large values of f_b and $n\tau$. This is due to the temperature dependence of these two parameters where a high value of T_1 is associated with the mirror reactor. This reactor also requires f_b of 4-5% for power break-even at an injection energy of 300 keV, and an f_b of 2 to 2.5% if the injection energy is reduced to 150 keV.
3. The Tokamak reactor exhibits the smallest f_b and $n\tau$ while the laser-fusion reactor shows the largest values of these parameters. f_b is less than 1% for power break-even in the Tokamak reactor and greater than 10% for the laser-fusion reactor.

The parameters Q, f_b, $n\tau$, S, and C required to yield overall plant efficiencies (η) of 0.30, 0.40, and 0.50 for each reactor concept are given in Table 53. The following points can be made:

1. The ratio Q/Q_c representing the degree of improvement in the plasma performance beyond power break-even must exceed three in order to yield an $\eta = 0.40$ with the advanced thermal converter. A plot of this ratio, for various values of η, as a function of the effective conversion efficiency η_e, is shown in Figure 143. This figure also indicates the range of η_e for the reference cases with the conventional thermal converter and with the advanced thermal converter. It clearly indicates the necessity of the development of advanced thermal converters if only moderate ratios of Q/Q_c are available.
2. Only the mirror and theta-pinch reactors can attain an overall plant efficiency of 40% for $\eta_{t_1} = 0.38$, but at the expense of large values of Q and f_b (f_b = 58% for the mirror and 23% for the theta-pinch reactor). This is a consequence of the direct conversion of a portion of the fusion energy in these devices.
3. The theta-pinch reactor exhibits the highest value of the supply energy ratio S and the lowest value of the circulating energy ratio C. It may, therefore, be characterized as requiring large amounts of supply energy and small amounts of circulating energy.

Some interesting and perhaps surprising conclusions emerge from the above detailed examination of the four D-T fusion reactor. It shows that the fusion energy multiplication factor Q (ratio of fusion energy to absorbed heating energy by the plasma)

Table 53

Q, f_b, $n\tau$, S and C for Given Overall Plant Efficiency η:
D-T Fusion Reactors Based on the Reference Cases
(Nozawa and Steiner 1974)

Quantity		*Laser Fusion Reactor*		*Mirror Fusion Reactor*		*Theta-Pinch Fusion Reactor*		*Tokamak Fusion Reactor*	
Plasma Temperature	T_2(keV)/T_1(keV)	50/5		200/200		10/6		18/6	
Effective Conversion Efficiency	η_e	0.38	0.57	0.438	0.593	0.436	0.609	0.38	0.57
Critical Q Value	Q_c	121.6	77.7	1.71	1.26	11.78	8.20	4.99	2.75
η = 0.30	Q	578	164	5.4	2.6	37.8	16.2	24	5.8
	f_b(%)	77	22	16	8	6.1	2.6	3.8	0.9
	$n\tau(10^{14}\ cm^{-3}\ sec)$	18	5.1	5.0	2.4	10.6	4.5	2.1	0.5
	S	0.3	1.0	0.8	1.8	5.86	13.6	0.62	2.6
	C	0.3	1.0	0.8	1.8	0.18	0.81	0.35	1.47
η = 0.40	Q		260	19.7	3.9	143	24		9.2
	f_b(%)		35	58	11	23	3.8		1.5
	$n\tau(10^{14}\ cm^{-3}\ sec)$		8.0	18	3.6	40	6.7		0.8
	S		0.5	0.1	0.8	1.16	6.9		1.22
	C		0.5	0.1	0.8	-0.13	0.34		0.69
η = 0.50	Q		633		8.0		45.8		22
	f_b(%)		85		24		7.4		3.6
	$n\tau(10^{14}\ cm^{-3}\ sec)$		19.5		7.4		12.8		2.0
	S		0.16		0.3		2.90		0.41
	C		0.16		0.3		0.06		0.23

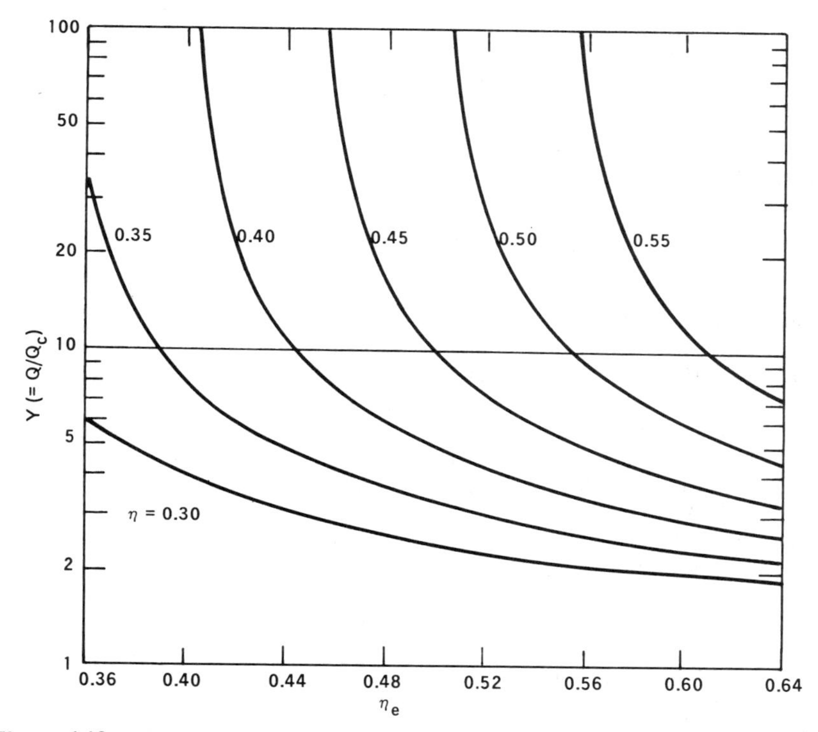

Figure 143. Y versus η_e for given overall efficiency η. A denotes conventional thermal converter and B denotes advanced thermal converter.

appears to be a more fundamental measure of fusion power achievement than the Lawson number $n\tau$. This is especially true in the pulsed systems where $n\tau$ is a time-averaged quantity. It also reveals that for a given overall plant efficiency: (1) the mirror reactor requires the lowest value of Q while the laser fusion requires the highest value, and (2) the Tokamak and theta-pinch reactors require the lowest values of fractional burn-up.

Because it represents the inherent energy handling and conversion capability of the system, the sensitivity of Q_c to variations in subsystem efficiencies is a useful measure of cost-benefit in the technological research and development of fusion power. Future progress in this field, as well as in plasma physics, will certainly yield more detailed and accurate information on the fusion plant and its components. When more precise values of the parameters become available, the above analysis will still prove useful in future assessments of fusion power requirements and implications.

REFERENCES

1. Booth, L. A. Los Alamos Scientific Laboratory Report LA-4858-MS.1 (1972).
2. Fries, R. P. *Nuclear Fusion, 13,* 247 (1973).
3. Futch, A. H., J. P. Holdren, J. Killeen, and A. A. Mirin. *Plasma Physics, 14,* 211 (1972).
4. Golovin, I. N., Y. N. Dnestrovsky, and D. P. Kostamarov. BNES Nuclear Fusion Reactors (Int. Conf. Culham), p. 194 (1969).
5. Krakowski, R. A., F. L. Ribe, T. A. Coultas, and A. J. Hatch. "An Engineering Design of a Reference Theta-Pinch Reactor," Argonne National Laboratory and Los Alamos Scientific Laboratory Report ANL-8019, LA-5336 (1974).
6. Moir, R. W. and W. L. Barr. *Nuclear Fusion, 13,* 35 (1973).
7. Nozawa, M. and D. Steiner. "An Assessment of the Power Balance in Fusion Reactors," USAEC Report ORNL-TM-4221 (1974).
8. Persiani, P. J., W. C. Lipinski, and A. J. Hatch. "Some Comments on the Power Balance Parameters Q and E as Measures of Performance for Fusion Power Reactors," Argonne National Laboratory Report ANL-7932 (1972).
9. Ribe, F. L. "Parameter Study of Long, Separated–Shock Theta-Pinch with Super Conducting Inductive-Energy Storage," Los Alamos Scientific Laboratory Report LA-4828-MS (1971).
10. Sweetman, D. R. BNES Nuclear Fusion Reactors (Int. Conf. Culham) p. 112 (1969).

APPENDIX 1

FREQUENTLY USED VECTOR RELATIONS

$\mathbf{A} \times \mathbf{B} = -\mathbf{B} \times \mathbf{A}$

$\mathbf{A}\cdot\mathbf{B} \times \mathbf{C} = \mathbf{B}\cdot\mathbf{C} \times \mathbf{A} = \mathbf{C}\cdot\mathbf{A} \times \mathbf{B} = (\mathbf{ABC})$

$\mathbf{A} \times (\mathbf{B} \times \mathbf{C}) = (\mathbf{A}\cdot\mathbf{C})\mathbf{B} - (\mathbf{A}\cdot\mathbf{B})\mathbf{C}$

$(\mathbf{A} \times \mathbf{B}) \times \mathbf{C} = (\mathbf{A}\cdot\mathbf{C})\mathbf{B} - (\mathbf{B}\cdot\mathbf{C})\mathbf{A}$

$(\mathbf{A} \times \mathbf{B}) \times (\mathbf{C} \times \mathbf{D}) = (\mathbf{ABD})\mathbf{C} - (\mathbf{ABC})\mathbf{D} = (\mathbf{ACD})\mathbf{B} - (\mathbf{BCD})\mathbf{A}$

$(\mathbf{A} \times \mathbf{B})\cdot(\mathbf{C} \times \mathbf{D}) = (\mathbf{A}\cdot\mathbf{C})(\mathbf{B}\cdot\mathbf{D}) - (\mathbf{A}\cdot\mathbf{D})(\mathbf{B}\cdot\mathbf{C})$

$(\mathbf{A}\cdot\mathbf{B})\mathbf{C} = \mathbf{A}\cdot(\mathbf{BC})$

$(\mathbf{A}\cdot\nabla)\mathbf{B} = \mathbf{A}\cdot(\nabla\mathbf{B})$

$\nabla(\phi + \psi) = \nabla\phi + \nabla\psi$

$\nabla(\phi\psi) = \phi\nabla\psi + \psi\nabla\phi$

$\nabla\cdot(\phi\mathbf{A}) = \phi\nabla\cdot\mathbf{A} + \mathbf{A}\cdot\nabla\phi$

$\nabla \times (\phi\mathbf{A}) = \phi\nabla \times \mathbf{A} + \nabla\phi \times \mathbf{A}$

$\nabla\cdot(\mathbf{A} \times \mathbf{B}) = \mathbf{B}\cdot\nabla \times \mathbf{A} - \mathbf{A}\cdot\nabla \times \mathbf{B}$

$\nabla \times (\mathbf{A} \times \mathbf{B}) = (\mathbf{B}\cdot\nabla)\mathbf{A} - (\mathbf{A}\cdot\nabla)\mathbf{B} + \mathbf{A}(\nabla\cdot\mathbf{B}) - \mathbf{B}(\nabla\cdot\mathbf{A})$

$\nabla(\mathbf{A}\cdot\mathbf{B}) = (\mathbf{A}\cdot\nabla)\mathbf{B} + (\mathbf{B}\cdot\nabla)\mathbf{A} + \mathbf{A} \times (\nabla \times \mathbf{B}) + \mathbf{B} \times (\nabla \times \mathbf{A})$

$\nabla \times \nabla\phi = 0$

$\nabla\cdot\nabla \times \mathbf{A} = 0$

$\nabla \times (\nabla \times \mathbf{A}) = \nabla(\nabla\cdot\mathbf{A}) - \nabla\cdot\nabla\mathbf{A}$

$\mathbf{r} = \mathbf{i}x + \mathbf{j}y + \mathbf{k}z$

$\nabla\cdot\mathbf{r} = 3$

$\nabla \times \mathbf{r} = 0$

$(\mathbf{A}\cdot\nabla)\mathbf{r} = \mathbf{A}$

$$\underset{\substack{\text{closed}\\ \text{volume}}}{\int} d\mathbf{r}\,\nabla\phi = \underset{\substack{\text{bounding}\\ \text{surface}}}{\int} d\mathbf{s}\,\phi$$

$$\underset{\substack{\text{closed}\\ \text{volume}}}{\int} d\mathbf{r}\,\nabla\cdot\mathbf{A} = \underset{\substack{\text{bounding}\\ \text{surface}}}{\int} d\mathbf{s}\cdot\mathbf{A}$$ (divergence, Gauss's theorem)

$$\int_{\text{closed volume}} d\mathbf{r}\, \nabla \times \mathbf{A} = \int_{\text{bounding surface}} d\mathbf{s} \times \mathbf{A}$$

$$\int_{\text{open surface}} d\mathbf{s} \times \nabla \phi = \oint_{\text{bounding circumference}} d\mathbf{l}\, \phi$$

$$\int_{\text{open surface}} d\mathbf{s} \cdot \nabla \times \mathbf{A} = \oint_{\text{bounding circumference}} d\mathbf{l} \cdot \mathbf{A} \qquad \text{(Stokes's theorem)}$$

APPENDIX II

PHYSICAL CONSTANTS AND CONVERSION FACTORS

Quantity	*Gaussian CGS*	*Rationalized MkS**
Length	1 cm (centimeter)	10^{-2} m (meter)
Mass	1 g (gram)	10^{-3} kg
Time	1 S (second)	1 S
Charge	1 esu or statcoulomb	$1/3 \times 10^{-9}$ C (coulomb)
Charge density	1 esu/cm^3	$1/3 \times 10^{-3}$ C/m^3
Current	1 statamp	$1/3 \times 10^{-9}$ A (ampere)
Current density	1 statamp/cm^2	$1/3 \times 10^{-5}$ A/m^2
Electric field	1 statvolt/cm	3×10^4 V/m
Potential	1 statvolt	300 V (volt)
Magnetic induction	1 G (gauss)	10^{-4} Wb (Weber)/m^2
Magnetic intensity	1 Oe (oerstead)	$1/4\pi \times 10^3$ A/m
Conductivity	1 S^{-1}	$1/9 \times 10^{-9}$ mho/m
Energy	1 erg	10^{-7} J (joule)
Inductance	1 CgS unit	9×10^{11} H (henry)
Capacitance	1 cm	$1/9 \times 10^{-11}$ F (farad)
Power	1 erg/sec	10^{-7} W (watt)
Electronic charge (e)	4.8×10^{-10} esu	1.6×10^{-19} C
Electronic mass (m_e)	9.1×10^{-28} g	9.1×10^{-31} kg
Planck's constant (h)	6.626×10^{-27} ergs-S	6.626×10^{-34} J-S
Boltzmann's constant	1.38×10^{-16} ergs/K	1.38×10^{-23} J/K

*For accurate work, factors of 3 should be replaced, except in exponents, by the speed of light $C = (2.997930 \pm 0.000003) \times 10^{10}$ cm/S.

APPENDIX III

MASS AND REST ENERGY OF PARTICLES OF INTEREST TO CTR

Particle	*Mass (kg x 10^{-27})*	*Rest Energy (MeV)*
e	9.108 x 10^{-4}	0.511
n	1.674	939.512
H^1 atom	1.673	938.730
H^2 atom	3.343	1876.017
H^3 atom	5.006	2809.272
$H_e{}^3$ atom	5.006	2809.250
$H_e{}^4$ atom	6.643	3728.189
$L_i{}^6$ atom	9.984	5602.735
$L_i{}^7$ atom	11.64	6534.995

FUNDAMENTAL CONSTANTS

Compiled by E. R. Cohen and B. N. Taylor under the auspices of the CODATA Task Group on Fundamental Constants. This set has been officially adopted by CODATA and is taken from J. Phys. Chem. Ref. Data, Vol. 2, No. 4, p. 663 (1973) and CODATA Bulletin No. 11 (December 1973).

Quantity	Symbol	Numerical Value *	Uncert. (ppm)	SI † ← Units →	cgs ‡
Speed of light in vacuum	c	299792458(1.2)	0.004	$m \cdot s^{-1}$	10^{2} $cm \cdot s^{-1}$
Permeability of vacuum	μ_0	4π		10^{-7} $H \cdot m^{-1}$	
		=12.5663706144		10^{-7} $H \cdot m^{-1}$	
Permittivity of vacuum, $1/\mu_0 c^2$	ϵ_0	8.854187818(71)	0.008	10^{-12} $F \cdot m^{-1}$	
Fine-structure constant, $[\mu_0 c^2/4\pi](e^2/\hbar c)$	α	7.2973506(60)	0.82	10^{-3}	10^{-3}
	α^{-1}	137.03604(11)	0.82		
Elementary charge	e	1.6021892(46)	2.9	10^{-19} C	10^{-20} emu
		4.803242(14)	2.9		10^{-10} esu
Planck constant	h	6.626176(36)	5.4	10^{-34} $J \cdot s$	10^{-27} $erg \cdot s$
	$\hbar = h/2\pi$	1.0545887(57)	5.4	10^{-34} $J \cdot s$	10^{-27} $erg \cdot s$
Avogadro constant	N_A	6.022045(31)	5.1	10^{23} mol^{-1}	10^{23} mol^{-1}
Atomic mass unit, $10^{-3} kg \cdot mol^{-1} N_A^{-1}$	u	1.6605655(86)	5.1	10^{-27} kg	10^{-24} g
Electron rest mass	m_e	9.109534(47)	5.1	10^{-31} kg	10^{-28} g
		5.4858026(21)	0.38	10^{-4} u	10^{-4} u
Proton rest mass	m_p	1.6726485(86)	5.1	10^{-27} kg	10^{-24} g
		1.007276470(11)		u	u
Ratio of proton mass to electron mass	m_p/m_e	1836.15152(70)	0.38		
Neutron rest mass	m_n	1.6749543(86)	5.1	10^{-27} kg	10^{-24} g
		1.008665012(37)	0.037	u	u
Electron charge to mass ratio	e/m_e	1.7588047(49)	2.8	10^{11} $C \cdot kg^{-1}$	10^{7} $emu \cdot g^{-1}$
		5.272764(15)	2.8		10^{17} $esu \cdot g^{-1}$
Magnetic flux quantum, $[c]^{-1}(hc/2e)$	Φ_0	2.0678506(54)	2.6	10^{-15} Wb	10^{-7} $G \cdot cm^2$
	h/e	4.135701(11)	2.6	10^{-15} $J \cdot s \cdot C^{-1}$	10^{-7} $erg \cdot s \cdot emu^{-1}$
		1.3795215(36)	2.6		10^{-17} $erg \cdot s \cdot esu^{-1}$
Josephson frequency-voltage ratio	$2e/h$	4.835939(13)	2.6	10^{14} $Hz \cdot V^{-1}$	
Quantum of circulation	$h/2m_e$	3.6369455(60)	1.6	10^{-4} $J \cdot s \cdot kg^{-1}$	$erg \cdot s \cdot g^{-1}$
	h/m_e	7.273891(12)		10^{-4} $J \cdot s \cdot kg^{-1}$	$erg \cdot s \cdot g^{-1}$
Faraday constant, $N_A e$	F	9.648456(27)	2.8	10^{4} $C \cdot mol^{-1}$	10^{3} $emu \cdot mol^{-1}$
		2.8925342(82)	2.8		10^{14} $esu \cdot mol^{-1}$
Rydberg constant, $[\mu_0 c^2/4\pi]^2(m_e e^4/4\pi\hbar^3 c)$	R_∞	1.097373177(83)	0.075	10^{7} m^{-1}	10^{5} cm^{-1}
Bohr radius, $[\mu_0 c^2/4\pi]^{-1}(\hbar^2/m_e e^2) = \alpha/4\pi R_\infty$	a_0	5.2917706(44)	0.82	10^{-11} m	10^{-9} cm
Classical electron radius, $[\mu_0 c^2/4\pi](e^2/m_e c^2) = \alpha^3/4\pi R_\infty$	$r_e = \alpha \bar{\lambda}_C$	2.8179380(70)	2.5	10^{-15} m	10^{-13} cm
Thomson cross section, $(8/3)\pi r_e^2$	σ_e	0.6652448(33)	4.9	10^{-28} m^2	10^{-24} cm^2
Free electron g-factor, or electron magnetic moment in Bohr magnetons	$g_e/2 = \mu_e/\mu_B$	1.0011596567(35)	0.0035		
Free muon g-factor, or muon magnetic moment in units of $[c](e\hbar/2m_\mu c)$	$g_\mu/2$	1.00116616(31)	0.31		
Bohr magneton, $[c](e\hbar/2m_e c)$	μ_B	9.274078(36)	3.9	10^{-24} $J \cdot T^{-1}$	10^{-21} $erg \cdot G^{-1}$
Electron magnetic moment	μ_e	9.284832(36)	3.9	10^{-24} $J \cdot T^{-1}$	10^{-21} $erg \cdot G^{-1}$
Gyromagnetic ratio of protons in H_2O	γ'_p	2.6751301(75)	2.8	10^{8} $s^{-1} \cdot T^{-1}$	10^{4} $s^{-1} \cdot G^{-1}$
	$\gamma'_p/2\pi$	4.257602(12)	2.8	10^{7} $Hz \cdot T^{-1}$	10^{3} $Hz \cdot G^{-1}$
γ'_p corrected for diamagnetism of H_2O	γ_p	2.6751987(75)	2.8	10^{8} $s^{-1} \cdot T^{-1}$	10^{4} $s^{-1} \cdot G^{-1}$
	$\gamma_p/2\pi$	4.257711(12)	2.8	10^{7} $Hz \cdot T^{-1}$	10^{3} $Hz \cdot G^{-1}$
Magnetic moment of protons in H_2O in Bohr magnetons	μ'_p/μ_B	1.52099322(10)	0.066	10^{-3}	10^{-3}
Proton magnetic moment in Bohr magnetons	μ_p/μ_B	1.521032209(16)	0.011	0^{-3}	10^{-3}
Ratio of electron and proton magnetic moments	μ_e/μ_p	658.2106880(66)	0.010		
Proton magnetic moment	μ_p	1.4106171(55)	3.9	10^{-26} $J \cdot T^{-1}$	10^{-23} $erg \cdot G^{-1}$
Magnetic moment of protons in H_2O in nuclear magnetons	μ'_p/μ_N	2.7927740(11)	0.38		
μ'_p/μ_N corrected for diamagnetism of H_2O	μ_p/μ_N	2.7928456(11)	0.38		
Nuclear magneton, $[c](e\hbar/2m_p c)$	μ_N	5.050824(20)	3.9	10^{-27} $J \cdot T^{-1}$	10^{-24} $erg \cdot G^{-1}$
Ratio of muon and proton magnetic moments	μ_μ/μ_p	3.1833402(72)	2.3		
Muon magnetic moment	μ_μ	4.490474(18)	3.9	10^{-26} $J \cdot T^{-1}$	10^{-23} $erg \cdot G^{-1}$
Ratio of muon mass to electron mass	m_μ/m_e	206.76865(47)	2.3		

Quantity	Symbol	Numerical Value *	Uncert. (ppm)	SI †	← Units → cgs ‡
Muon rest mass	m_μ	1.883566(11)	5.6	10^{-28} kg	10^{-25} g
		0.11342920(26)	2.3	u	u
Compton wavelength of the electron, $h/m_e c = \alpha^2/2R_\infty$	λ_C	2.4263089(40)	1.6	10^{-12} m	10^{-10} cm
	$\bar{\lambda}_C = \lambda_C/2\pi = \alpha a_0$	3.8615905(64)	1.6	10^{-13} m	10^{-11} cm
Compton wavelength of the proton, $h/m_p c$	$\lambda_{C,p}$	1.3214099(22)	1.7	10^{-15} m	10^{-13} cm
	$\bar{\lambda}_{C,p} = \lambda_{C,p}/2\pi$	2.1030892(36)	1.7	10^{-16} m	10^{-14} cm
Compton wavelength of the neutron, $h/m_n c$	$\lambda_{C,n}$	1.3195909(22)	1.7	10^{-15} m	10^{-13} cm
	$\bar{\lambda}_{C,n} = \lambda_{C,n}/2\pi$	2.1001941(35)	1.7	10^{-16} m	10^{-14} cm
Molar volume of ideal gas at s.t.p.	V_m	22.41383(70)	31	10^{-3} $m^3 \cdot mol^{-1}$	10^3 $cm^3 \cdot mol^{-1}$
Molar gas constant, $V_m p_0/T_0$ ($T_0 \equiv 273.15$ K; $p_0 \equiv 101325$ Pa $\equiv$ 1atm)	R	8.31441(26)	31	$J \cdot mol^{-1} \cdot K^{-1}$	10^7 $erg \cdot mol^{-1} \cdot K^{-1}$
		8.20568(26)	31	10^{-5} $m^3 \cdot atm \cdot mol^{-1} \cdot K^{-1}$	10 $cm^3 \cdot atm \cdot mol^{-1} \cdot K^{-1}$
Boltzmann constant, R/N_A	k	1.380662(44)	32	10^{-23} $J \cdot K^{-1}$	10^{-16} $erg \cdot K^{-1}$
Stefan-Boltzmann constant, $\pi^2 k^4/60\hbar^3 c^2$	σ	5.67032(71)	125	10^{-8} $W \cdot m^{-2} \cdot K^{-4}$	10^{-5} $erg \cdot s^{-1} \cdot cm^{-2} \cdot K^{-4}$
First radiation constant, $2\pi hc^2$	c_1	3.741832(20)	5.4	10^{-16} $W \cdot m^2$	10^{-5} $erg \cdot cm^2 \cdot s^{-1}$
Second radiation constant, hc/k	c_2	1.438786(45)	31	10^{-2} $m \cdot K$	$cm \cdot K$
Gravitational constant	G	6.6720(41)	615	10^{-11} $m^3 \cdot s^{-2} \cdot kg^{-1}$	10^{-8} $cm^3 \cdot s^{-2} \cdot g^{-1}$
Ratio, kx-unit to ångström, $\Lambda = \lambda(\text{Å})/\lambda(\text{kxu})$; $\lambda(\text{Cu}K\alpha_1) \equiv 1.537400$ kxu	Λ	1.0020772(54)	5.3		
Ratio, Å* to ångström, $\Lambda^* = \lambda(\text{Å})/\lambda(\text{Å}^*)$; $\lambda(\text{W}K\alpha_1) \equiv 0.2090100$ Å*	Λ^*	1.0000205(56)	5.6		

ENERGY CONVERSION FACTORS AND EQUIVALENTS

Quantity	Symbol	Numerical Value *	Units	Uncert. (ppm)
1 kilogram ($kg \cdot c^2$)		8.987551786(72)	10^{16} J	0.008
		5.609545(16)	10^{29} MeV	2.9
1 Atomic mass unit ($u \cdot c^2$)		1.4924418(77)	10^{-10} J	5.1
		931.5016(26)	MeV	2.8
1 Electron mass $m_e \cdot c^2$)		8.187241(42)	10^{-14} J	5.1
		0.5110034(14)	MeV	2.8
1 Muon mass ($m_\mu \cdot c^2$)		1.6928648(96)	10^{-11} J	5.6
		105.65948(35)	MeV	3.3
1 Proton mass ($m_p \cdot c^2$)		1.5033015(77)	10^{-10} J	5.1
		938.2796(27)	MeV	2.8
1 Neutron mass ($m_n \cdot c^2$)		1.5053738(78)	10^{-10} J	5.1
		939.5731(27)	MeV	2.8
1 Electron volt		1.6021892(46)	10^{-19} J	2.9
			10^{-12} erg	2.9
	1 eV/h	2.4179696(63)	10^{14} Hz	2.6
	1 eV/hc	8.065479(21)	10^5 m^{-1}	2.6
			10^3 cm^{-1}	2.6
	1 eV/k	1.160450(36)	10^4 K	31
Voltage-wavelength conversion, hc		1.986478(11)	10^{-25} $J \cdot m$	5.4
		1.2398520(32)	10^{-6} $eV \cdot m$	2.6
			10^{-4} $eV \cdot cm$	2.6
Rydberg constant	$R_\infty hc$	2.179907(12)	10^{-18} J	5.4
			10^{-11} erg	5.4
		13.605804(36)	eV	2.6
	$R_\infty c$	3.28984200(25)	10^{15} Hz	0.075
	$R_\infty hc/k$	1.578885(49)	10^5 K	31
Bohr magneton	μ_B	9.274078(36)	10^{-24} $J \cdot T^{-1}$	3.9
		5.7883785(95)	10^{-5} $eV \cdot T^{-1}$	1.6
	μ_B/h	1.3996123(39)	10^{10} $Hz \cdot T^{-1}$	2.8
	μ_B/hc	46.68604(13)	$m^{-1} \cdot T^{-1}$	2.8
			10^{-2} $cm^{-1} \cdot T^{-1}$	2.8
	μ_B/k	0.671712(21)	$K \cdot T^{-1}$	31
Nuclear magneton	μ_N	5.505824(20)	10^{-27} $J \cdot T^{-1}$	3.9
		3.1524515(53)	10^{-8} $eV \cdot T^{-1}$	1.7
	μ_N/h	7.622532(22)	10^6 $Hz \cdot T^{-1}$	2.8
	μ_N/hc	2.5426030(72)	10^{-2} $m^{-1} \cdot T^{-1}$	2.8
			10^{-4} $cm^{-1} \cdot T^{-1}$	2.8
	μ_N/k	3.65826(12)	10^{-4} $K \cdot T^{-1}$	31

* Note that the numbers in parentheses are the one standard-deviation uncertainties in the last digits of the quoted value computed on the basis of internal consistency, that the unified atomic mass scale $^{12}C \triangleq 12$ has been used throughout, that u=atomic mass unit, C=coulomb, F=farad, G=gauss, H=henry, Hz=hertz=cycle/s, J=joule, K=kelvin (degree Kelvin), Pa=pascal=$N \cdot m^{-2}$, T=tesla (10^4 G), V=volt, Wb=weber=$T \cdot m^2$, and W=watt. In cases where formulas for constants are given (e.g., R_∞), the relations are written as the product of two factors. The second factor, in parentheses, is the expression to be used when all quantities are expressed in cgs units, with the electron charge in electrostatic units. The first factor, in brackets, is to be included only if all quantities are expressed in SI units. We remind the reader that with the exception of the auxiliary constants which have been taken to be exact, the uncertainties of these constants are correlated, and therefore the general law of error propagation must be used in calculating additional quantities requiring two or more of these constants.

† Quantities given in u and atm are for the convenience of the reader; these units are not part of the International System of Units (SI).

‡ In order to avoid separate columns for "electromagnetic" and "electrostatic" units, both are given under the single heading "cgs Units." When using these units, the elementary charge e in the second column should be understood to be replaced by e_m or e_s, respectively.

INDEX